Neurotoxicity of Industrial and Commercial Chemicals

Volume II

Editor

John L. O'Donoghue, V.M.D., Ph.D.
Pathology Group Leader
Health and Environment Laboratories
Eastman Kodak Company
Rochester, New York

CRC Press, Inc.
Boca Raton, Florida

Library of Congress Cataloging in Publication Data
Main entry under title:

Neurotoxicity of industrial and commercial chemicals.

Includes bibliographies and indexes.
1. Nervous system--Diseases. 2. Neurotoxic agents.
3. Industrial toxicology. I. O'Donoghue, John L.

This book represents information obtained from authentic and highly regarded sources. Reprinted material is quoted with permission, and sources are indicated. A wide variety of references are listed. Every reasonable effort has been made to give reliable data and information, but the author and the publisher cannot assume responsibility for the validity of all materials or for the consequences of their use.

All rights reserved. This book, or any parts thereof, may not be reproduced in any form without written consent from the publisher.

Direct all inquiries to CRC Press, Inc., 2000 Corporate Blvd., N.W., Boca Raton, Florida, 33431.

© 1985 by CRC Press, Inc.
International Standard Book Number 0-8493-6454-X (v. 1)
International Standard Book Number 0-8493-6455-8 (v. 2)

Library of Congress Card Number 84-4261
Printed in the United States

FOREWORD

The purpose of this book on neurotoxicity is to collect and integrate information on the neurotoxicity of chemicals used in industry or having commercial value to aid in making choices to reduce neurotoxic risks. Toxicologists, industrial hygienists, occupational physicians, and others interested in the neurotoxic effects of chemicals now have a readily available source of information. Chemicals reported to cause a variety of effects on the nervous system are reviewed. These effects include behavioral changes, electrophysiologic effects, encephalopathy, peripheral neuropathy, neuroteratogenesis, neurooncologic effects, and neurochemical alterations. Exposure data, clinical manifestations, neuropathology, experimental neurology, metabolism, and structure activity correlates have been integrated into one source. This information is presented by the anatomical and functional areas of the nervous system affected and also for the first time by chemical classes with neurotoxic effects. Extensive use of tables and references provides the reader with a previously unavailable source of information.

John L. O'Donoghue

PREFACE

Increased concern about subtle effects of occupational and environmental exposures to chemicals has led to an increased interest in neurotoxicological information since neurologic deficits in sensory perception, movement, and thought processes have profound implications for the individual and society. The objective of this book is to concisely present information on the neurotoxicity of industrial and commercial chemicals for use by toxicologists, industrial hygienists, occupational physicians, neuroscientists, and others needing a readily available information source.

The first chapters review the effects of chemicals on the fundamental building blocks of the nervous system: the neuron, the axon, the myelin sheath, and myelinating cells: the oligodendroglia and Schwann cell. The effects of chemicals on integrated nervous system function or behavior are covered in Chapter 3. In Chapter 3, many chemicals reported to produce a neurologic effect have been collated from standard toxicology texts. Chapter 4 reviews chemical and biological interactions which modify the neurotoxicity of several substances.

The remaining chapters in Volumes I and II bring together information on the neurotoxicity of specific chemicals and classes of chemicals. The intent of this type of approach is to draw structure-activity relationships important to understanding neurotoxicity. Whenever possible, related substances, both negative and positive for neurotoxicity, have been included in the reviews. Information on structure-activity relationships has two important uses. The first is to help the toxicologist to make judgments of the probability of neurotoxicity within a family of substances to guide the safety testing process. The second potential use is to the neuroscientist because an accurate definition of the structural features of a chemical necessary for neurotoxicity defines the nervous system receptor site(s) responsible for neurotoxicity. Identification of neurotoxin receptors is essential for a full understanding of the pathogenesis of nervous system dysfunction. For many chemicals, understanding of structure-activity is primitive.

Substances reviewed are those of past, present, or potentially future use in industry or commerce which have also been reported to be neurotoxic. In some instances, the information on neurotoxicity is contradictory or weak and such substances may in fact not be neurotoxic. Chemicals used strictly as therapeutic drugs or only experimentally generally have been excluded from review.

The choice of which chemicals to review as neurotoxicants may at first seem straight forward, but in fact there are various opinions on what is a neurotoxin. For instance, any substance producing a neurologic effect may be considered neurotoxic. In the extreme then, all substances would likely be classified as neurotoxins since at very high or lethal doses, terminal seizures or depression of neurologic activity are usual. For the purposes of this book, those chemicals resulting in specific effects on the nervous system at sublethal doses have been included. Many other substances are tabulated in Volume I, Chapter 3.

In the preparation of a book there are many people whose help is necessary but who are often unrecognized. Appreciation is particularly due to Jean Ambrose and Maimie Reitano for their assistance in collecting data and to Tracey Coogan, Beverly Morris, and Marcia Swanger for preparing the manuscripts. There are also many other people who, through their concern and inspiration, also play a necessary role. I am particularly indebted to Drs. John T. McGrath, Allan Kelly, Sheldon Steinberg, and Clarence J. Terhaar, and also to my family who has always been there when needed most.

John L. O'Donoghue

THE EDITOR

John L. O'Donoghue, V.M.D., Ph.D., is Pathology Group Leader of the Health and Environment Laboratories at the Eastman Kodak Company in Rochester, New York.

Dr. O'Donoghue obtained his training at the University of Pennsylvania in Philadelphia, receiving the V.M.D. degree in 1970 and the Ph.D. in 1979. He served as U.S.P.H.S. Trainee in Pathology at the University of Pennsylvania, Pathologist (1974 to 1979) and Pathology Group Leader (1979 to present) of the Toxicology Section in the Health and Environment Laboratories at Eastman Kodak, and Instructor of Laboratory Animal Medicine (1975 to 1982), Assistant Professor of Laboratory Animal Medicine (1982 to present), and Assistant Professor of Toxicology (1982 to present) at the University of Rochester School of Medicine and Dentistry.

Dr. O'Donoghue is a member of the Wildlife Disease Association, the American Veterinary Medical Association, the Society of Toxicologic Pathologists, the Electron Microscopy Society of America, the American Association for the Advancement of Sciences, the American College of Veterinary Toxicologists, the Amereican Society for Veterinary Clinical Pathology, and the American Society of Neuropathologists. He is also a member of the honorary society Phi Zeta, and has received an award from this society as well as one from Sigma Xi. Dr. O'Donoghue was Student Research Fellow (1967 to 1969) and Post-Doctoral Research Fellow (1970 to 1974) at the University of Pennsylvania, is an International Member of the United States OECD Ad hoc Expert Group on Neurotoxicity, and is Chairman of the Neurotoxicology Expert Group, a division of the CMA.

CONTRIBUTORS

W. Kent Anger, Ph.D.
Chief
Neurobehavioral Research Section
Division of Biomedical and Behavioral
 Science
National Institute for Occupational Safety
 and Health
Cincinnati, Ohio

Joseph Arezzo, Ph.D.
Associate Professor
Departments of Neuroscience, Neurology,
 and Pathology
Albert Einstein College of Medicine
Bronx, New York

Cinda S. Davis, Ph.D.
Lecturer
Program in Toxicology
Neurotoxicology Research Laboratory
School of Public Health
University of Michigan
Ann Arbor, Michigan

Ian D. Duncan, B.V.M.S., Ph.D., M.R.C.V.S.
Associate Professor
Department of Medical Sciences
School of Veterinary Medicine
University of Wisconsin
Madison, Wisconsin

Barry L. Johnson, Ph.D.
Director
Division of Biomedical and Behavioral
 Science
National Institute for Occupational Safety
 and Health
Cincinnati, Ohio

Martin K. Johnson, Ph.D.
Senior Scientist
MRC Toxicology Unit
Medical Research Council
Carshalton, Surrey, England

Gary V. Katz, Ph.D.
Technical Associate
Health and Environment Laboratories
Eastman Kodak Company
Rochester, New York

David O. Marsh, M.D.
Professor of Clinical Neurology and
 Toxicology
Department of Neurology
School of Medicine and Dentistry
University of Rochester
Rochester, New York

Rudy J. Richardson, Sc.D.
Professor of Toxicology
Program in Toxicology
Neurotoxicology Research Laboratory
School of Public Health
University of Michigan
Ann Arbor, Michigan

Herbert H. Schaumburg, M.D.
Professor, Vice-Chairman
Departments of Neuroscience, Neurology,
 and Pathology
Albert Einstein College of Medicine
Bronx, New York

Peter S. Spencer, Ph.D., M.R.C.Path.
Professor of Neuroscience
Director, Institute of Neurotoxicology
Albert Einstein College of Medicine
Bronx, New York

TABLE OF CONTENTS

Volume I

Chapter 1
Chemicals Causing Disease of Neurons and Their Processes 1
P. S. Spencer, J. Arezzo, and H. Schaumburg

Chapter 2
Toxic Myelinopathies 15
Ian D. Duncan

Chapter 3
Chemicals Affecting Behavior 51
W. Kent Anger and Barry L. Johnson

Chapter 4
Chemical and Biological Interactions Affecting Neurotoxicity 149
Gary V. Katz

Chapter 5
The Neurotoxicity of Mercury and Lead 159
David O. Marsh

Chapter 6
Metals and Metalloids Other than Mercury and Lead 171
Gary V. Katz

Chapter 7
Carbon Monoxide, Inorganic Nitrogenous Compounds, and Phosphorus 193
John L. O'Donoghue

Index 205

Volume II

Chapter 1
Organophosphorus Compounds 1
Cinda S. Davis, Martin K. Johnson, and Rudy J. Richardson

Chapter 2
Cyanide, Nitriles, and Isocyanates 25
John L. O'Donoghue

Chapter 3
Carbon Disulfide and Organic Sulfur-Containing Compounds 39
John L. O'Donoghue

Chapter 4
Alkanes, Alcohols, Ketones, and Ethylene Oxide 61
John L. O'Donoghue

Chapter 5
Aliphatic Halogenated Hydrocarbons, Alcohols, and Acids and Thioacids 99
John L. O'Donoghue

Chapter 6
Aromatic Hydrocarbons 127
John L. O'Donoghue

Chapter 7
Phenol and Related Substances 139
John L. O'Donoghue

Chapter 8
Cyclic Halogenated Hydrocarbons and Related Substances 155
John L. O'Donoghue

Chapter 9
Acrylamide and Related Substances 169
John L. O'Donoghue

Chapter 10
Miscellaneous Organic Nitrogen and Aromatic Compounds 179
John L. O'Donoghue

Index 197

Chapter 1

ORGANOPHOSPHORUS COMPOUNDS

C. S. Davis, M. K. Johnson, and R. J. Richardson

TABLE OF CONTENTS

I. Introduction 2
    A. Acute Effects 2
    B. Delayed Neurotoxicity 2

II. Identification of the Axonopathic Receptor 3

III. Nature of Axonopathic Inhibitors: Aging and Protection 3

IV. Neurotoxicity Testing 5
    A. Single-Dose Tests 5
    B. Multi-Dose Tests 6

V. Predictions of Neurotoxicity 6

VI. Conclusions 7

References 23

I. INTRODUCTION

A. Acute Effects

Organophosphorus compounds (OPs) can have several diverse effects on both the central and peripheral nervous systems. In this chapter, we will survey only briefly the acute effects, and concentrate on the more insidious delayed neurotoxicity.

The most extensively studied acute action of some OP compounds, including some insecticides and the so-called nerve gases, is the inhibition of acetylcholinesterase, one of the hydrolyzing enzymes of the neurotransmitter acetylcholine. The biochemical mechanism and treatment of this poisoning have been reviewed extensively elsewhere.[1-3] Signs and symptoms in mammals are attributable to accumulation of acetylcholine in cholinergic synapses in the nervous system, peripheral ganglia, and neuromuscular junctions. Initial effects due to overstimulation of glands and muscles include miosis (pin-point pupils), lacrimation, salivation, sweating (especially at sites of dermal contact with the OP), and muscular twitching, fasciculation, and convulsions. Late effects due to loss of voluntary muscle control and fatigue of the cholinergic synapses and neuromuscular junctions include involuntary urination and defecation, weakness, and flaccid paralysis. Death is usually due to respiratory paralysis — a combined central and peripheral effect. The entire spectrum of effects due to acetylcholinesterase inhibition can be produced within a few minutes from the time of exposure. Prophylaxis or treatment consists of first blocking the effects of excess acetylcholine with atropine, and reactivating inhibited acetylcholinesterase with agents such as pralidoxime (2-PAM).

Certain 4-alkyl derivatives of 1-phospha-2,6,7-trioxabicyclo[2.2.2]-octane-1-oxide produce convulsive seizure and death within a few minutes after i.p. injection into mice. Inhibition of brain acetylcholinesterase is not involved.[4] While the biochemical mechanism is still uncertain, antagonism of γ-aminobutyrate (GABA)[5] and elevation of cyclic guanosine 3',5'-monophosphate (cyclic GMP)[6] have been implicated.

A subacute effect of certain organophosphorus compounds such as paraoxon (diethyl-4-nitrophenyl phosphate) is a dose-dependent necrosis in rat skeletal muscle. Pathological change is initiated at the motor end-plate and, by 24 hr, a generalized breakdown of muscle fiber architecture is evident.[7] A critical loss of acetylcholinesterase activity (85%) over a specified period (2 hr) is an obligatory correlate in the development of necrosis.[8]

B. Delayed Neurotoxicity

Some organophosphorus esters are classified as axonopathic. They can cause a delayed neurotoxicity beginning with cramping, tingling, ataxia, and weakness in the lower limbs, progressing to paralysis, with possible upper limb involvement in severe cases. These effects can be produced by a single sufficiently large dose, but do not occur immediately. There is always a delay that averages 1 to 2 weeks in many animals and 2 to 4 weeks in man between exposure and appearance of signs and symptoms. Pathologically, the neuropathy is characterized as a central-peripheral distal axonopathy. This means that the distal region of axons in the central and peripheral nervous system have undergone degeneration. In the case of OP-induced central-peripheral distal axonopathy, degeneration occurs largely in spinal cord tracts and peripheral nerves in hindlimbs. Demyelination occurs as a secondary event after the primary axonopathy. The lesions are symmetrical, and involve both sensory and motor pathways. Many species, including cat, hen, water buffalo, and man, are susceptible, but some rodents, such as rat and guinea pig, seem highly resistant. Generally, the young of a susceptible species are resistant to single, but not repeated, doses. The adult chicken has become the accepted test species. The initiation of this effect does not involve the inhibition of acetylcholinesterase, but rather the phosphorylation of a receptor protein for axonopathic OPs. This protein has esteratic activity which fortuitously provides us with a convenient method for its assay and has given rise to the name neurotoxic esterase (NTE). For reviews concerning clinical and morphological characteristics of organophosphorus-induced delayed neurotoxicity as well as biochemical characteristics of NTE, see References 9 to 11.

II. IDENTIFICATION OF THE AXONOPATHIC RECEPTOR

Axonopathic OPs were hypothesized to phosphorylate a specific receptor in neural tissue, quite likely an esterase-like site in a protein. In order to identify the specific target, brain tissue from the hen (a susceptible species) was incubated with [^{32}P]-DFP, an axonopathic agent. This compound was expected to bind to the axonopathic OP receptor, but it was also known to phosphorylate other esterases, including acetylcholinesterase, known not to be involved in the production of neuropathy. Therefore, nonspecific binding sites were blocked by incubating with nonaxonopathic OPs. The difference in radiolabeling of proteins between tissue preincubated with nonaxonopathic OPs and those preincubated with a combination of nonaxonopathic and axonopathic OPs gave an indication of the amount of specific axonopathic OP receptor site contained in the tissue.[12] Subsequently, this specific labeling has been localized to a single band on SDS-polyacrylamide gel electrophoresis.[13,14]

To show that this receptor site was an esterase, advantage was taken of the fact that the rate of phosphorylation of an esterase by an OP compound is greatly reduced when a substrate is present. Several esters were tested for their ability to reduce [^{32}P]-DFP labeling of the axonopathic OP receptor protein. Phenyl phenylacetate was effective and subsequently shown to be hydrolyzed by the axonopathic OP-sensitive protein.[15] This protein was then named neurotoxic esterase (NTE), reflecting the fact that the organophosphorylation of this protein could be conveniently monitored by assaying its esterase activity. In subsequent evaluations of approximately 50 substrates, the most effective has proved to be phenyl valerate.[16]

Optimum NTE assay conditions for brain tissue have been described by Johnson;[17] however, 6 mg of brain tissue are required, not 0.6 mg as erroneously stated there. Since other enzymes in nervous tissue are also able to utilize phenyl valerate as a substrate, the presence of neurotoxic esterase can be detected only after selective inhibition of these similar but irrelevant enzymes. This is done by preincubation of brain homogenate with appropriate concentrations of a nonaxonopathic inhibitor such as paraoxon. After the esterases sensitive to nonaxonopathic compounds are inhibited, the neurotoxic esterase activity can be determined using an axonopathic inhibitor such as mipafox (N,N'-diisopropyl phosphorodiamidofluoridate). A small amount of esterase activity, insensitive to both these axonopathic and nonaxonopathic inhibitors remain under these conditions, although much of it is sensitive to other OP esters.

Neurotoxic esterase is inhibited in vivo by axonopathic compounds. Inhibition in the brain and spinal cord of a dosed hen following a single acute dose has to reach at least a value of approximately 75% before ataxia and paralysis develop.[9] This critical value may be nearer 50% in the case of subacute or chronic dosing.[11]

III. NATURE OF AXONOPATHIC INHIBITORS: AGING AND PROTECTION

As more compounds were tested for potential neurotoxicity, it became apparent that several nonaxonopathic compounds could inhibit neurotoxic esterase; these included certain phosphinates, sulfonates, and carbamates. However, paralysis did not result. Thus, it appeared that mere loss of NTE catalytic activity was not sufficient to produce a neuropathy and the relevance of NTE to delayed neurotoxicity seemed questionable. The crucial experiment that provided an answer to this problem was conducted as follows: Hens were pretreated with inhibitory, yet nonaxonopathic, compounds. During the interval of NTE inhibition by

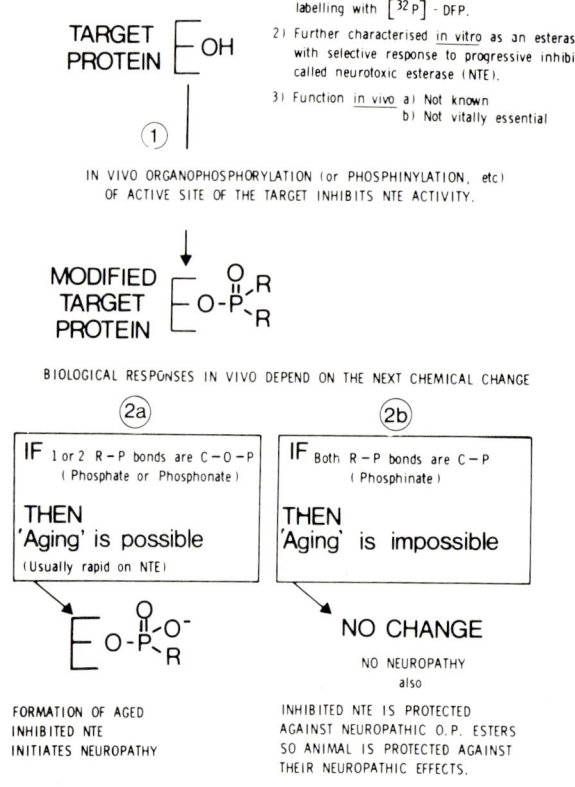

FIGURE 1. Alternative chemical modifications (1 and 2a or 2b) of the target protein with their pathological correlates. (Reproduced by permission of the publisher from Johnson, M. K., *Rev. Biochem. Toxicol.*, 4, 141. Copyright 1982 by Elsevier Science Publishing Co., Inc.)

nonaxonopathic compounds, the animals were given known axonopathic doses of organophosphorus compounds. These animals failed to develop neuropathy.[18-20] Thus, the nonaxonopathic NTE inhibitors protected the target from phosphorylation by the axonopathic inhibitors and produced a protected state in the birds. Protection has since been demonstrated against a variety of axonopathic OP compounds and in several species.[21] This is strong confirmation of the relevance of NTE to organophosphorus-induced delayed neurotoxicity.

Inhibitors of NTE can now be put into two groups: Those in Group A (phosphates, phosphonates, or phosphoramidates) are axonopathic, while those in Group B (carbamates, sulfonyl fluorides, and phosphinates) are not axonopathic, but prior administration of any Group B inhibitor is protective against the delayed neurotoxic action of subsequently administered Group A compounds. The essential difference between these two groups is the ability of Group A compounds to undergo a second time-dependent reaction called "aging" (loss or transfer of a side-chain from the OP) subsequent to becoming bound to the esterase active site (Figure 1). The inhibited enzyme is thus converted to a modifed form with an ionized acidic group on the phosphorus atom. Recently, the aging of OP-inhibited NTE has been established.[22,23]

IV. NEUROTOXICITY TESTING

Prescribed test protocols for organophosphorus pesticides have been introduced both in the U.S. and in England for assessment of delayed neurotoxicity.[24,25] In the U.S. it is required that adult hens, at least 10 per dose level, protected by atropine alone or atropine in combination with a suitable oxime reactivator, receive a maximum tolerated dose of the test substance and then be observed daily for at least 21 days. Some animals would be dosed at lower levels. All clinical signs of toxicity, including behavioral abnormalities, ataxia, and paralysis, are recorded. The animals are then sacrificed and a histopathological exam is performed. The difficulty with these procedures is that they are limited to yes-or-no decisions with the results only applicable to the exact dosing conditions employed. It is difficult to determine whether a negative result means totally negative or not-quite-positive.

Knowledge of NTE has transformed the situation for toxicity testers.[11] Assays of NTE 1 to 2 days after any chosen in vivo dose of a test compound indicate the percentage inhibition relative to control animals and thereby the degree of phosphorylation of the primary target. It is important to remember, however, that measurements of NTE inhibition correlate with the clinical or histopathological state of dosed birds 7 to 14 days later, not on the day of assay. In fact, during this delay period and subsequently, resynthesis of NTE can be considerable.[19] Given sufficient time following a single effective dose of an axonopathic OP, NTE levels in brain and spinal cord can approach control levels in frankly ataxic or paralyzed birds. The test is predictive of later events; we can think of inhibition of NTE as something that initiates the pathological process and recovers while the susceptible neurons degenerate.

In any neurotoxicity testing, particularly at high doses, it is desirable to consider the nature of the compound being administered. If the compound has anticholinesterase properties, then prophylaxis with physostigmine plus atropine can markedly raise the maximum tolerated dose. Reactivator oximes such as 2-PAM can also be considered. If the compound is devoid of anticholinesterase activity, then efficiency of absorption, rather that acute toxicity, controls the dose. It is probably inefficient to dose more than 1 g/kg at one time.[11] Poorly soluble compounds may be absorbed more readily if they are dissolved or suspended in such things as vegetable oils or various glycols. Another way to maximize absorption is to administer moderate doses several times a day for, say, 5 consecutive days. (See triethylphenyl phosphate isomers in Table 1.*) Some OP esters that are absorbed or metabolized very slowly, such as 2,4,5-trichlorophenyl ethyl phenylphosphonothioate, show a much greater inhibition of brain NTE two days after a single dose, rather than one day after, as is normally performed.[26]

Finally, in the case of two dimethyl phosphates which are very unstable in vivo, inhibition in spinal cord has been known to lag behind that in brain.[27] In these cases, the brain level would be an overly sensitive indicator of axonopathic potential, so spinal cord assays might be considered as a cross-check with new classes of OPs. It is important to note that the brain assay has never resulted in a false negative finding (with compounds classified as group A inhibitors). It seems that for most compounds reported, the inhibition in spinal cord lags by 5 to 15% less than brain.[11,21]

A. Single-Dose Tests

For severe clinical signs of ataxia and/or paralysis to be seen in hens 14 days after dosing, at least 70 to 80% of the NTE activity in brain and spinal cord must be inhibited within 1 to 40 hr after dosing. Great savings of time, chemicals, and hens are possible if a preliminary study is done to look at NTE activity levels in the brains of a few dosed birds in order to

---

* All tables are presented at the end of the text.

establish appropriate dosages for a combined NTE and clinical test.[45] Lengthy negative tests at inappropriate doses can be avoided in this manner.

The dose-related nature of NTE inhibition is obvious in Table 1. For the first three compounds, all inhibitory at even the lowest doses tested, it is readily apparent which would be the most likely dose to produce a clinical response. From this information, an experimental protocol could be determined, using both NTE assay and clinical observations. In the case of the three isomeric tri-ethylphenyl phosphates, an additional benefit of preliminary NTE testing becomes apparent. The 2-isomer has been reported negative in single or multi-dose tests by several groups.[26] However, the degree of NTE inhibition after one dose of 1200 mg/kg prompted one of us (MKJ) to try four daily doses with the positive result shown. Repeated dosing of the 3-isomer gave no further change in inhibition. This correlates with the negative clinical effect at this dose level and also strongly suggests that even more persistent dosing would be equally ineffective.

B. Multi-Dose Tests

This topic has been reviewed lately.[11,28] As we have seen, divided doses of poorly absorbed neurotoxic compounds actually may be more potent than a single massive dose (Table 1). However, there appears to be a clear threshold where further subdivisions of dose appear to be tolerable for long periods of time.[29-30] In chronic dosing, as for the single-dose test, there is a delay between initiation and expression of neuropathy. Thus, with daily doses, the inhibition on any particular day will only correlate properly with the effect 10 to 14 days later. It is essential, therefore, to continue observation for 2 to 3 weeks after dosing ceases in order to correlate dose and effect and to perform NTE assays at times which contribute most to the interpretation of chronic feeding experiments.

It is now clear that in a chronic dose situation, inhibition of NTE and resynthesis, balanced by degradation of the enzyme, brings about an equilibrium point of residual activity. The equilibrium point and the time needed to come to it will depend on the dose (as a fraction of a single effective dose) and the persistence of active material in the body. Most multi-dose studies reported so far[31-36] with a variety of axonopathic compounds, indicate that clinical neuropathy is only seen when a high-point of inhibition is obtained. When daily dosing was maintained for 3 to 10 weeks, the threshold of persistent NTE inhibition which correlates with clinical neuropathy seems to be about 45 to 50% in spinal cord or 50 to 60% in brain.[33,35,36] This is not far short of the inhibition needed to cause neuropathy after a single dose. Only one study claims that neuropathy can be associated with prolonged low levels of inhibition.[37] These results could not be duplicated,[31] and the discrepancy has yet to be resolved.

V. PREDICTIONS OF NEUROTOXICITY

A large amount of published test data has been earlier reviewed and is summarized here in Tables 2 through 9. NTE data are included whenever available. All data listed are from Reference 26, except where noted.

Some rules for predicting neurotoxicity have been suggested.[11,26]

For alkyl or mixed alkyl/aryl esters, with the general structure R^1R^2P (O or S)X, factors which increase delayed neurotoxicity potential when compared with acute toxicity potential include:

1. Choice of phosphonates or phosphoramidates rather than analogous phosphates.
2. Increase in chain length (especially up to 4 or 5 carbon atoms) or hydrophobicity of R^1 and R^2.
3. A leaving group X which does not sterically hinder approach to the active site of NTE.

Factors which decrease the comparative potential are

1. The converse of (1) to (3) above.
2. Choice of R or X groups which are very bulky or nonplanar. (This may counter the effects of (2) above as in the case of higher alkyl aryl phenylphosphorothioates.)
3. Choice of a nitrophenyl group at X.
4. Choice of comparatively more hydrophilic X groups (oximes or heterocyclics).
5. Choice of thioether linkages at X.

For triaryl phosphates, the following generalizations can be made:

1. Triaryl phosphates must be activated before they can phosphorylate NTE.
2. If the *ortho*-alkyl group has at least one hydrogen on the α-carbon, cyclic derivatives can be obtained. These derivatives are often highly neurotoxic.
3. If the *ortho*-substituted ring contains further substitution, neurotoxicity is reduced. Further substitution in other rings does not substantially reduce neurotoxicity.
4. Isomers having only one *ortho*-substituent have higher neurotoxicity than the symmetrical tri-*ortho*-ester.
5. As the *ortho*-substituent becomes larger, neurotoxicity declines.
6. At the *para*-position, a substituent requires two hydrogen atoms on the α-carbon in order to produce a potentially inhibitory (α-oxo) metabolite.
7. Substituents at the *meta*-position may be metabolized but do not yield inhibitory products.
8. Methyl groups in the *meta*- or *para*-position reduce toxicity.
9. Symmetrical tri-*para*-esters are more active than mixed esters with some substituents in the *meta*-position and some in the *para*.

VI. CONCLUSIONS

Only recently has the importance of determining biochemical mechanisms of toxicity been realized. In the case of organophosphorus-induced delayed neuropathy, the initial steps in the biochemical mechanism, the phosphorylation and aging of NTE, have been identified, although we are still a long way from understanding the process that follows aging and the subsequent degeneration of the axon. Currently, much investigative work is focused on this aspect.

NTE is a potentially useful adjunct to other toxicological endpoints in evaluation of OPs for delayed neurotoxicity. A suggested protocol for approaching evaluation of a new OP ester in a simple coherent fashion is given in the report of the Delayed Neurotoxicity Workshop.[21,45] However, some points still need to be resolved. These also are discussed in Reference 21. Thus the concept of a threshold level of inhibition for initiation by chronic dosing deserves more attention. Factors which may influence this threshold level need identification. Procedures and materials used in these studies, such as the NTE assay and strain of hen, need to be standardized. Thresholds, both in the acute and chronic sense, need to be established for species that are sensitive to a single dose, other than the hen. Nonetheless, it is to be hoped that with future development of new organophosphorus compounds, NTE assays will be performed as routinely as acetylcholinesterase assays, and that the information gained will be appropriately and completely utilized.

Table 1
DOSE-RELATED EFFECTS ON HEN BRAIN NTE AND ON CLINICAL RESPONSE[a]

Compound	Dose (mg/kg)	NTE activity remaining (%)	Ataxia seen
2,2-Dichlorovinyl di-*n*-pentyl phosphate	0.5 s.c.	58	—
	2.0 s.c.	9	+
2,4,5-Trichlorophenyl ethyl ethylphosphonothioate (trichloronate)	100 oral	76	—
	250 oral	35	—
	310 oral	14	+
Leptophos oxon	10 oral	67	—
	25 oral	57	—
	60 oral	11	+
Tri-4-ethylphenyl phosphate	250 oral	8	+
Tri-2-ethylphenyl phosphate	1200 oral	58	—
	3 × 700 oral	24	—
	4 × 1200 oral	13	+
Tri-3-ethylphenyl phosphate	1250 oral	84—91	—
	11 × 1200 oral	96	—

[a] NTE assays were performed 17 to 24 hr after dosing; surviving birds were observed for 3 weeks.

Data from Johnson, M. K., *Arch. Toxicol.*, 34, 259, 1975.

Table 2
FLUORIDES

Phosphorofluoridates

$$R_1-O, \quad R_3 \atop R_2-O \diagdown P \diagdown F$$

Cmpd. No.	R_1	R_2	R_3	In Vivo dose (i.m.) (mg/kg)	Ataxia
1	CH_3	CH_3	O	30	+
2	C_2H_5	C_2H_5	O	0.75	+
3	C_2H_5	C_2H_5	S	5	+
4	nC_3H_7	nC_3H_7	O	0.25	+
5	$isoC_3H_7$	$isoC_3H_7$	O	0.3	+[a]
6	nC_4H_9	nC_4H_9	O	0.5	+
7	$isoC_4H_9$	$isoC_4H_9$	O	1.5	+
8	$secC_4H_9$	$secC_4H_9$	O	1.5	+[a]
9	nC_5H_{11}	nC_5H_{11}	O	2.5	+
10	$C_3H_7CH(CH_3)$	$C_3H_7CH(CH_3)$	O	2.5	+
11	Cyclohexyl	Cyclohexyl	O	2.5	+
12	C_6H_5	C_6H_5	O	800	—
13	$2CH_3C_6H_4$	$2CH_3C_6H_4$	O	800	—

Table 2 (continued)
FLUORIDES

Cmpd. No.	R_1	R_2	R_3	In Vivo dose (i.m.) (mg/kg)	Ataxia
14	C_2H_5	C_3H_7	O	1.0	+
15	$C_6H_5(CH_2)_2$	$C_6H_5(CH_2)_2$	O	81[b]	+[c]
16	$C_6H_5(CH_2)_4$	$C_6H_5(CH_2)_4$	O	275[b]	−[c]
17	$C_6H_5(CH_2)_5$	$C_6H_5(CH_2)_5$	O	64[b]	−[c]
18	$C_6H_5(CH_2)_7$	$C_6H_5(CH_2)_7$	O	31[b]	−[c]

Phosphonofluoridates

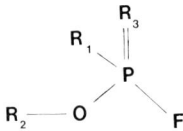

19	CH_3	C_2H_5	O	3	+
20	CH_3	$isoC_3H_7$	O	5 × 0.2	+
21	C_2H_5	$isoC_3H_7$	O	1.0	+
22	C_2H_5	$isoC_4H_9$	O	7 × 0.4	+
23	$isoC_3H_7$	CH_3	O	5	+
24	CH_3	Pinacolyl	O	2.2[b]	−[c]

Phosphinofluoridates

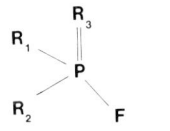

25	C_2H_5	C_2H_5	O	5	−
26	nC_3H_7	nC_3H_7	O	5	−
27	$isoC_3H_7$	$isoC_3H_7$	O	5	−
28	nC_4H_9	nC_4H_9	O	2.5	−

Phosphorodiamidofluoridates

29	CH_3	CH_3	O	15	+
30	C_2H_5	C_2H_5	O	3	+
31	nC_3H_7	C_3H_7	O	0.25	+
32	nC_4H_9	nC_4H_9	O	0.10	+[a]
33	nC_5H_{11}	nC_5H_{11}	O	2.5	+
34	nC_9H_{19}	nC_9H_{19}	O	100	+
35	$isoC_3H_7$	$isoC_3H_7$	O	25	+[a]

Table 2 (continued)
FLUORIDES

Cmpd. No.	R_1	R_2	R_3	In Vivo dose (i.m.) (mg/kg)	Ataxia
36	isoC$_4$H$_9$	isoC$_4$H$_9$	O	1.0	+
37	Cyclohexyl	Cyclohexyl	O	5	+
38	Phenyl	Phenyl	O	10	+
39	4-tolyl	4-tolyl	O	20	+
40	2-tolyl	2-tolyl	O	100	+

Note: Data are from Reference 26 unless indicated otherwise.

[a] In vitro, the I_{50} for DFP (cmpd. 5) was 0.7 μM, <0.1 μM for Butafox (cmpd. 32), and 7 μM for Mipafox (cmpd. 35).
[b] In vivo dose, mmol/kg.
[c] Data from Reference 39.

Table 3
PHOSPHATES AND PHOSPHOTHIOATES

$$R_1-O-\underset{\underset{O-R_2}{|}}{\overset{\overset{X}{\|}}{P}}-R_3$$

Cmpd. No.	R_1	R_2	R_3	X	In vivo dose (mg/kg)		Ataxia	Activity of NTE remaining %	In vitro I_{50} (μM)
41	CH_3	CH_3	$Cl_2C=CHO$	O	100	s.c.	—	10[a]	70
42	C_2H_5	C_2H_5	$Cl_2C=CHO$	O	18	s.c.	+	25	2
43	nC_3H_7	nC_3H_7	$Cl_2C=CHO$	O	2.0	s.c.	+	5	0.05
44	nC_5H_{11}	nC_5H_{11}	$Cl_2C=CHO$	O	2.0	s.c.	+	9	0.003
45	C_6H_5	C_6H_5	$Cl_2C=CHO$	O	2.5	i.v.	+	17	0.007
46	2,4,6-$(CH_3)_3C_6H_2$	2,4,6-$(CH_3)_3C_6H_2$	$Cl_2C=CHO$	O	10	s.c.	—	100	>100
47	4-tertBu-C_6H_4	4-tertBu-C_6H_4	$Cl_2C=CHO$	O	15	i.v.	+	13	0.8
48	2-isoPro-4-MeC$_6H_3$	2-isoPro-4-MeC$_6H_3$	$Cl_2C=CHO$	O					>100
49	$ClCH_2CH_2$	$ClCH_2CH_2$	$Cl_2C=CHO$	O	25	s.c.	+		
50	$ClCH_2CH_2$	$ClCH_2CH_2$	$Cl_2C=CHO$	O	5	s.c.	+		
51	CH_3	CH_3	$4NO_2C_6H_4O$	O	12	s.c.	—	100	b
52	CH_3	CH_3	$4NO_2C_6H_4O$	S	3 × 10	i.v.	—		
53	C_2H_5	C_2H_5	$4NO_2C_6H_4O$	O	15	s.c.	—	88	>100
54	C_2H_5	C_2H_5	$4NO_2C_6H_4O$	S	3 × 20	s.c.	—		
55	nC_3H_7	nC_3H_7	$4NO_2C_6H_4O$	O	25	s.c.	—	62	100
56	$isoC_3H_7$	$isoC_3H_7$	$4NO_2C_6H_4O$	O	15	s.c.	—	99	b
57	$isoC_3H_7$	$isoC_3H_7$	$4NO_2C_6H_4O$	S	3 × ?		—		
58	nC_4H_9	nC_4H_9	$4NO_2C_6H_4O$	O	30	s.c.	—	52	6
59	nC_5H_{11}	nC_5H_{11}	$4NO_2C_6H_4O$	O	22	i.v.	+		2
60	ClC_2H_4	ClC_2H_4	$4NO_2C_6H_4O$	O	100	oral	+	8	10
61	ClC_2H_4	C_2H_5	$4NO_2C_6H_4O$	O	20	s.c.	—		
62	CH_3	CH_3	$ClCH=CHO$	O	110	i.v.	—	16,20,28[a]	60

Table 3 (continued)
PHOSPHATES AND PHOSPHOTHIOATES

Cmpd. No.	R_1	R_2	R_3	X	In vivo dose (mg/kg)		Ataxia	Activity of NTE remaining %	In vitro I_{50} (μM)
63	C_2H_5	C_2H_5	ClCH=CHO	O	18	s.c.	+	25	3
64	CH_3	CH_3	ClC_2H_4O	O	50	i.v.	−	65	>100
65	C_2H_5	C_2H_5	ClC_2H_4O	O	30	i.v.	−	62	>100
66	nC_4H_9	nC_4H_9	ClC_2H_4O	O	10	i.v.	−	88	3
67	CH_3	CH_3	CH≡CH_2O	O	60	i.v.	−	91	b
68	CH_3	CH_3	1-(4-chlorophenyl thiovinyl) O	O	50	s.c.	−	100	60
69	CH_3	CH_3	1-(2,4-dichlorophenyl) 2-chlcrovinyl-O	O	60	i.v.	−	84	60
70	CH_3	CH_3	$2,5Cl_2 4IC_6H_2O$	O	30	i.v.	−	100	20
71	CH_3	CH_3	$2,5Cl_2 4BrC_6H_2O$	S	400	oral	−	100	
72	CH_3	CH_3	$2,4,5Cl_3C_6H_2O$	S	1600	s.c.	−		
73	CH_3	CH_3	$3Cl 4NO_2 C_6H_3O$	S	1200	s.c.	−		
74	CH_3	CH_3	S-1,2-dicarbethoxyethyl-thio	S	1000	s.c.	−	85[b]	
75	CH_3	CH_3	3-methyl-4-acetyl-phenyl	S	1000	oral	−		
76	CH_3	CH_3	$C_6H_5C_2H_4O$	O	60	i.v.	−	100	
77	CH_3	CH_3	$C_2H_5SC_2H_4O$	O	60	i.v.	−	100	
78	C_2H_5	C_2H_5	$C_2H_5SC_2H_4O$	S	20	s.c.	−		
79	C_2H_5	C_2H_5	$4ClC_6H_4SCH_2S$	S	2 × 500	s.c.	+		
80	C_2H_5	C_2H_5	2-isopropyl-4-methyl-pyrimidyl-6-0	S	10	s.c.	−		3
81	ClC_2H_4	ClC_2H_4	$2,3,5Cl_3C_6H_2O$	O	500	oral	+		
82	ClC_2H_4	ClC_2H_4	7-(3-chloro-4-methyl-coumariny1)-O	O	3000	oral	+		
83	$ClCH_2CH_2CH_2$	$ClCH_2CH_2CH_2$	7-(3-chloro-4-methyl-coumariny1)-O	O	400	oral	−		
84	$CH_3CHClCH_2$	$CH_3CHClCH_2$	7-(3-Chloro-4-methylcoumariny1)-O	O	2000	oral	−		
85	nC_4H_9	nC_4H_9	nC_4H_9O	O	184 × 2	p.o.	−[c]		
86	C_6H_5	C_6H_5	Isodecyl-O	O	120	p.o.[d]	−[c]		
87	C_6H_5	C_6H_5	Ethylhexyl-O	O	120	p.o.[d]	−[c]		
88	nC_4H_9	nC_4H_9	C_6H_5O	O	1.34 × 2	p.o.	−[c]		
89	CH_3	CH_3	CH_3O	O	50 × 10	i.p.	−[c]		
90	CH_3	CH_3	CH_3O	S	50 × 10	i.p.	−[c]		
91	CH_3	CH_3	CH_3S	O	20 × 10	i.p.	−[c]		

#								
92	CH₃ClCHCH₂	CH₃ClCHCH₂	CH₃CHClCH₂O	O	13,200	oral	—[f]	97
93	ClCH₂CH₂	ClCH₂CH₂	ClCH₂CH₂O	O	14,200	oral	—[f]	70
94			CH₃	3CH₃,4NO₂C₆H₃O	O			>102[g]
95			CH₃	3CH₃,4NO₂C₆H₃O	S	500	oral	—	92[g]
96			CH₃	α-Ethoxycarbonylbenzyl-S	O			103[g]
97			CH₃	4CNC₆H₄O	O			>119[g]
98			CH₃	4CNC₆H₄O	S	20	oral	—	100[g]

Note: Data are from reference 26 unless indicated otherwise.

[a] Massive doses of dichlorvos (cmpd. 46) and its 2-chlorovinyl analogue (cmpd. 67) caused high inhibition of brain but not spinal cord NTE without ataxia. See Reference 27 for discussion.
[b] Compounds caused < 20% inhibition in vitro at nominal concentration of 100 μM.
[c] From Reference 38.
[d] Cumulative dose.
[e] From Reference 41.
[f] From Reference 42.
[g] From Reference 34.
[h] From Reference 11.

Table 4
PHOSPHONATES AND PHOSPHONOTHIOATES

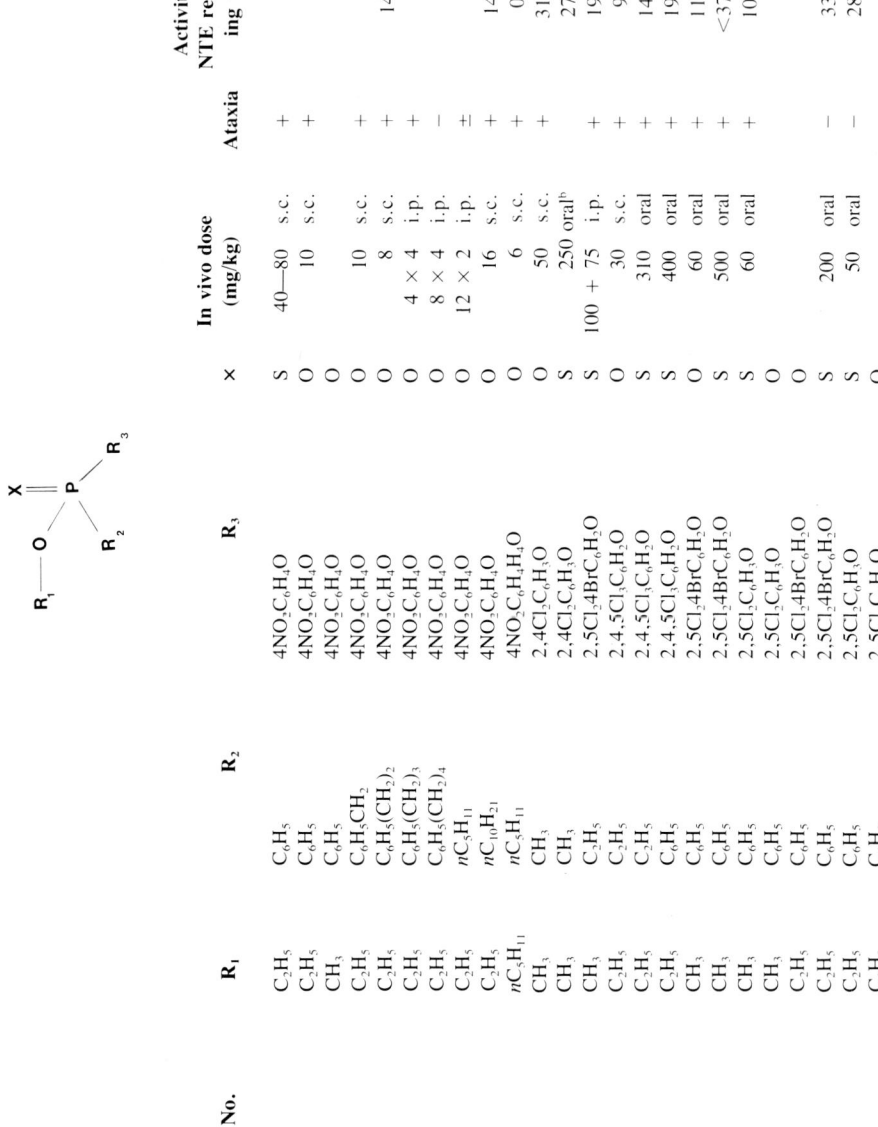

Cmpd. No.	R_1	R_2	R_3	X	In vivo dose (mg/kg)		Ataxia	Activity of NTE remaining %	In vitro I_{50} (μM)
99	C_2H_5	C_6H_5	$4NO_2C_6H_4O$	S	40—80	s.c.	+		(2.14^a)
100	C_2H_5	C_6H_5	$4NO_2C_6H_4O$	O	10	s.c.	+		(6.81^a)
101	CH_3	C_6H_5	$4NO_2C_6H_4O$	O					2
102	C_2H_5	$C_6H_5CH_2$	$4NO_2C_6H_4O$	O	10	s.c.	+		0.3
103	C_2H_5	$C_6H_5(CH_2)_2$	$4NO_2C_6H_4O$	O	8	s.c.	+	14	0.5
104	C_2H_5	$C_6H_5(CH_2)_3$	$4NO_2C_6H_4O$	O	4 × 4	i.p.	+		
105	C_2H_5	$C_6H_5(CH_2)_4$	$4NO_2C_6H_4O$	O	8 × 4	i.p.	−		
106	C_2H_5	nC_5H_{11}	$4NO_2C_6H_4O$	O	12 × 2	i.p.	±		1.4
107	C_2H_5	$nC_{10}H_{21}$	$4NO_2C_6H_4O$	O	16	s.c.	+	14	0.2
108	nC_5H_{11}	nC_5H_{11}	$4NO_2C_6H_4H_2O$	O	6	s.c.	+	0	0.06
109	CH_3	CH_3	$2,4Cl_2C_6H_3O$	O	50	s.c.	+	31	15
110	CH_3	CH_3	$2,4Cl_2C_6H_3O$	S	250	oral^b	+	27	NS^c
111	CH_3	C_2H_5	$2,5Cl_2 4BrC_6H_2O$	S	100 + 75	i.p.	+	19	2
112	C_2H_5	C_2H_5	$2,4,5Cl_3C_6H_2O$	O	30	s.c.	+	9	
113	C_2H_5	C_2H_5	$2,4,5Cl_3C_6H_2O$	S	310	oral	+	14	NS^c
114	C_2H_5	C_2H_5	$2,4,5Cl_3C_6H_2O$	S	400	oral	+	19^d	>100
115	CH_3	C_6H_5	$2,5Cl_2 4BrC_6H_2O$	O	60	oral	+	11	0.4^c
116	CH_3	C_6H_5	$2,5Cl_2 4BrC_6H_2O$	S	500	oral	+	<37	$NS^{a,c}$
117	CH_3	C_6H_5	$2,5Cl_2C_6H_3O$	S	60	oral	+	10	>100^d
118	CH_3	C_6H_5	$2,5Cl_2C_6H_3O$	O					0.204^{a,c}
119	C_2H_5	C_6H_5	$2,5Cl_2 4BrC_6H_2O$	O					0.077^{a,c}
120	C_2H_5	C_6H_5	$2,5Cl_2 4BrC_6H_2O$	S	200	oral	−	33^{a,c,f}	
121	C_2H_5	C_6H_5	$2,5Cl_2C_6H_3O$	S	50	oral	−	28^{a,c,f}	
122	C_2H_5	C_6H_5	$2,5Cl_2C_6H_3O$	O					0.059^{a,c}

No.	R₁	R₂	R₃	S/O	Dose	Route			
123	CH₃	C₆H₅	CH₃O	s	300	oral	−	100	
124	2,5Cl₂,4BrC₆H₂	C₆H₅	2,5Cl₂,4BrC₆H₂O	s	50	oral	−	100	NS[c]
125	CH₃	CH₃O	Cl₃CCH(OH)	O	200 + 100	s.c.		32	NS[c]
126	ClCH₂CH₂	C₂H₅	Cl₃C=CHO	O	112.5	s.c.	+	19	>100
127	isoC₃H₇	CH₃	2,5Cl₂,4BrC₆H₂O	O	100	i.p.	+	18	0.05
128	C₂H₅	C₂H₅	2thiopheneC(CN)=NO	s	75	oral	+	91	
129	C₂H₅	4ClC₆H₄	2,5Cl₂,4BrC₆H₂O	s	100	oral	−		
130	C₂H₅	C₆H₅	C₂H₅O	s	250	oral	−[g]		
131	C₂H₅	C₆H₅	C₂H₅O	O	500	oral	+[g]		
132	2,5Cl₂,4BrC₆H₂	C₆H₅	C₆H₅S	s	200	oral	+[g]		
133	C₂H₅	C₆H₅	C₆H₅S	O	500	oral	−[g]		
134	CH₃	4ClC₆H₄	C₆H₅O	O	200	oral	+[g]		
135	CH₃	C₆H₅	CH₃O	O	50 × 10	i.p.	−[h]		
136	CH₃	C₆H₅	CH₃S	O	50 × 10	i.p.	−[h]		
137	CH₃	C₆H₅	CH₃O	s	50 × 10	i.p.	−[h]		
138	CH₃	C₆H₅	CH₃O	O	50 × 10	i.p.	−[h]		
139	C₂H₅	C₆H₅	C₂H₅O	s	50 × 10	i.p.	−[h]		
140	C₂H₅	C₆H₅	C₂H₅O	O	50 × 10	i.p.	−[h]		
141	C₂H₅	C₆H₅	C₆H₅S	O	20 × 5	i.p.	−[h]		
142	C₆H₅	C₆H₅	4CNC₆H₄O	s	303	oral	+	10[i]	
143	CH₃	C₆H₅	4CNC₆H₄O	O					4.63[a]
144	CH₃	C₆H₅	4CNC₆H₄O	O					19.0[a]
145	CH₃	C₆H₅	2,4,5Cl₃C₆H₂O	O					0.104[a]
146	C₂H₅	C₆H₅	2,4,5Cl₃C₆H₂O	O					0.090[a]
147	CH₃	C₆H₅	2,4Cl₂C₆H₃O	O					1.11[a]
148	C₂H₅	C₆H₅	2,4Cl₂C₆H₃O	O					0.422[a]

Note: Data are from Reference 26 unless otherwise indicated.

[a] From Reference 34.
[b] Death occurred several days after dosing; insufficient compound for retesting.
[c] No significant inhibition in vitro at nominal concentration of 100 μM.
[d] Assayed 2 days after dosing.
[e] For a discussion and literature review on neurotoxic potential of leptophos and its analogues, see Reference 11.
[f] A near threshold response.
[g] From Reference 43.
[h] From Reference 41.
[i] From Reference 44.

Table 5
PHOSPHINATES[a]

$$R_1\text{-}P(=O)(R_2)\text{-}X$$

Cmpd. No.	R	X	In vivo dose (mg/kg)	Activity of NTE remaining %	Ataxia	In vitro I_{50} (μM)
149	C_2H_5	$4NO_2C_6H_4O$	10 s.c.	100	–	100
150	nC_4H_9	$4NO_2C_6H_4O$	10 s.c.	20	–	0.3
151	nC_5H_{11}	$4NO_2C_6H_4O$	16 s.c.	10	–	0.02
152	nC_5H_{11}	$Cl_2C=CHO$	10 s.c.	17	–	0.06

Note: Data from Reference 26.

[a] These compounds which are capable of inhibiting NTE are not neurotoxic but protective.

Table 6
PYROPHOSPHATES AND RELATED ESTERS[a]

$$R_1R_2P(=O)\text{-}O\text{-}P(=O)R_3R_4$$

Cmpd. No.	R_1	R_2	R_3	R_4	s.c. dose (mg/kg)	Ataxia
153	C_2H_5O	C_2H_5O	C_2H_5O	C_2H_5O	5 × 20	–
154	C_2H_5O	CH_3	nC_3H_7O	CH_3	10	–
155	nC_4H_9O	CH_3	nC_4H_9O	CH_3	10	–
156	C_2H_5O	CH_3	C_2H_5O	C_2H_5	10	–
157	$isoC_3H_7O$	CH_3	$isoC_3H_7O$	CH_3	10	–
158	$isoC_3H_7O$	$isoC_3H_7O$	$isoC_3H_7O$	$isoC_3H_7O$	3 × 100	–
159	C_2H_5O	CH_3	C_2H_5O	CH_3	10	–
160	$(CH_3)_2N$	$(CH_3)_2N$	$(CH_3)_2N$	$(CH_3)_2N$	160	–
161	$isoC_3H_7NH$	$isoC_3H_7NH$	$isoC_3H_7NH$	$isoC_3H_7NH$	300 (oral)	–

Note: Data from Reference 26.

[a] None of the compounds tested were neurotoxic. Only TEPP (cmpd. 153) and IsoOmpa (cmpd. 161) have been tested against NTE in vivo and neither caused much inhibition.

Table 7
CYCLIC SALIGENIN ESTERS

Cmpd. No.	R	O or S	i.p. dose (mg/kg)	Ataxia
162	CH_3O	O	12	−
163	CH_3O	S	100[a]	−
164	C_6H_5O	O	2	+
165	$2CH_3C_6H_4O$	O	5	+
166	$3CH_3C_6H_4O$	O	2	+
167	$4CH_3C_6H_4O$	O	0.5	+
168	$3,5(CH_3)_2C_6H_3O$	O	8	+
169	$2ClC_6H_4O$	O	25	+
170	C_2H_5	O	2	−
171	$ClCH_2$	O	25	−
172	C_6H_5	O	200	+
173	C_6H_5	S	100	+
174	$(CH_3)_2N$	O	10	−

Note: Data from Reference 26, except where noted.

[a] From Reference 34; brain NTE inhibition was 72%, near to the value usually associated with a positive clinical response of delayed neuropathy.

Table 8
MISCELLANEOUS ESTERS

Cmpd. No.	R	Dose (mg/kg)	Ataxia	Activity of NTE remaining (%)
175	$(C_2H_5O)_2POCl$	100 i.m.	−	
176	$(C_2H_5O)_2POCN$	50 i.m.	−	
177	$(C_2H_5O)_2POOC_2H_5$	10 i.m.	−	
178	$(C_2H_5O)_2POS(CH_2)_2N(C_2H_5)_2$	20 i.m.	−	
179	$(C_3H_7O)_2POO(2CH_3C_6H_4)$	10 i.m.	−	
180	$(isoC_3H_7O)_2PON_3$	5 i.m.	+	
181	$(isoC_3H_7O)_2PONHC_6H_5$	50 i.m.	−	
182	$(nC_4H_9O)_2POCl$	20 i.m.	−	
183	$(isoC_4H_9O)_2POCl$	20 i.m.	−	
184	$(2CH_3C_6H_4O)_2POCH_3$	50 i.m.	−	
185	$\begin{array}{c} C_2H_5O \\ \diagdown \\ POF \\ \diagup \\ (CH_3)_2N \end{array}$	5 i.m.	+	
186	$\begin{array}{c} C_2H_5O \\ \diagdown \\ POCN \\ \diagup \\ (CH_3)_2N \end{array}$	3.5 i.m.	−	34[a]
187	$\begin{array}{c} C_2H_5O \\ \diagdown \\ PON_3 \\ \diagup \\ (CH_3)_2N \end{array}$	5 × 3 i.m.	−	
188	$[(CH_3)_2N]POF$	5 i.m.	−	99
189	$(C_6H_5NH)_2POOCH=CCl_2$	130 s.c.	+	12
190	$(C_2H_5S)_3PO$	10 × 100 i.p.	−	
191	$(nC_3H_7S)_3PO$	10 × 5 i.p.	+	
192	$(nC_4H_9S)_3PO$	1110 s.c.	+	23
193	$(nC_4H_9S)_3P$	10 × 100 i.p.	+	
194	$(nC_5H_{11}S)_3PO$	7 × 200 i.p.	−	
195	$(nC_6H_{13}S)_3PO$	10 × 300 i.p.	−	
196	$(nC_8H_{17}S)_3PO$	10 × 100 i.p.	−	
197	$\begin{array}{c} CH_2-O- \\ \vert \quad\;\; \vert \\ (C_2H_5)_2CCH_2OPSCl \end{array}$	250 oral	−	91

Note: Data from Reference 26, except where noted.

[a] From Reference 39.

TABLE 9
TRIARYL PHOSPHATES

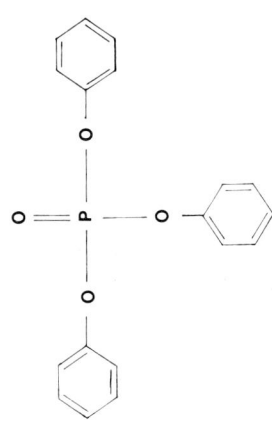

Cmpd. No.	Substituents on phenyl ring			Oral dose (mg/kg)	Interval between successive doses (days)	NTE activity remaining one day after last dose	Neurotoxic response
	Ring 1	Ring 2	Ring 3				
198	None	None	None	700		100	—
199	2CH₃	None	None	120		7	+
200	2CH₃	2CH₃	2CH₃	250		4	+
201	2CH₃	2CH₃	3CH₃	100			+
202	2CH₃	2CH₃	4CH₃	100			+
203	2CH₃	3CH₃	3CH₃	50			+
204	2CH₃	4CH₃	4CH₃	50—100			+
205	2CH₃	3CH₃	4CH₃	50			+
206	2CH₃	3,5-di-CH₃	3,5-di-CH₃	1000		5	+
207	3CH₃	3CH₃	3CH₃	25 × 250	1		+
208	4CH₃	4CH₃	4CH₃	1200		103	—
209	3CH₃	3CH₃	4CH₃	2500			—
210	3CH₃	4CH₃	4CH₃	2500			—

TABLE 9 (continued)
TRIARYL PHOSPHATES

No.	R₁	R₂	R₃	Dose	n	%	Effect
211	2C₂H₅	2C₂H₅	2C₂H₅	4 × 1200	2	13	+
212	2C₂H₅	2C₂H₅	4CH₃	1000			+
213	2C₂H₅	4CH₃	4CH₃	50			+
214	2C₂H₅	3C₂H₅	3C₂H₅	50		<18	+
215	3C₂H₅	3C₂H₅	3C₂H₅	11 × 1200	1	96ᶜ	−
216	3C₂H₅	None	None	6 × 1200	1		−
217	4C₂H₅	4C₂H₅	4C₂H₅	200—300		8	+
218	4C₂H₅	3C₂H₅	3C₂H₅	5 × 500	1		−
219	4C₂H₅	4CH₃	4CH₃	5 × 500	1		−
220	4C₂H₅	4CH₃	4CH₃	7 × 1000	1		−
221	4C₂H₅	4CH₃CO	4CH₃CH(OH)	50 i.v.		15	+
222	4C₂H₅	4C₂H₅	4CH₃CO	100 i.v.		15	+
223	4-α-hydroxy-ethyl	4-α-hydroxy-ethyl	4-α-hydroxy-ethyl	50 i.m.			+
224	4CH₃CO	4CH₃CO	4CH₃CO	25—100			+
225	4CH₃CO	4CH₃CO	4CH₃CO	ca 100		6	+
226	4CH₃CO	None	None	1000			−
227	4C₂H₅	2-n-propyl	2-n-propyl	100			+
228	2C₂H₅	3,5-di-CH₃	3,5-di-CH₃	1000			+
229	4C₂H₅	3,5-di-CH₃	3,5-di-CH₃	7 × 1000			−
230	2-n-C₃H₇	2-n-C₃H₇	2-n-C₃H₇	1000		76	+/−
231	2-n-C₃H₇	4CH₃	4CH₃	4 × 500	1		−
232	2-n-C₃H₇	4C₂H₅	4C₂H₅	100		67	+
233	2-n-C₃H₇	3,5-di-CH₃	3,5-di-CH₃	1000		86	−
234	4-n-C₃H₇	4-n-C₃H₇	4-n-C₃H₇	1200		100	−
235	2-iso-C₃H₇	2-iso-C₃H₇	2-iso-C₃H₇	4 × 1000	1	15	+/−
236	2-iso-C₃H₇	2-iso-C₃H₇	None	1200		10	+
237	2-iso-C₃H₇	None	None	1200		100	−
238	3-iso-C₃H₇	3-iso-C₃H₇	3-iso-C₃H₇	1000		100	−
239	4-iso-C₃H₇	4-iso-C₃H₇	4-iso-C₃H₇	1000		100	−

No.	R	R'	Dose	%		
240	4-iso-C₃H₇	4-iso-C₃H₇	1000	100	—	—
241	4-iso-C₃H₇	None	1000	96	—	—
242	2-sec-C₃H₉	None	3 × 1200		—	+/−
243	4-sec-C₃H₉	None	1200	97	—	—
244	2-tert-C₃H₉	None	1200	93	—	—
245	4-tert-C₃H₉	None	1000	90	—	—
246	4-tert-C₃H₉	None	1000	100	—	—
247	4-tert-C₃H₉	4-tert-C₄H₉	450	100	—	—
248	2,2-di-CH₃	2,2-di-CH₃	12[b]		—	+
249	2,3-di-CH₃	2,3-di-CH₃	40 × 1000		—	+
250	2,4-di-CH₃	2,4-di-CH₃	8 × 2500[d]		—	—
251	2,5-di-CH₃	2,5-di-CH₃	18 × 2500		—	—
252	2,6-di-CH₃	2,6-di-CH₃	18 × 2500		—	—
253	3,4-di-CH₃	3,4-di-CH₃	18 × 2500		—	—
254	3,5-di-CH₃	3,5-di-CH₃	18 × 2500		—	—
255	2-CH₃,4-C₂H₅	2-CH₃,4-C₂H₅	5 × 500		—	—
256	2,3-di-CH₃	3,5-di-CH₃	50 × 900		—	+
257	2,4-di-CH₃	3,5-di-CH₃	28 × 900		—	—
258	2,5-di-CH₃	3,5-di-CH₃	50 × 900		—	+
259	2,6-di-CH₃	3,5-di-CH₃	30 × 900		—	+
260	3,4-di-CH₃	3,5-di-CH₃	50 × 900		—	—
261	2,4-di-CH₃	3,5-di-CH₃	13 × 900[a]		—	—
262	2,6-di-CH₃	3,5-di-CH₃	50 × 900		—	—
263	2Cl	2Cl	1000		—	—
264	2Cl	None	1000		—	—
265	2Cl	None	1000		—	—
266	2OCH₃	2OCH₃	3000		—	—
267	2-Phenyl	2-Phenyl	1000		—	—
268	H	H	120[a]		—	−/+
269	H	H	120[a]		—	−/+

Note: Data are from Reference 26 unless indicated otherwise.

[a] Cumulative dose, g/kg.
[b] From Reference 40.
[c] From Reference 11.
[d] Minimum effective dose is probably less since symptoms appeared only 3 to 4 days after last dose.

REFERENCES

1. **Heath, D. F.**, *Organophosphorus Poisons*, Pergamon Press, New York, 1961.
2. **O'Brien, R. D.**, *Insecticides, Action and Metabolism*, Academic Press, New York, 1967.
3. **Eto, M.**, *Organophosphorus Pesticides: Organic and Biological Chemistry*, CRC Press, Boca Raton, Fla., 1974.
4. **Bellet, E. M. and Casida, J. E.**, Bicyclic phosphorus esters: high toxicity without cholinesterase inhibition, *Science*, 182, 1135, 1973.
5. **Bowery, N. G., Collins, J. F., and Hill, R. G.**, Bicyclic phosphorus esters that are potent convulsants and GABA antagonists, *Nature (London)*, 261, 601, 1976.
6. **Coult, D. B., Howells, D. J., and Smith, A. P.**, Cyclic nucleotide concentrations in the brains of mice treated with the convulsant bicyclic organophosphate, 4-isopropyl-2,6,7-trioxa-1-phosphabicyclo[2.2.2]octane, *Biochem. Pharmacol.*, 28, 193, 1979.
7. **Wecker, L. and Dettbarn, W. D.**, Paraoxon-induced myopathy: muscle specificity and acetylcholine involvement, *Exp. Neurol.*, 51, 281, 1976.
8. **Wecker, L., Kiauta, T., and Dettbarn, W. D.**, Relationship between acetylcholinesterase inhibition and the development of a myopathy, *J. Pharmacol. Exp. Ther.*, 206, 97, 1978.
9. **Johnson, M. K.**, The delayed neuropathy caused by some organophosphorus esters: mechanism and challenge, *Crit. Rev. Toxicol.*, 3, 289, 1975.
10. **Davis, C. S. and Richardson, R. J.**, Organophosphorus compounds, in *Experimental and Clinical Neurotoxicology*, Spencer, P. S. and Schaumburg, H. H., Eds., Williams & Wilkins, Baltimore, 1980, chap. 36.
11. **Johnson, M. K.**, The target for initiation of delayed neurotoxicity by organophosphorus esters: biochemical studies and toxicological applications, *Rev. Biochem. Toxicol.*, 4, 141, 1982.
12. **Johnson, M. K.**, A phosphorylation site in brain and the delayed neurotoxic effect of some organophosphorus compounds, *Biochem. J.*, 111, 487, 1969.
13. **Williams, D. G. and Johnson, M. K.**, Gel-electrophoretic identification of hen brain neurotoxic esterase, labelled with tritiated di-isopropyl phosphorofluoridate, *Biochem. J.*, 199, 323, 1981.
14. **Huggins, D. J. and Richardson, R. J.**, unpublished observations, 1982.
15. **Johnson, M. K.**, The delayed neurotoxic effect of some organophosphorus compounds. Identification of the phosphorylation site as an esterase, *Biochem. J.*, 114, 711, 1969.
16. **Johnson, M. K.**, Structure-activity relationship for substrates and inhibitors of hen brain neurotoxic esterase, *Biochem. Pharmacol.*, 24, 797, 1975.
17. **Johnson, M. K.**, Improved assay of neurotoxic esterase for screening organophosphates for delayed neurotoxicity potential, *Arch. Toxicol.*, 37, 113, 1977.
18. **Johnson, M. K.**, Organophosphorus and other inhibitors of brain "neurotoxic esterase" and the development of delayed neuropathy in hens, *Biochem. J.*, 120, 523, 1970.
19. **Johnson, M. K.**, The primary biochemical lesion leading to the delayed neurotoxic effects of some organophosphorus esters, *J. Neurochem.*, 23, 785, 1974.
20. **Johnson, M. K. and Lauwerys, R.**, Protection by some carbamates against the delayed neurotoxic effects of di-isopropyl phosphorofluoridate, *Nature (London)*, 222, 1066, 1969.
21. **Johnson, M. K. and Richardson, R. J.**, Biochemical endpoints: neurotoxic esterase assay, in *Proc. Delayed Neurotoxicity Workshop*, Urbana, Ill., 1982; *Neurotoxicology*, 4, 311, 1983.
22. **Clothier, B. and Johnson, M. K.**, Rapid aging of neurotoxic esterase after inhibition by di-isopropyl phosphorofluoridate, *Biochem. J.*, 177, 549, 1979.
23. **Clothier, B. and Johnson, M. K.**, Reactivation and aging of neurotoxic esterase inhibited by a variety of organophosphorus esters, *Biochem. J.*, 185, 739, 1980.
24. *Fed. Regist.*, 43, No. 163, Sec. 163.81-7, August 22, 1978.
25. Pesticides Branch, Document No. B5, U.K. Ministry of Agriculture, F. F., London, 1979.
26. **Johnson, M. K.**, Organophosphorus esters causing delayed neurotoxic effects: mechanism of action and structure-activity studies, *Arch. Toxicol.*, 34, 259, 1975.
27. **Johnson, M. K.**, The anomalous behavior of dimethyl phosphates in the biochemical test for delayed neurotoxicity, *Arch. Toxicol.*, 41, 107, 1978.
28. **Hollingshaus, J. G. and Fukuto, T. R.**, The effect of chronic exposure to pesticides on delayed neurotoxicity, in *The Effects of Chronic Exposure to Pesticides on Animal Systems*, Chambers, J. E. and Yarbrough, J., Eds., Raven Press, New York, in press.
29. **Barnes, J. M.**, Assessing hazards from prolonged and repeated exposure to low doses of toxic substances, *Br. Med. Bull.*, 31, 196, 1975.
30. **Abou-Donia, M. B. and Preissig, S.**, Delayed neurotoxicity from continuous low-dose oral administration of leptophos to hens, *Toxicol. Appl. Pharmacol.*, 38, 595, 1976.

31. **Johnson, M. K. and Lotti, M.,** Delayed neurotoxicity caused by chronic feeding of organophosphates requires a high-point of inhibition of neurotoxic esterase, *Toxicol. Lett.*, 5, 99, 1980.
32. **Lotti, M. and Johnson, M. K.,** Repeated small doses of a neurotoxic organophosphate: monitoring of neurotoxic esterase in brain and spinal cord, *Arch. Toxicol.*, 45, 263, 1980.
33. **Hussain, M. A. and Oloffs, P. C.,** Neurotoxic effects of leptophos in chickens and rats following chronic low level feeding, *J. Environ. Sci. Health B*, 14, 367, 1979.
34. **Okhawa, H., Oshita, H., and Miyamoto, J.,** Comparison of inhibitory activity of various organophosophorus compounds against acetylcholinesterase and neurotoxic esterase of hens with respect to delayed neurotoxicity, *Biochem. Pharmacol.*, 29, 2721, 1980.
35. **Sprague, G. L. and Bickford, A. A.,** Effect of multiple diisopropyl fluorophosphate injections in hens: a behavioral, biochemical and histological investigation, *J. Toxicol. Environ. Health*, 8, 973, 1981.
36. **Sprague, G. L., Sandvik, L. L., and Bickford, A. A.,** Time course for neurotoxic esterase inhibition in hens given multiple diisopropyl fluorophosphate injections, *Neurotoxicology*, 2, 523, 1981.
37. **Olajos, E. J., DeCaprio, A. P., and Rosenblum, I.,** Central and peripheral neurotoxic esterase activity and dose response relationship in adult hens after acute and chronic oral administration of diisopropyl fluorophosphate, *Ecotoxicol. Environ. Saf.*, 2, 383, 1978.
38. **Abou-Donia, M. B.,** Organophosphorus ester-induced delayed neurotoxicity, *Annu. Rev. Pharmacol. Toxicol.*, 21, 511, 1981.
39. **Gordon, J. J., Inns, R. H., Johnson, M. K., Leadbeater, L., Maidment, M. P., Upshall, D. G., Cooper, G. H., and Rickard, R. L.,** The delayed neuropathic effects of nerve agents and some other organophosphorus compounds, *Arch. Toxicol.*, 52, 71, 1983.
40. **Johannsen, F. R., Wright, P. L., Gordon, D. E., Levinskas, G. J., Radue, R. W., and Graham, P. R.,** Evaluation of delayed neurotoxicity and dose-response relationships of phosphate esters in the adult hen, *Toxicol. Appl. Pharmacol.*, 41, 291, 1977.
41. **Hollingshaus, J. G., Armstrong, D., Toia, R. F., McCloud, L., and Fukoto, T. R.,** Delayed toxicity and delayed neurotoxicity of phosphorothioate and phosphonothioate esters, *J. Toxicol. Environ. Health*, 8, 619, 1981.
42. **Sprague, G. L., Sandvik, L. L., Brookins-Hendricks, M. J., and Bickford, A. A.,** Neurotoxicity of two organophosphorus ester flame retardants in hens, *J. Toxicol. Environ. Health*, 8, 507, 1981.
43. **Hollingshaus, J. G., Nishioka, T., March, R. B., and Fukoto, T. R.,** Effect of impurities on the delayed neurotoxicity of O-(4-bromo-2,5-dichlorophenyl) O-ethyl phenylphosphonothioate administered orally to hens, *J. Agric. Food Chem.*, 29, 593, 1981.
44. **Soliman, S. A., Curley, A., Farmer, J., and Novak, R.,** *In vivo* inhibition of chicken brain acetylcholinesterase and neurotoxic esterase in relation to the delayed neurotoxicity of leptophos and cyanofenphos, *J. Environ. Pathol. Toxicol. Oncol.*, in press.
45. **Johnson, M. K.,** Delayed neurotoxicity tests of organophosphorus esters: a proposed protocol integrating Neuropathy Target Esterase (NTE) assays with behaviour and histopathology tests to obtain more information more quickly from fewer animals, in *Proc. Int. Conf. on Environ. Hazards of Agrochemicals in Developing Countries*, Alexandria, November 1983, Vol. 1, El-Sebae, A. H., Ed., University of Alexandria Press, Egypt, 1984, 474.

Chapter 2

CYANIDE, NITRILES, AND ISOCYANATES

John L. O'Donoghue

TABLE OF CONTENTS

I.	Introductory Comments	26
II.	Malononitrile	26
	A. Introduction	26
	B. Neurotoxicity	26
III.	β-Dimethylaminopropionitrile	27
	A. Synonyms	27
	B. Introduction	27
	C. Human Neurotoxicity	27
	D. Experimental Neurotoxicity	28
	E. Neuropathology	29
	F. Other Effects	29
IV.	β,β'-Iminodipropionitrile	29
	A. Synonyms	29
	B. Introduction	29
	C. Experimental Neurotoxicity	29
	D. Neuropathology	30
	E. Metabolism	30
	F. Mechanism of Action	30
	G. Other Effects	30
	H. Other Chemicals Producing Similar Effects	30
V.	Acrylonitrile	31
	A. Synonyms	31
	B. Introduction	31
	C. Human Neurotoxicity	31
	D. Experimental Neurotoxicity	31
	E. Neuropathology	32
	F. Metabolism	32
	G. Other Effects	32
VI.	2,4-Hexadiene-1-Nitrile	33
	A. Synonyms	33
	B. Neurotoxicity	33
	C. Neuropathology	33
	D. Other Effects	33

VII. p-Bromophenylisothiocyanate 34
    A. Synonyms 34
    B. Introduction 34
    C. Neurotoxicity 34

VIII. Toluene Diisocyanate 34

References.. 35

I. INTRODUCTORY COMMENTS

Cyanide as hydrogen cyanide gas or simple salts of sodium, potassium, or calcium is well recognized as a highly toxic substance. Most intoxications with these substances result from accidental exposures or suicide attempts. Cyanide is acutely toxic by its binding to the iron in cytochrome oxidase, inactivating this enzyme which is essential for cellular respiration.

Acute symptoms of cyanide poisoning may include dizziness, weakness, and headache prior to collapse. Neuropathologic lesions following acute cyanide poisoning are few and involve acute vascular changes such as congestion. In those individuals who survive the acute phase of intoxication, hypoxic-ischemic lesions may be seen in the brain. These have been reviewed by Brierley.[1] Similar lesions may occur after nitroprusside poisoning and other materials which induce a profound coma.[2] With chronic poisoning by cyanogenetic glucosides in plants such as cassava, and laetrile from almond pits, a different type of neurotoxicity involving axonal pathology is reported.[3,4] Sodium cyanate, used to treat sickle-cell anemia, also is reported to produce axonal damage in man[5] and myelin damage experimentally. Doherty et al.[7] have reported neural tube defects resulting in exencephaly and encephalocoele in hamsters exposed *in utero* to sodium cyanide infused between the sixth and ninth day of gestation.

Aliphatic nitriles are metabolized releasing cyanide in vivo and acute poisoning by nitriles may have some of the characteristics of cyanide poisoning including inhibition of cytochrome oxidase. The full spectrum of nitrile toxication, however, is not due solely to cyanide.[8] The neurotoxicity of nitriles has not been linked to cyanide production and is unlikely to be, since each nitrile covered in this chapter produces a different type of neurotoxicity. There is also no evidence currently to link isocyanates with neurotoxic action to cyanide metabolism.

II. MALONONITRILE

A. Introduction

Malononitrile ($C{\equiv}N-CH_2-C{\equiv}N$) is a chemical intermediate in the manufacture of thiamin, anticancer drugs, acrylic fibers, photosensitizers, and dyes, and is used as a lubricating oil additive.[9]

B. Neurotoxicity

Malononitrile has a unique effect on the nervous system which is not related to the formation of cyanide ions. Hyden and Hartelius[10] injected rabbits intravenously with malononitrile and found that nucleoprotein synthesis was stimulated in the brain, spinal cord, and other areas of the nervous system. Based on animal studies, malononitrile was administered to patients with mental illness in an effort to stimulate cerebral nucleoprotein synthesis

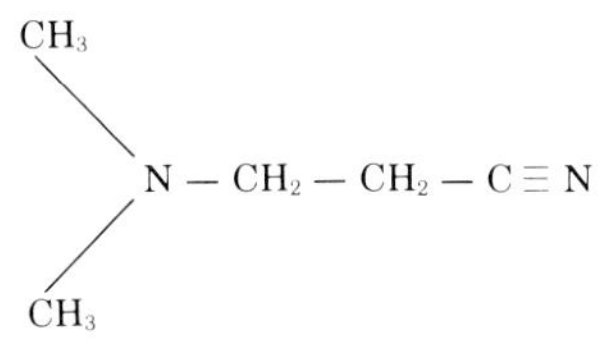

$$\text{(CH}_3\text{)}_2\text{N} - \text{CH}_2 - \text{CH}_2 - \text{C} \equiv \text{N}$$

FIGURE 1. β-Dimethylaminopropionitrile.

and reverse their illness.[10] Necrosis of neurons in the corpus striatum and demyelination of the optic tracts occurred in rats given multiple injections of malononitrile repetitively over a 1- to 2-day period.[11] Other toxicity studies have been reviewed in Reference 1.

Eberts et al.[12] identified a chromophore in commercial samples of aqueous malononitrile which they identified as a malononitrile dimer, 1,1,3-tricyano-2-amino-1-propene. Formation of the dimer increased during storage particularly in alkaline solutions. The dimer increases neural protein and ribonucleic acid and may be responsible for the activity of commercial grades of malononitrile.[12,13]

III. β-DIMETHYLAMINOPROPIONITRILE

A. Synonyms

3-(Dimethylamino)propanenitrile
3-(Dimethylamino)propionitrile
N,N-Dimethylamino-3-propionitrile
β-N-Dimethylaminopropionitrile
3-(N,N-Dimethylamino)propionitrile
DMAPN

B. Introduction

β-Dimethylaminopropionitrile (DMAPN) has had limited commercial use as a catalyst for flexible polyurethane foams and as a component of acrylamide-based grouting gels. It is also found as an intermediate in the production of dimethylaminopropylamine and some pharmaceuticals.[14] Following neurotoxicity in two polyurethane molding plants in 1978, the manufacturer removed DMAPN from the polyurethane catalyst market.

Early toxicity studies on DMAPN were concerned with its structural similarity to β-aminopropionitrile and the production of osteolathyrism. A group of six rats fed 0.35% DMAPN for 56 days did not develop observable skeletal abnormalities and gained weight normally.[14] Feeding 0.3% DMAPN to two pregnant rats beginning on the 17th day of gestation was not maternal or fetotoxic.[15] A collagen fragility assay in which 0.054 mM of DMAPN was injected into 14-day-old chicken eggs did not show any evidence suggestive of interference in collagen formation.[16] Methyl substitution of the amine group on DMAPN appears to block the occurrence of osteolathyrism.[14-16]

C. Human Neurotoxicity

In 1978, two incidences of neurotoxicity related to DMAPN exposure occurred in a very short time.[18,19] In both situations, DMAPN made up 95% of a polyurethane foam catalyst, NIAX-catalyst ESN. The remaining 5% of the catalyst was primarily *bis*-2-dimethyl-aminoethylether and <1% acrylonitrile and dimethylamine.

Keogh et al.[18] reported an epidemic of neurogenic urinary retention from a company, employing 141 and 75 people at 2 different plants, manufacturing polyurethane foam products. In the larger plant, in which the exposure to DMAPN was greater, 85% of the employees had symptoms of urinary retention. Symptomatology related to DMAPN included urinary retention, sexual dysfunction (including difficulty getting and maintaining erection, decreased libido, and rarely impotence), irritability, insomnia, headaches, muscle weakness, and paresthesias. Cystometrograms performed on four individuals with at least moderate urinary retention had abnormal findings. Symptoms of peripheral neuropathy included decreased sensation (pain, temperature, and light touch) over the distal region of the legs and hands and areas innervated by the lower sacral dermatomes and decreased muscle strength in the muscles of the hand and feet. Abnormalities in vision, upper respiratory irritation, dermatitis, and pulmonary abnormalities also occurred but appeared to be related to the presence of other chemicals including N-ethylmorpholine and toluene diisocyanate.

Exposure to DMAPN was through direct skin contact with foam cushions containing DMAPN and by inhalation of DMAPN vapor released from the foam following curing. Both routes may have contributed to neurotoxicity.

Kreiss et al.[19] described neurogenic bladder abnormalities occurring in 104 (63%) of 166 employees engaged in the production of polyurethane foam automobile seat cushions.[19] The catalyst used was the same as that reported by Keogh et al.[18] The onset of bladder dysfunction in this plant followed the introduction of the DMAPN catalyst and the number of employees affected increased as DMAPN use increased. The first case occurred during the first month of exposure, but later a few people were affected within their first week of exposure to DMAPN. The latency from exposure to onset of urinary symptoms could have been only a few days. There was no association between handling foam containing DMAPN and urinary symptoms. The most likely route of exposure to DMAPN was by inhalation.[19,21]

Clinical symptomatology was similar to the cases described by Keogh et al.[18] Affected people had difficulty urinating, lost the urge to urinate, urinated infrequently, lost uretheral sensation, and/or had abdominal discomfort. Some people reported hand pressure on the abdomen or performed a Valsalva maneuver to initiate or maintain urine flow. Two required surgery for urinary retention. Sexual dysfunction was also reported. Both men and women were affected similarly. Cystoscopic examination of the bladder and urine analyses showed no differences between affected and unaffected employees.

Symptoms of peripheral neuropathy were limited to numbness of the arms in 13 people. Seven of eight symptomatic people examined neurologically had evidence of peripheral neuropathies in the legs. The majority of cases had a sensory neuropathy involving decreased vibration, pinprick, proprioception, and light touch, hyperesthesia, and numbness. The remaining cases had a combined sensorimotor neuropathy with decreases or absent ankle jerks. Evidence of decreased nerve conduction velocity and abnormal cystometrograms were reported.

With cessation of use of the DMAPN catalyst, improvement was rapidly observed in nearly all cases. Repeat examinations two years later showed that those people with the most severe symptomatology had persistent urological, sexual and neurological abnormalities.[22]

DMAPN produces an unusual and so far unique pattern of neurotoxicity in that urological abnormalities are very prominent while peripheral nerve function is affected little. In toxic neuropathies the reverse is the usual situation.

D. Experimental Neurotoxicity

Experimental animal data on the neurotoxic effects of DMAPN or the NIAX catalyst ESN are limited to relatively short term exposures.[23] Tests on *bis*-dimethylaminoethyl ether, which made up about 5% of the catalyst, produced loss of body weight and general malaise in rats and mice and polyuria in rats but no evidence of neurotoxicity. Both DMAPN and the catalyst produced tonic convulsions, tremors, akinesia, cardiovascular abnormalities, and loss of the micturation reflex in rats dosed with 0.25 mℓ/kg i.p., 5 days/5 week for two weeks. Mice receiving the same dose had convulsions and tremors. Doses as low as 0.01 mℓ/kg produced loss of the micturation reflex in rats.

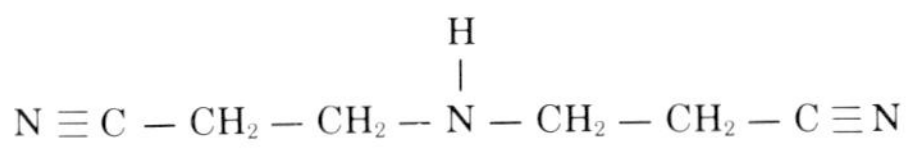

$$N \equiv C - CH_2 - CH_2 - \overset{\overset{H}{|}}{N} - CH_2 - CH_2 - C \equiv N$$

FIGURE 2. β,β′-Iminodipropionitrile.

E. Neuropathology

Rats exposed to 0.5% DMAPN in their drinking water for 2 to 9 months showed electron microscopic evidence of enlarged distal motor and spindle axons with neurofilament abnormalities in the soleus and plantar interosseous muscles.[20] Silver-cholinesterase stains showed enlargement of soleus motor nerve terminals.[24]

Biopsies from an individual severely affected by DMAPN showed histological evidence of mild denervation atrophy in the gastrocnemius muscle, slight reduction in the population of myelinated and unmyelinated axon in the sural nerve, and early degenerative axonal changes in the sural nerve.[25]

F. Other Effects

Jaeger et al.[26] reported that 2.0 mℓ/kg i.p. of the NIAX catalyst ESN, produced excitation and death in rats but reduction of the dose to 0.31 or 0.62 mℓ/kg orally administered twice resulted in acute bladder lesions. The lesions consisted of mass transmural edema, ulceration, acute inflammation, and in some animals hemorrhagic necrosis of the bladder wall. Similar effects have not been observed in people with NIAX catalyst ESN-related urinary dysfunction.[18,19]

IV. β,β′-IMINODIPROPIONITRILE

A. Synonyms

3,3′-Iminodipropionitrile
3,3′-Imino-*bis*-propanenitrile
Bis(β-cyanoethyl)amine
Bis-β-aminopropionitrile
Imino-β-β-dipropionitrile

N,N-*Bis*(2-cyanoethyl)amine
Di(2cyanoethyl)amine
2-Cyano-N-(2-cyanoethyl)ethanamine
Aminodipropionitrile
IDPN

B. Introduction

β,β′-Iminodipropionitrile (IDPN) is a clear, colorless liquid used in the preparation of certain pharmaceuticals and in synthetic chemistry. No reports of human toxicity due to IDPN have been reported, although the literature contains many studies of IDPN neurotoxicity in laboratory animals as models for hyperactivity and human diseases which produce proximal axonal damage such as seen in amyotrophic lateral sclerosis.

C. Experimental Neurotoxicity

IDPN toxicity has been described in rats, mice, dogs, cats, monkeys, fish, brine shrimp, and insects.[27-30] In rats and mice, 1 to 2 g/kg of IDPN by i.p. or s.c. injection or by gavage in a single dose or divided and given over several days produces a characteristic group of signs within 48 hr of the last dose. Signs include excitement, increased activity and alertness, head twitching, shaking or chorea-like movements, circling, retropulsion, and backward somersaults. These abnormalities are generally regarded as irreversible since they may be present a year or more after onset.[28] Rats placed on diets containing 0.3% IDPN progress through the hyperexcitable phase and after the eighth week develop decreased activity, ataxia, loss of the placing reaction, and paresis or paralysis, especially of the hindlimbs.[31]

In dogs and cats, IDPN produces severe ataxia, depression, anorexia, and retropulsion, but not hyperactivity.[32]

D. Neuropathology

The characteristic histologic lesion produced by IDPN is a focal fusiform swelling or ballooning of the proximal axonal segment. Electron microscopy has shown these swellings contain neurofilaments and axonal organelles.[28] Early reports referred to the axonal balloons as "ghost cells" or necrotic neurons, but more recent studies have clearly shown them to be axonal.[28,31] Affected axons generally are from larger neurons, especially spinal motor neurons, neurons of the brain stem reticular formation, red nuclei, medial and lateral vestibular nuclei, dorsal root ganglia, ganglia and nuclei of the cranial nerves, and possibly Purkinje cells.[27,28,31] Axonal ballooning is accompanied by secondary demyelination and attenuation of the axon distal to the swelling.[28,33-35] Gliosis may occur in spinal ventral roots and the root exit zone in chronic intoxications.[33] Degeneration of the distal axon is infrequently reported but may occur more commonly than is recognized.[35]

E. Metabolism

IDPN is absorbed by oral, dermal, inhalation, and parenteral routes. Following a dose of 2 g/kg of IDPN s.c., rats excreted cyanoacetic acid, β-alanine, and β-aminopropionitrile in their urine.[36] I.p. injections of ethionine (240 mg/kg i.p. for 5 days) prevented signs of IDPN toxicity but did not inhibit the formation of urinary metabolites.[36] Thyroxine also inhibits IDPN neurotoxicity.[37]

Thuillier and Nakajima, as reported by Chou and Hartmann,[28] found that replacement of the amine in IDPN by an oxygen or sulfur resulted in the absence of IDPN-like signs.

F. Mechanism of Action

IDPN was originally studied because of its structural similarity to β-aminopropionitrile (β-APN). Like β-APN, IDPN produces lesions of osteolathyrism, possibly due to its metabolism to β-APN.[27,36] It seems likely that IDPN undergoes biological interactions in the nervous system which are different from those which occur with collagen since β-APN, a more potent osteolathyrogenic agent, does not produce neurotoxicity like IDPN, and IDPN can produce neurotoxicity without osteolathyrism.[27]

Within the axon, IDPN specifically interferes with slow axonal transport of neurofilaments. This results in their proximal accumulation within the axon.[38] Fast transport in both proximal and distal directions are not affected by IDPN.[38] Griffin and Price have recently reviewed this area of neurotoxicity.[33]

G. Other Effects

IDPN produces osteolathyrism in the rat but is less active than β-aminopropionitrile.[27] Lesions produced by IDPN and associated with osteolathyrism include fetotoxicity, poor body weight gain, and abnormal bone growth.[27,36] Blindness due to ocular hemorrhage, lens anomalies, and reduction in the size of the thymus also occur.[40-41]

H. Other Chemicals Producing Similar Effects

Chemicals which produce IDPN-like signs in laboratory animals include methyl, ethyl, or propyl substituted chloroethylamines, sodium acetylarsanilate, 3,3-dimethyl-1-phenyltriazene, dimethylamino-hexose-reductone, piperidino-hexose-reductone, and lysergic acid diethyl-amide.[28] Each compound producing this syndrome contains a dialkyl or heterocyclic amine.[28]

V. ACRYLONITRILE

A. Synonyms 2-Propenenitrile
Vinyl cyanide
Cyanoethylene

B. Introduction

Acrylonitrile ($H_2C=CH-C\equiv N$) is a very high volume (10^{10} lb/year) chemical intermediate used in the synthesis of acrylic and modacrylic fibers, resins with styrene and butadiene, synthetic rubber, and other chemicals. It is also used as a pest fumigant. It is a clear, colorless to light yellow, volatile liquid.

C. Human Neurotoxicity

The toxicity, properties, and uses of acrylonitrile have been reviewed by Miller and Villaume.[42] Acute human poisonings have been associated with house fumigation and accidental exposure in chemical laboratories and manufacturing plants. Common symptoms include ocular and nasal irritation, nausea, vomiting, headache, difficulty breathing, visual disturbances, tremors, incoordination, and convulsions.[42-44] Recovery can be rapid and complete.[44] Lethal exposures are associated with incomplete ventilation following house fumigation.[42]

Chronic human exposure in the workplace has generally not been associated with neurotoxicity. Sakurai et al.[45] studied 102 workers exposed to acrylonitrile for 5 years. The highest concentration was 4.2 ppm. No health effects were attributed to the exposures. Agayeva, as quoted by Miller and Villaume,[42] reported decreases in "epinephrine-like substances" and elevated epinephrine levels in workers chronically exposed to acrylonitrile. The author attributed these effects and changes in cardiovascular parameters to central nervous system effects due to acrylonitrile.

O'Berg[46] reported an excess number of respiratory cancers in a population of 1345 male chemical workers exposed to acrylonitrile. Ott et al.[47] found no increase in the cancer rate among a group of 100 men exposed to >1 ppm acrylonitrile but exposures were <20 years. Brain tumors in the working population due to acrylonitrile have not been reported, but the latency period covered in these studies is close to the expected minimum time necessary for induction of neoplasia in man.

D. Experimental Neurotoxicity

Acute toxicity in laboratory animals and man appears similar. Hyperexcitability, salivation, vomiting, cyanosis, difficulty breathing, incoordination, transitory paralysis, and dilation and paralysis of the pupil occur at lethal or near lethal doses.[42,43,48-50]

Svirbely and Floyd, as quoted by Quast et al.,[51] reported that female rats receiving 500 ppm of acrylonitrile in their drinking water unexpectedly developed hindlimb weakness approximately 16 to 19 weeks after weaning a second litter.

Chronic exposure to acrylonitrile has resulted in central nervous system tumors in laboratory rats. Maltoni et al.[52] found three gliomas in the brains of rats inhaling 20 or 40 ppm of acrylonitrile 4 hr/day, 5 days/week for 52 weeks. The low number of tumors in the acrylonitrile rats and the high incidence of gliomas in control rats from a companion study prevented a conclusion as to the carcinogenicity of acrylonitrile.

Quast et al.[53] reported the occurrence of astrocytomas in male and female rats inhaling acrylonitrile at 20 or 80 ppm or drinking water containing 35, 100, and 300 ppm. Astrocytoma incidence was statistically significant at all levels except 20 ppm by inhalation. The tumors were clinically silent except that some animals ingesting acrylonitrile became paralyzed.

The earliest tumors were found in rats examined between 7 and 12 months. In addition to astrocytomas, perivascular inflammatory cell cuffing, focal gliosis, and focal and multifocal glial cell proliferation suggestive of early tumor formation were observed. Other tumors in the central nervous system were not statistically increased.

Teratogenicity has been reported in rats ingesting 25 mg/kg/day or inhaling 80 ppm of acrylonitrile.[50] Malformations included missing vertebrae and short trunks and tails presumably due to missing vertebrae. Cerebral and spinal cord abnormalities were not found.

Willhite et al.[54] induced exencephaly, encephalocoeles, and rib abnormalities in the progeny of hamsters given 80 to 120 mg/kg of acrylonitrile by i.p. injection. Propionitrile, the saturated analogue of acrylonitrile, produced similar effects.

E. Neuropathology

Brieger et al.[49] described the lesions resulting from exposure of rats, monkeys, and dogs to 50, 75, or 100 ppm of acrylonitrile as anoxic changes. Neurons, particularly in the cerebral cortex, were shrunken and hyperchromatic.

Cerebral and spinal inflammatory vascular changes, and perivascular cuffing occur in chronic acrylonitrile exposure and are accompanied by gliosis with foci of transformed or altered cells preceding the onset of tumor, or astrocytoma formation.[51-53]

F. Metabolism

Acrylonitrile is absorbed by oral, respiratory, and dermal routes. Liquid or, more commonly, vapor exposures produce significant irritation which may enhance absorption.

While there are differing opinions on the importance of the cyanide ion to acrylonitrile toxicity, cyanide may be found in the blood following near lethal doses of acrylonitrile and increased thiocyanate, a metabolite of cyanide, is commonly found in the blood and urine of test animals and man. Similarities between the clinical course of cyanide and acute acrylonitrile poisoning, and improvement from acrylonitrile intoxication following treatment for cyanide suggest that cyanide may play a role in acute acrylonitrile intoxication.

The unsaturated bond and the close proximity between the unsaturated bond and a reactive group (CN) must be important to the toxicity of acrylonitrile. Differing toxicity between propionitrile, the saturated analogue of acrylonitrile, and acrylonitrile and the induction of neoplasms by both acrylonitrile and vinyl chloride support this opinion.

Biotransformation of acrylonitrile is affected by the species involved, the route of administration, the activity of microsomal enzymes, and the presence of inhibitors.[55,56] Acrylonitrile is strongly bound in the blood. Binding and cyanoethylation reactions strongly influence subsequent metabolism.[55] Tissue binding is also very strong. Protein and nonprotein sulfhydryls are reduced to 15% and 50% of control values in the liver and brain following a dose of 100 mg/kg acrylonitrile.[57] Pyruvate and lactate levels increase in the brain following sulfhydryl binding.[57]

Both cysteine and phenobarbital are protective against acute acrylonitrile toxicity.[57,58] Other thiols including glutathione and especially diethyl dithiocarbamate inhibit acrylonitrile tissue binding in vitro although they do not change thiocyanate excretion.[55,59] Aroclor 1254 pretreatment enhances the production of cyanide levels from acrylonitrile threefold and enhances central nervous system dysfunction.[56]

Major urinary excretory products of acrylonitrile, in addition to thiocyanate, are conjugates with glutathione either directly or through the formation of an epoxide.[60,61]

G. Other Effects

Acrylonitrile may cause severe burns and respiratory irritation. Hepatic dysfunction, anemia, and possibly renal dysfunction have been reported.[42,44,45]

VI. 2,4-HEXADIENE-1-NITRILE

A. Synonyms   Sorbonitrile

B. Neurotoxicity

This liquid ($CH_3-CH=CH-CH=CH-C\equiv N$) has very limited use as an intermediate in synthetic chemistry. No human or animal toxicity data were available in the literature for this material. During an acute screening study, dramatic signs of central nervous system toxicity occurred.[62]

Groups of four male and female rats were given 400, 800, 1600, or 3200 mg/kg of 2,4-hexadiene-1-nitrile by gavage. Beginning 1 hr after dosing, generalized weakness, rough hair coats, and ptosis were observed. One female in the 1600 mg/kg group survived, but all others receiving 1600 or 3200 mg/kg died within 24 to 48 hr.

By the second day, rats given 800 mg/kg developed gross ataxia, tremors, and difficulty posturing. Both males and females eventually were unable to stand or posture unless allowed to grasp a perforated surface. On a smooth surface, rats rocked from side to side alternately hyperextending the upper most limbs. The rats were easily aroused by any sensory input, i.e., sound, touch, movement. Visual and tactile placing, flexor, and righting reflexes were intact but overriden during hyperextension. Bowel and bladder function were normal.

Rats in the 400 mg/kg group and those 800 mg/kg rats which did not show obvious signs of neurotoxicity initially, subsequently (11 days after dosing) developed a very exaggerated hindlimb gait. They moved their hindlimbs in a high stepping, overextended, prancing gait. Hindlimb strength was reduced and falls were observed.

A second group of 15 female rats were given 800 mg/kg of 2,4-hexadiene-1-nitrile and killed serially. Twenty-four hours later, a portion of the group showed increased respiratory rate and depth, central nervous system depression, bright red blood, vasodilatation, and death. Clinically, the rats appeared to be undergoing cyanide poisoning. Rats with milder clinical signs survived and developed neurological abnormalities as originally observed.

C. Neuropathology

Morphological changes were strictly confined to the inferior olivary nuclear complex (ION) of the brainstem. All areas of the ION were involved, but some sparing of the medial accessory nucleus was observed. Extent of ION damage correlated well with the clinical signs. Severely affected animals showed complete loss of ION neurons. The pattern of degeneration was bilaterally symmetrical with no vascular association.

The earliest lesions showed shrunken, hyperchromatic neurons. Later lesions showed clumping or dispersion of Nissel and nuclear chromatin and finally cytolysis. Damage was confined to individual neurons, and normal neurons could be found next to completely degenerate ones. In Bodian preparations, small axons near degenerating neurons occasionally were fragmented or had knobby surfaces, but otherwise, the remaining central nervous system appeared unaffected. Secondary damage to the cerebellum, which receives afferent fibers from the ION, was not apparent. Histochemical stains for cytochrome oxidase, which is inhibited by cyanide, did not show a reduction except in ION neurons that were obviously degenerate. Therefore, it appears unlikely that neuronal degeneration is linked to cytochrome oxidase inhibition due to cyanide.

The high specificity of 2,4-hexadiene-1-nitrile for the ION is not unique since 3-acetylpyridine also shows a similar affinity in the rat and monkey.[63]

D. Other Effects

Within 24 hr of a single dose of 800 mg/kg of 2,4-hexadiene-1-nitrile, severe necrosis of thymic cortical lymphocyte occurs and is followed by atrophy of all regions of the thymus.[62] 400 mg/kg produces less severe damage.

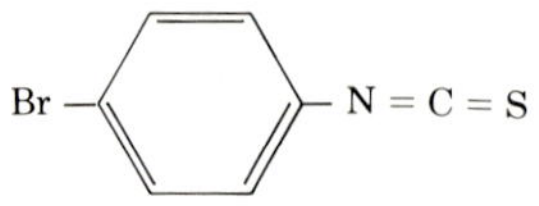

FIGURE 3. *p*-Bromophenylisothiocyanate.

VII. *p*-BROMOPHENYLISOTHIOCYANATE

A. Synonyms
4-Bromophenylisothiocyanate
1-Bromo-4-isothiocyanatobenzene
Bromobenzene-4-iso thiocyanatobenzene
Bromobenzene-4-isothiocyanate
p-bromophenyl ester of isothiocyanic acid

B. Introduction

p-Bromophenyl isothiocyanate has been used to treat human dermatomycoses and has been studied for use as a veterinary anthelmintic. No industrial uses have been reported for this compound but related materials, *p*-bromophenylisocyanate and phenylisothiocyanate, are used as intermediates in chemical syntheses.

C. Neurotoxicity

Lessel and Towlerton[64] briefly reported that single oral doses of 150 or 200 mg/kg of *p*-bromophenylisothiocyanate produced ataxia and paraplegia in 8-week-old unweaned lambs. A 2-week delay was present between the dose and the onset of ataxia. Six-month-old weaned lambs, rabbits, cats, and dogs receiving 200 mg/kg and rats and hens receiving 400 mg/kg did not show neurotoxicity. The lack of neurotoxicity in weaned lambs is thought to be due to dilution in the rumen or detoxification of *p*-bromophenylisothiocyanate by rumen organisms.[64,65]

Light microscopic examination revealed swollen, degenerating axons, degeneration of myelin sheaths, and degeneration and lysis of neurons. The lumbar spinal cord was most severely affected.

VIII. TOLUENE DIISOCYANATE

Neurological symptoms were reported in 23 firemen heavily exposed to toluene diisocyanate during a polyurethane manufacturing plant fire.[66] Symptoms included euphoria, ataxia, and loss of consciousness with long lasting mild subjective symptoms of personality change, irritability, depression, and memory difficulties.

REFERENCES

1. **Brierley, J. B.,** Cerebral Hypoxia, in *Greenfield's Neuropathology,* 3rd ed., Blackwood, W. and Corsellis, J. A. N., Eds., Year Book Medical Publishers, Chicago, 1976, 43.
2. **Kim, Y. H., Foo, M., and Terry, R. D.,** Cyanide encephalopathy following therapy with sodium nitroprusside: report of a case, *Arch. Pathol.,* 106, 392, 1982.
3. **Montgomery, R. D.,** Cyanogenetic Glucosides, in *Handbook of Clinical Neurology: Intoxications of the Nervous System,* Vol. 36, Pt. 1, Vinken, P. J. and Bruyn, G. W., Eds., Elsevier/North Holland, Amsterdam, 1979, 515.
4. **Schmidt, E. S., Newton, G. W., Sanders, S. M., Lewis, J. P., and Conn, E. E.,** Laetrile toxicity studies in dogs, *JAMA,* 239, 943, 1978.
5. **Ohnishi, A., Peterson, C. M., and Dyck, P. J.,** Axonal degeneration in sodium cyanate-induced neuropathy, *Arch. Neurol.,* 33, 530, 1975.
6. **Tellez-Nagel, I., Korthals, J. K., Vlassara, H. V., and Cerami, A.,** An ultrastructural study of chronic sodium cyanate-induced neuropathy, *J. Neuropathol. Exp. Neurol.,* 36, 352, 1977.
7. **Doherty, P. A., Ferm, V. H., and Smith, R. P.,** Congenital malformations induced by infusion of sodium cyanide in the golden hamster, *Toxicol. Appl. Pharmacol.,* 64, 456, 1982.
8. **Ahmed, A. E. and Farooyui, Y. H.,** Comparative toxicities of aliphatic nitriles, *Toxicol. Lett.,* 12, 157, 1982.
9. National Institute for Occupational Safety and Health, Occupational Exposure to Nitriles, Department of Health, Education, and Welfare, Publ. No. 78-212, Washington, D.C., 1978.
10. **Hyden, H. and Hartelius, H.,** Stimulation of the nucleoprotein-production in the nerve cells by malononitrile and its effects on psychic functions in mental disorders, *Acta Psychiatr. Neurol Suppl.,* 48, 1, 1948.
11. **Hicks, S. P.,** Brian metabolism *in vivo.* II. The distribution of lesions caused by azide, malononitrile, plasmocid and nitrophenol poisoning in rats, *Arch. Pathol.,* 50, 545, 1950.
12. **Eberts, F. S., Jr., Slomp, G., and Johnson, J. L.,** 1,1,3-Tricyano-2-amino-1-propene (U-9189), a biologically active component of aqueous solutions of malononitrile, *Arch. Biochem. Biophys.,* 95, 305, 1961.
13. **Dhindsa, K. S. and Enesco, H. E.,** Radioautographic study of the effect of malononitrile dimer on RNA synthesis in mice, *Acta Anat.,* 100, 44, 1978.
14. Current Intelligence Bulletin No. 26, NIAX catalyst ESN. A Mixture of Dimethylaminopropionitrile and bis[2-(dimethylamino)-ethyl] ether, Department of Health, Education, and Welfare (NIOSH), Publ. No. 78-157, 1978, 1.
15. **Bachhuber, T. E., Lalich, J. J., Angevine, D. M., Schilling, E. D., and Strong, F. M.,** Lathyrus factor activity of beta-aminopropionitrile and related compounds, *Proc. Soc. Exp. Biol. Med.,* 89, 294, 1955.
16. **Stamler, F. W.,** Reproduction in rats fed *Lathyrus* peas or aminonitriles, *Proc. Soc. Exp. Biol. Med.,* 90, 294, 1955.
17. **Levene, C. I.,** Structural requirements for lathyrogenic agents, *J. Exp. Med.,* 114, 295, 1961.
18. **Keogh, J. P., Pestronk, A., Werthheimer, D., and Moreland, R.,** An epidemic of urinary retention caused by dimethylaminopropionitrile, *JAMA,* 243, 746, 1980.
19. **Kreiss, K., Wegman, D. H., Niles, C. A., Siroky, M. B., Krane, R. J., and Feldman, R. G.,** Neurological dysfunction of the bladder in workers exposed to dimethylaminopropionitrile, *JAMA,* 243, 741, 1980.
20. Product Information on NIAX-catalyst ESN, Union Carbide Corporation, New York, 1978.
21. **White, G. L.,** Health Hazard Evaluation Determination Report No. HE 78-68-546, Lear Siegler, Inc., Marblehead, Mass., National Institute for Occupational Safety and Health, Cincinnati, Health Hazard and Technical Assistance Branch, 1978.
22. **Baker, E. L., Christiani, D. C., Wegman, D. H., Siroky, M., Niles, C. A., and Feldman, R. G.,** Follow-up studies of workers with bladder neuropathy caused by exposure to dimethylaminopropionitrile, *Scand. J. Work Environ. Health,* Suppl. 4, 54, 1981.
23. **Gad, S. C., McKelvey, J. A., and Turney, R. A.,** NIAX catalyst ESN: subchronic neuropharmacology and neurotoxicology, *Drug Chem. Toxicol.,* 2, 223, 1979.
24. **Pestronk, A., Keogh, J. P., and Griffin, J. W.,** Dimethylaminopropionitrile intoxication: a new industrial neuropathy, *Neurology,* 29, 540, 1979.
25. **Pestronk, A., Keogh, J. P., and Griffin, J. W.,** Dimethylaminopropionitrile, in *Experimental and Clinical Neurotoxicology,* Spencer, P. S. and Schaumburg, H. H., Eds., Williams & Wilkins, Baltimore, 1980, 422.
26. **Jaeger, R. J., Plugge, H., and Szabo, S.,** Acute urinary bladder toxicity of a polyurethane foam catalyst mixture: a possible new target organ for a propionitrile derivative, *J. Environ. Pathol. Toxicol.,* 4, 555, 1980.

27. **Bachhuber, T. E., Lalich, J. J., Angevine, D. M., Schilling, E. D., and Strong, F. M.,** Lathyrus factor activity of Beta-aminopropionitrile and related compounds, *Proc. Soc. Exp. Biol. Med.*, 89, 294, 1955.
28. **Chou, S.-M. and Hartmann, H. A.,** Axonal lesions and waltzing syndrome after IDPN administration in rats, *Acta Neuropathol.*, 3, 428, 1964.
29. **Schott, K., Schneider, G., and Oepen, H.,** A pharmacological and behavioral study on β,β'-iminodipropionitrile (IDPN)-treated brine shrimp, *Artemia salina L., Toxicol. Lett.*, 6, 231, 1980.
30. **Delay, J., Pichot, P., Thuillier, J., and Marquiset, J. P.,** Action de l'aminodipropionitrile sur le comportement moteur de la souris blanche, *C. R. Soc. Biol.*, 146, 533, 1952.
31. **Hartmann, H. A., Lalich, J. J., and Akert, K.,** Lesions in the anterior motor horn cells of rats after administration of bis-β-cyanoethylamine, *J. Neuropathol. Exp. Neurol.*, 17, 298, 1958.
32. **Tews, J. K., Kopf, G. M., and Stone, W. E.,** Effects of β,β'-iminodipropionitrile on cerebral constituents and on cortical electric activity, *Int. J. Neuropharmacol.*, 7, 29, 1968.
33. **Griffin, J. W. and Price, D. L.,** Proximal axonopathies induced by toxic chemicals, in *Experimental and Clinical Neurotoxicology*, Spencer, P. S. and Schaumburg, H. H., Eds., Williams & Wilkins, Baltimore, 1980, 161.
34. **Fukuhara, N. and Chou, S.-M.,** IDPN-induced acute neuraxonal dystrophy, with special reference to ultrastructural changes of the neuromuscular junction, *Clin. Neurol.*, 15, 500, 1975.
35. **Griffin, J. W., Gold, B. G., Cork, L. C., Price, D. L., and Lowndes, H. E.,** IDPN neuropathy in the cat: coexistence of proximal and distal axonal swellings, *Neuropathol. Appl. Neurobiol.*, 8, 351, 1982.
36. **Williams, S., Brownlow, E. K., and Heath, H.,** Metabolism of β,β'-iminodipropionitrile in the rat, *Biochem. Pharmacol.*, 19, 2277, 1970.
37. **Vivanco, F., Ramos, F., and Jimenez-Diaz, C.,** Determination of γ-aminobutyric acid and other free amino acids in whole brains of rats poisoned with β,β'-iminodipropionitrile and α,γ-diaminobutyric acid with, or without, administration of thyroxine, *J. Neurochem.*, 13, 1461, 1966.
38. **Griffin, J. W., Hoffman, P. N., Clark, A. W., Carroll, P. T., and Price, D. L.,** Slow axonal transport of neurofilament proteins: impairment by β,β'-iminodipropionitrile administration, *Science*, 202, 633, 1978.
39. **Stamler, F. W.,** Reproduction in rats fed *Lathyrus* pea or aminonitriles, *Proc. Soc. Exp. Biol. Med.*, 90, 294, 1955.
40. **Sackler, A. M., Weltman, A. S., and Schwartz, R.,** Behavioral and endocrine effects of aminodipropionitrile (ADPN) in male mice, *Pharmacol., Biochem. Behav.*, 1, 191, 1973.
41. **Wolff, G., Kunze, E., Rodden, A., and Oepen, H.,** Loss of auditory startle-reflex in the iminodipropionitrile (IDPN) treated rat, *Life Sci.*, 20, 1163, 1977.
42. **Miller, L. M. and Villaume, J. E.,** Investigation of Selected Potential Environmental Contaminants: Acrylonitrile, Final Report to the U.S. Environmental Protection Agency, Washington, D.C., 1978.
43. **Wilson, R. H., Hough, G. V., and McCormick, W. E.,** Medical problems encountered in the manufacture of American-made rubber, *Ind. Med.*, 17, 199, 1948.
44. **Sartorelli, E.,** Acute poisoning due to acrylonitrile, *Med. Lav.*, 57, 184, 1966.
45. **Sakurai, H., Onodera, M., Utsunomiya, T., Minakuchi, H., Iwai, H., and Matsumura, H.,** Health effects of acrylonitrile in acrylic fibre factories, *Br. J. Ind. Med.*, 35, 219, 1978.
46. **O'Berg, M. T.,** Epidemiologic study of workers exposed to acrylonitrile, *J. Occup. Med.*, 22, 245, 1980.
47. **Ott, M. G., Kolesar, R. C., Scharnweber, H. C., Schneider, E. J., and Venable, J. R.,** A mortality survey of employees engaged in the development or manufacture of styrene-based products, *J. Occup. Med.*, 22, 445, 1980.
48. **Dudley, H. E., Sweeney, T. R., and Miller, J. W.,** Toxicology of acrylonitrile (vinyl cyanide). II. Studies of effects of daily inhalation, *J. Ind. Hyg. Toxicol.*, 24, 255, 1942.
49. **Brieger, H., Rieders, F., and Hodes, W. A.,** Acrylonitrile: spectro-photometric determination, acute toxicity, and mechanism of action, *Arch. Ind. Hyg. Occup. Med.*, 6, 128, 1952.
50. **Murray, F. J., Schwetz, B. A., Nitschke, K. D., John, J. A., Norris, J. M., and Gehring, P. J.,** Teratogenicity of acrylonitrile given to rats by gavage or by inhalation, *Food Cosmet. Toxicol.*, 16, 547, 1978.
51. **Quast, J. F., Wade, C. E., Humiston, C. G., Carreon, R. M., Hermann, E. A., Park, C. N., and Schwetz, B. A.,** A two-year Toxicity and Oncogenicity Study with Acrylonitrile Incorporated in the Drinking Water of Rats, Final Report to the Chemical Manufacturers Association, Washington, D.C., 1980.
52. **Maltoni, C., Ciliberti, A., and DiMaio, V.,** Carcinogenicity bioassays on rats of acrylonitrile administered by inhalation and by ingestion, *Med. Lav.*, 68, 401, 1977.
53. **Quast, J. F., Schwetz, D. J., Balmer, M. F., Gushow, T. S., Park, C. N., and McKenna, M. J.,** A Two-year Toxicity and Oncogenicity Study with Acrylonitrile Following Inhalation Exposure of Rats, Final Report to the Chemical Manufacturers Association, Washington, D.C., 1980.
54. **Willhite, C. C., Ferm, V. H., and Smith, R. P.,** Teratogenic effects of aliphatic nitriles, *Teratology*, 23, 317, 1981.

55. **Gut, I., Nerudorá, J., Kopecký, J., and Holećek, V.**, Acrylonitrile biotransformation in rats, mice, and Chinese hamsters as influenced by the route of administration and by phenobarbital, SKF 525-A, cysteine, dimercaprol, or thiosulfate, *Arch. Toxicol.*, 33, 151, 1975.
56. **Ahmed, A. E. and Patel, K.**, Acrylonitrile: *in vivo* metabolism in rats and mice, *Drug. Metab. Disp.*, 9, 219, 1981.
57. **Hashimoto, K. and Kawai, R.**, Effect of acrylonitrile on sulfhydryls and pyruvate metabolism in tissues, *Biochem. Pharmacol.*, 21, 635, 1972.
58. **Bondarev, G. I., Stassenkova, K. P., Vissarionova, V. Y., and Zemlyanskaya, T. A.**, Experimental study of the dependence of acrylic acid nitrile toxicity on feeding, *Gig. Tr. Prof. Zabol.*, 8, 63, 1981.
59. **Peter, H. and Bolt, H. M.**, Irreversible protein binding of acrylonitrile, *Xenobiotica*, 11, 51, 1981.
60. **Langvardt, P. W., Putzig, C. L., Braun, W. H., and Young, J. D.**, Identification of the major urinary metabolites of acrylonitrile in the rat, *J. Toxicol. Environ. Health*, 6, 273, 1980.
61. **van Bladeren, P. J., Delbressine, L. P., Hoogeterp, J. J., Beaumont, A. H., Breimer, D. D., Seutter-Berlage, F., and van der Gen, A.**, Formation of mercapturic acids from acrylonitrile, crotononitrile, and cinnamonitrile by direct conjugation and via an intermediate oxidation process, *Drug Metab. Disp.*, 9, 246, 1981.
62. Eastman Kodak Company, unpublished data, Rochester, N.Y., 1983.
63. **Anderson, W. A. and Flumerfelt, B. A.**, A light and electron microscopic study of the effects of 3-acetylpyridine intoxication on the inferior olivary complex and cerebellar cortex, *J. Comp. Neurol.*, 190, 157, 1980.
64. **Lessel, B. and Towlerton, R. G.**, Neurotoxicity of *p*-bromophenyl isothiocyanate, *Food Cosmet. Toxicol.*, 5, 741, 1967.
65. **Cavanagh, J. B.**, Toxic substances and the nervous system, *Br. Med. Bull.*, 25, 268, 1969.
66. **LeQuesne, P. M., Axford, A. T., McKerrow, C. B., and Jones, A. P.**, Neurological complications after a single severe exposure to toluene di-isocyanate, *Br. J. Ind. Med.*, 33, 72, 1976.

Chapter 3

CARBON DISULFIDE AND ORGANIC SULFUR-CONTAINING COMPOUNDS

John L. O'Donoghue

TABLE OF CONTENTS

I. Carbon Disulfide 40
    A. Synonyms 40
    B. Introduction 40
    C. Human Neurotoxicity 40
    D. Experimental Neurotoxicity 41
    E. Neuropathology 42
    F. Toxicokinetics 42
    G. Mechanisms of Action 43
    H. Effects on Other Organs 44

II. 2,4-Dithiobiuret 44
    A. Introduction 44
    B. Neurotoxicity 44

III. Disulfiram 45
    A. Synonyms 45
    B. Introduction 45
    C. Human Neurotoxicity 45
    D. Experimental Neurotoxicity and Neuropathology 46
    E. Metabolism 46
    F. Mechanism of Action 46
    G. Other Effects 47

IV. Tetramethylthiuramdisulfide 47
    A. Synonyms 47
    B. Introduction 47
    C. Neurotoxicity 47
    D. Neuropathology 47
    E. Mechanism of Action 48
    F. Other Effects 48

V. Other Dithiocarbamate Compounds 48

VI. Thiophene 48
    A. Synonyms 48
    B. Introduction 48
    C. Human Neurotoxicity 48
    D. Experimental Neurotoxicity 50
    E. Neuropathology 50
    F. Metabolism 50
    G. Other Effects 51

VII. Pyridinethione Derivatives 51
       A.   Introduction 51
       B.   Neurotoxicity 51
       C.   Neuropathology 52
       D.   Metabolism 52
       E.   Mechanism of Action 54
       F.   Other Effects of Pyridinethiones 54

References 54

I. CARBON DISULFIDE

A. Synonyms Carbon bisulfide
Dithiocarbonic anhydride

B. Introduction

Carbon disulfide (CS_2) in its pure form is a clear, colorless, volatile, flammable liquid with an ethereal odor, but common technical grades have an unpleasant odor due to the presence of other sulfur-containing materials. It has been an important chemical since about 1880 when it was first produced in commercial amounts.[1] Recent estimated world production is over one million metric tons.[1]

The main uses of CS_2 are in the production of regenerated cellulose fibers (viscose rayon) and films (cellophane) and as a raw material in the production of carbon tetrachloride. In the U.S. approximately a third of the CS_2 is used for rayon, a third for carbon tetrachloride, and 13% for cellophane.[1] Other uses include grain and soil fumigation, parasiticide for horses, polymerization inhibitor for vinyl chloride, an ingredient in organic solvents, adhesives, electroplating solutions, corrosion inhibitors, pesticides and resins, and in the production of rubber accelerators.

Simpson, who had introduced the use of chloroform for surgery, also studied the anesthetic properties of CS_2 for surgery but abandoned it because of its acute toxicity.[2] When the India rubber trade started to expand in the 1850s, the first reports of occupational illness due to CS_2 began appearing in the French medical literature.[3] Until the early 1900s, exposure to carbon disulfide in European industries producing cold vulcanized rubber was widespread resulting in frequent cases of acute psychosis due to carbon disulfide. Laudenheimer, in 1899, is credited with recognizing that CS_2 exposure could be controlled inexpensively thus reducing exposure from hundreds of ppm to 30 ppm.[4] In the U.S., CS_2 intoxication reports were few until the introduction of the viscose rayon industry. Following a study of health problems in Pennsylvania viscose rayon workers in 1938, standards for CS_2 exposure levels were first proposed for the American workplace.[5]

Primary concern for the neurotoxicity of carbon disulfide remains centered on the health of viscose rayon workers particularly workers in spinning operations or workers repairing equipment where exposures to CS_2 and hydrogen sulfide are the highest. It has been estimated that 20,000 American workers are potentially exposed to CS_2.[4]

C. Human Neurotoxicity

The toxicity of CS_2 has been reviewed by several authors.[4-11] Early reports of CS_2 toxicity in the rubber and viscose rayon industries emphasized the psychological and central nervous system damage following exposure to relatively high levels (several hundred ppm) of CS_2. Fortunately, these exposures are rare today and reports of acute CS_2 are very infrequent.[12-15] Major psychiatric symptoms associated with CS_2 encephalopathy include rapid changes in mood including mania and suicidal tendencies, extreme irritability, uncontrollable anger, memory loss, insomnia, bad dreams, and decreased or loss of libido.[8] In some respects, the acute symptomatology is similar to that produced by methyl halides such as methyl bromide (Chapter 5).

Poor hygiene conditions and long work hours led to the recognition of chronic carbon disulfide intoxication, which may produce obvious peripheral neuropathy as well as encephalopathy. Vigliani[16] reported symptoms of headache, vertigo, gastrointestinal disturbances, and signs of polyneuropathy, including decrease or absent tendon reflexes, weakness in the legs, difficulty walking, and leg pain in workers exposed to 144 to 321 ppm with peaks of up to 800 ppm in a World War II era plant.

More recent workplace studies have concentrated on subtle neurological and electrophysiological effects of CS_2 generally between 30 and 50 ppm.[17-27] Several problems exist in trying to associate CS_2 levels with the presence or absence of effects. Frequently, the people exposed have worked for many years with varying CS_2 so that studies done today at lower levels of CS_2 are confounded by previous higher exposures. Usually the actual exposures are not known or are not known accurately. This is especially true when equipment failure or other factors cause temporary increases in CS_2 levels. Reports rarely mention hydrogen sulfide exposure but exposure to both CS_2 and hydrogen sulfide are common and hydrogen sulfide may increase the hazard of exposure to CS_2.[4,28,29]

Of great concern is the continuing presence of encephalopathic abnormalities at reduced exposure levels. Hänninen[18] performed a series of psychological tests on a group of apparently normal men exposed to CS_2 at approximately 20 to 45 ppm for 5 to 20 years. As a group, these men showed poor visual performance, impaired dexterity, difficulty in coordination, and decreased originality and spontaneity. Abnormal electrophysiological abnormalities and reduction in nerve conduction velocities have been reported in studies of subclinical cases of CS_2 neurotoxicity occurring at exposures as low as 20 ppm.[17,20,24] The present suggested threshold limit value of occupational exposure is 10 ppm in the U.S. and in several European countries. This value was recommended in the U.S. to protect against cardiovascular toxicity due to CS_2 but is expected to also protect against neurotoxicity.[32]

D. Experimental Neurotoxicity

Experimental neurotoxicity studies have been conducted to confirm findings made in case studies of man and to explore mechanisms of actions rather than to collect data to assess safe exposure levels. Rats, mice, rabbits, cats, and dogs have been intoxicated with CS_2. Rats appear to be the most resistant and dogs the most sensitive.[33-35]

Lewey[34] exposed a group of nine dogs to 400 ppm of CS_2 8 hr/day, 5 days/week. Beginning between the fifth and eighth weeks the dogs developed limb weakness followed by ataxia, jerking, choreiform movements, loss of position sense, and behavioral abnormalities including aggression, excitability, apprehension, and later somnolence and apathy to surroundings.

Lukáš[36] exposed rats to approximately 190 ppm of CS_2 6 hr/day, 5 days/week for 6 months without clinical or electromyographic (EMG) evidence of neurotoxicity. Concentrations of 385 ppm or greater produced signs of neuropathy but the onset was altered by diet especially copper, zinc, and vitamin B6 supplementation.[36,37] Frantík[38] did not produce motor impairment in rats exposed to 48 ppm of CS_2 5 days/week for 40 weeks but did produce a neuropathy after 18 weeks of exposure at 385 ppm or 8 weeks at 770 ppm. Knoblock et al.[39] found peripheral nerve conduction velocity reduction at levels that did not result in clinical evidence of neuropathy. Rats exposed to approximately 290 ppm of CS_2 5 hr/day, 6 days/week for

6 months had reduced conduction velocity in the sciatic and tibial nerves. Recovery following 12 months of exposure was incomplete 6 months following the last exposure. Clinical and/or electrophysiological evidence of peripheral neuropathy has been reported following repeated CS_2 exposures greater than 500 ppm, usually 750 ppm.[40-43]

E. Neuropathology

Pathologic changes occur in both the central and peripheral nervous systems in chronic CS_2 poisoning. Despite large numbers of cases of CS_2 poisoning, there are relatively few reports on morphologic changes in man. Quensel in 1904 described diffuse neuronal degeneration in the brain and spinal cord and acute vascular damage in human CS_2 poisoning.[44] Later Abe described neuronal and myelin degeneration and gliosis, and Alpers and Lewey described neuronal degeneration, axonal degeneration, and sclerosis of cerebral vasculature in human autopsy material.[44]

Early experimental studies by Lewey[34] reported thickening of cerebral vessels as well as axonal degeneration in dogs severely intoxicated with CS_2, and Ferraro et al.[35] also reported marked vascular damage and neuronal degeneration in cats. Some clinical case reports on CS_2 encephalopathy have emphasized the role of CS_2-related cerebrovascular damage in producing encephalopathy (for reviews see Davidson and Feinleib[8] and Tolonen[10]), but it is clear from experimental studies that CS_2 can directly affect the central and peripheral nervous system resulting in axonal damage which fits the pattern of a central-peripheral distal axonopathy.

In the long spinal tracts including the spinocerebellar tracts and the fasciculus gracilis and the peripheral nerves of both the rabbit[45] and the rat,[41,45-48] the most prominent findings are large or "giant" focal swellings of axons in a "dying-back" or distal pattern.[37,45-49] The swellings contain masses of neurofilaments and a reduced number of neurotubules. Swollen mitochondria and accumulations of vacuoles, tubules, glycogen granules, or amorphous or granular material may be seen.

Juntunen et al.[50] found degeneration and complete destruction of the preterminal portion of the myoneural junction in rats showing a clinical polyneuropathy after CS_2 exposure. The post synaptic membrane appeared unaltered although denervation or neurogenic muscle atrophy has been described.

These more recent studies have not demonstrated significant neuronal, myelin, or vascular damage as did earlier studies, but rather they reveal primarily axonal damage which morphologically is very similar to that produced by n-hexane and methyl n-butyl ketone (see Chapter 4).

F. Toxicokinetics

Although the major route of CS_2 exposure is considered to be by inhalation, CS_2 may also be absorbed through the skin and by the gastrointestinal tract. The contribution to toxicity of skin absorption in the workplace is unknown. Dutkiewicz and Baranowska[51] calculated a human dermal absorption rate of 0.232 to 0.780 mg/cm^2/hr and reported that relatively less CS_2 was exhaled unchanged following dermal absorption as compared to pulmonary absorption. Cohen and colleagues[52] found that rabbits exposed to very high levels of CS_2 (1500 ppm) vapor for 3 hr absorbed CS_2 through the skin. No CS_2 was detected in exhaled air at exposures of 150 ppm. In unusual conditions, CS_2 may be directly toxic to cutaneous nerves.[53]

Because CS_2 is soluble in lipids but relatively insoluble in water, it reaches equilibrium in the body rapidly. An air-blood equilibrium is reached in 1.5 to 2 hr and tissue equilibrium in about 5 hr.[54] In blood, CS_2 is partitioned so that twice as much is present in red blood cells as in plasma.[54]

Elimination of CS_2 is slower than uptake. In 1 hr there is approximately a 25% reduction in blood levels.[55] In 2 hr, blood levels approach zero[55] while complete clearance takes 3 to 8 hr,[50] although traces may be detected for 80 hr.[54] The uptake, distribution, and elimination of CS_2 depend to a large extent on the method of exposure, i.e., pulmonary, dermal, or parenteral[54] and the exposure concentration.[8]

Only a small amount of residual CS_2 is excreted via the lungs, amounting to about 5%.[55] A small amount, about 0.05%, is excreted by the kidneys.[55] Metabolites in the urine include inorganic sulfates,[56] 5-mercaptothiazolidone,[57] and thiourea.[57]

CS_2 is widely distributed throughout the body although, because of its solubility, accumulation is greatest in organs with high lipid content,[58] especially the brain and peripheral nerves.[54]

Within tissues, CS_2 is found in free and bound forms. The ratio of free to bound CS_2 varies with the lipid content of the tissue. Generally, high lipid tissue content corresponds to higher free CS_2 levels.[58] Except for fat and blood, 40 to 90% of tissue CS_2 is found as nonvolatile acid labile metabolites.[58]

CS_2 is highly reactive and once absorbed reacts with nucleophilic compounds such as those containing amino, sulfhydryl, or hydroxyl groups. Important substrates for this reaction include amino acids, steroidal hormones, and catecholamines. Depending on the nucleophilic group, dithiocarbamates, trithiocarbonic acids, or xanthogenic acids are formed.[59] Dithiocarbamates formed by the reaction of CS_2 and amino acids may be excreted in the urine or metabolized back to CS_2 and amino acids, or isothiocyanates, or hydrogen sulfide which are excreted as sulfates.[8]

G. Mechanisms of Action

High reactivity of CS_2 with a variety of biologically important molecules makes it very likely that several mechanisms are involved in its neurotoxicity. A number of mechanisms have been explored but none has been generally accepted.

Dithiocarbamates formed by CS_2 are chelators of copper and zinc. Rabbits show a reduction in spinal cord copper but a slight elevation of zinc on exposure to CS_2 and increasing dietary copper and zinc levels lead to increased spinal copper and zinc levels and protection from the effects of CS_2 on body weight.[60] Addition of dietary copper and zinc also has a protective effect on rats exposed to CS_2.[37,61] Kotas and colleagues[62] showed that copper levels were increased but zinc levels did not change in the peripheral nerves of rats exposed to CS_2. The in vitro inhibition of dopamine-β-hydroxylase by synthesized dithiocarbamate derivatives of amino acids was reversed by increasing copper levels.[63] Equimolar concentrations of copper ions and dithiocarbamate inhibitors were without effect on dopamine-β-hydroxylase activity. This led McKenna and DiStefano[63] to propose that CS_2 might react with endogenous amino acids in catecholamine storage sites to form dithiocarbamate derivatives which bind copper thus inactivating dopamine-β-hydroxylase which is copper dependent. Other important enzyme systems such as cytochrome oxidases,[64,65] tyrosinase, monoamine oxidases, carbonic anhydrase, and alcohol dehydrogenase,[66] which require copper or zinc as cofactors, may also be affected by similar mechanisms.

CS_2 also reacts with vitamin B6, pyridoxamine, forming pyridoxamine dithiocarbamic acid, which acts as an inhibitor of pyridoxal requiring enzymes including transaminases and amino oxidases.[59] It has been suggested that vitamin B6 deficiency may be partially responsible for CS_2 neurotoxicity. However, supplementation of animal diets with vitamin B6 increases the exposure time needed for CS_2-induced neurotoxicity but does not prevent neurotoxicity.[36,37,67]

Deficiencies of thiamine secondary to hepatotoxicity have also been proposed as possible mechanisms of CS_2 action.[34] Thiamine supplementation delayed the onset of neurotoxicity in CS_2-exposed dogs and decreased the severity of neuropathological lesions.[34]

Inhibition of glycolysis in brain slices was investigated by DeMeio and Brieger.[68] They found that oxygen consumption was decreased in vitro but not in vivo. Other in vitro work showed that cytochrome oxidase and respiration were inhibited in rat brain by CS_2.[8] In vitro inhibition of α-glycerophosphate dehydrogenase (GPDH), both crystalline and enzyme isolates from brain and peripheral nerve, by CS_2 has been postulated to interfere with neural energy production.[69] Inactivation of GPDH, a major enzyme involved in axonal transport, by CS_2 and 2,5-hexanedione may be the reason for the similarities in their peripheral nervous system lesions.[69]

The work of Tarkowski and colleagues[70] points out the importance of dose level in determining the mechanism of action of CS_2. When rats were exposed to CS_2 at approximately 770 ppm continuously for 12 hr, brain adenosine triphosphate (ATP) increased and adenosine diphosphate (ADP) and monophosphate (AMP) levels decreased while mitochondria were morphologically altered suggesting that CS_2 uncoupled mitochondrial oxidative phosphorylation. When the continuous exposure was at lower levels, approximately 257 ppm for 10 months, there were no effects on ATP, ADP, AMP, or mitochondria. The authors suggested that at very high levels CS_2 has a nonspecific effect similar to anesthetic agents.

High binding of CS_2 in axons has led to the suggestion that neurofilaments may directly be damaged by CS_2.[71,72]

H. Effects on Other Organs

CS_2 is a metabolic inhibitor of many enzyme processes which result in specific and nonspecific effects on many organs. The most studied target site after the nervous system is the cardiovascular system, particularly the heart[8] and arteries.[8,10,73,74] In both experimental animals[33,75,76] and man,[77,78] abnormalities in the testes and sperm have been described. Effects on the female reproductive tract[4,6,79] and the developing fetus[80,81] are more controversial. Reports of liver, kidney, bone marrow,[8] endocrine,[8,82] and gastrointestinal effects[6] are less frequent.

II. 2,4-DITHIOBIURET

A. Introduction

2,4-Dithiobiuret (thioimidodicarbonic diamide, [$H_2NCNHCNH_2$] with S=C and C=S groups) is a stable, colorless, crystalline solid used as a rubber accelerator, plasticizer, and an intermediate in the synthesis of resins, insecticides, and rodenticides. It can be used to delay flower wilting and has some activity as a promoter of plant root growth.[83]

B. Neurotoxicity

Astwood et al.[84] discovered that 2,4-dithiobiuret was a cumulative neurotoxin while investigating the antithyroid activity of analogues of thiourea. Single doses of 20 to 50 mg were lethal to rats. Large single doses did not produce paralysis while five daily doses of <0.5 mg by s.c. injection or in the drinking water produced a fatal progressive paralysis. At a concentration of 0.002% in the drinking water, rats appeared normal for 2 to 4 days then developed hindlimb weakness which ascended until only the head, neck, and respiratory muscles were not paralyzed. If continued, animals died of respiratory paralysis but if the water concentration was reduced to 0.001 to 0.002%, the animals remained paralyzed for weeks. Even with muscle atrophy, the paralysis was reversible when 2,4-dithiobiuret exposure was stopped.

No pathological changes were identified in the brain, spinal cord, dorsal root ganglia, or peripheral nerves. Brain cholinesterase levels were unaffected. Administration of pilocarpine, prostigmine, atropine, epinephrine, ephedrin, vitamin A and B complexes, and biotin were without effect.[84]

Seifter et al.[85] injected rabbits with 7 to 100 mg/kg and gave rats water containing 0.002% 2,4-dithiobiuret producing a flaccid paralysis after 5 to 8 doses (rats) and 1 to 12 days (rabbits). Skeletal muscle Cl, Na, K, P, and creatinine, and plasma CO_2, Na, Ca, and K were unaffected.[85] Rabbits had elevated blood glucose, difficulty urinating, and on autopsy had enlarged urinary bladders.

Peterson and colleagues injected rats i.p. with 1 mg/kg/day until rotarod performance failed (5 to 7 days).[86,87] Phenobarbital delayed the onset of rotarod failure but did not alter the acute toxicity of 2,4-dithiobiuret for mice.

Atchison and Peterson[87] gave daily injections of 1 mg/kg of dithiobiuret. Cumulative doses of 4.9, 7.1, and 8.0 mg/kg, respectively, impaired rotarod performance, righting reflexes, or resulted in death of male rats. Female rats and older animals were more sensitive to dithiobiuret.[87] The no-effect level was 0.125 mg/kg daily for 52 days.[88]

Concurrent exposure to diethyldithiocarbamate, d-penicillamine, or disulfiram delayed onset of neurotoxicity from 5 to 6 days to 21 to 26 days.[87] The paralysis induced by dithiobiuret was reversed after 5 days treatment with disulfiram.[89]

Fragmentation and swelling of axons were found microscopically in the recovery phase but not at the onset of motor failure.[86]

The available literature suggests that 2,4-dithiobiuret is a cumulative neurotoxin that may produce axonal impairment. A defect in nerve terminal release of acetylcholine in response to nerve stimulation may be a primary or secondary effect of 2,4-dithiobiuret.[90]

III. DISULFIRAM

A. Synonyms Tetraethylthioperoxydicarbonic diamide
Tetraethylthiuram disulfide
Bis(diethylthiocarbamoyl) disulfide
Antabuse®

B. Introduction

Disulfiram [$(C_2H_5)_2$-N-C-S-S-C-N-$(C_2H_5)_2$] is a solid used in the rubber industry as an accelerator and vulcanizer. It is also used as a fungicide, seed disinfectant, and medication. As a medication, it is used in the treatment of alcoholism.[91,92] People receiving disulfiram who also consume ethanol either as a beverage, or an ingredient in foods, or cosmetics, develop severe signs and symptoms referred to as the "Antabuse®" or "acetaldehyde syndrome". The syndrome includes vasodilation of the face and later the entire body, headache, tachycardia, respiratory difficulty, vomiting, weakness, and hypotension. The cause of this reaction is thought to be disulfiram-induced inhibition of ethanol metabolism leading to the build-up of blood acetaldehyde levels although acetaldehyde administration alone does not reproduce the syndrome.

C. Human Neurotoxicity

In reviewing the literature, no reports were found of workers developing neurotoxicity due to industrial exposure to disulfiram, although some have shown the "acetaldehyde syndrome". It is very likely that the "acetaldehyde syndrome" provides the same type of negative reinforcement to reduce chronic disulfiram exposure as it does for ethanol in the treatment of alcoholism.

The most common form of disulfiram neurotoxicity is a polyneuropathy affecting the legs and, less severely, the arms.[93-101] When proximal muscles are affected first, the neuropathic process may appear to be a myopathy rather than a neuropathy.[95]

Polyneuropathy occurs after 5 to 6 months of disulfiram therapy, even at common therapeutic doses of 250 to 500 mg/day. Symptoms include weakness, muscle cramps, decreased tendon reflexes in the legs, and abnormal sensation in a glove and stocking distribution. Weakness in the upper limbs may involve intrinsic muscles of the hand.[95]

Symptoms of central nervous system (CNS) toxicity occur less frequently than polyneuropathy and boh may occur in the same individual.[94,102-107] Acute disulfiram toxicity following massive doses may produce seizures or coma while chronic intoxication may lead to psychiatric illness. Optic neuritis[108,112] and cerebral hemorrhage[113] occur at lower rates. Recovery from polyneuropathy, psychiatric changes, and optic neuritis is usually very good but may require several months.

D. Experimental Neurotoxicity and Neuropathology

Ueno et al.[114] gave 100 mg/kg of disulfiram i.p. to rats for periods of 1, 2, 4, 6, and 8 weeks to study disulfiram effects on the CNS. Although decreased spontaneous activity, anorexia, diarrhea, and transient decreased weight gain occurred, cerebral neurons were normal until the 4th week. By 6 weeks cerebrocortical and thalamic neurons were showing degenerative effects. Anzil et al.[115] found that rats given 500 mg of disulfiram weekly became incoordinated. Four to five weeks of exposure resulted in neuronal lipofuscin deposition and chromatolysis and swelling of terminal axons. A model of local disulfiram neuropathy has also been described.[116]

Although early studies by Child and Crump[117] in laboratory animals indicated demyelination was produced by disulfiram, more recent reports emphasize the presence of axonal degeneration, especially distally. Thus, disulfiram appears to produce a central-peripheral distal axonopathy with focal axonal enlargement similar to that produced by carbon disulfide.[93,96,97,99,101,115]

E. Metabolism

Disulfiram is rapidly absorbed by intestinal, s.c., and i.p. routes. Distribution of disulfiram to many organs including lungs, kidney, liver, and brain and metabolism to diethyldithiocarbamate, the methyl ester and glucuronide of diethyldithiocarbamate, diethylamine, CS_2, and sulfates is rapid.[119] CS_2 excretion through the lungs reaches a peak in 1 hr in dogs given disulfiram s.c.[120] Plasma levels of disulfiram, diethyldithiocarbamate, and the methyl ester of diethyldithiocarbamate peak in 15 min and disulfiram is cleared from the plasma in 45 min.[119] Brain levels of disulfiram and metabolites are also rapidly eliminated from the brain.[119] A major portion of a dose of disulfiram is found in the kidneys as diethyldithiocarbamate-glucuronide and sulfate which are presumably excreted in the urine.[119] A portion of a dose of disulfiram or an active metabolite remains for a prolonged period of time, up to 6 days.[121,122] People receiving disulfiram may show signs of the "acetaldehyde syndrome", on exposure to alcohol, 4 to 5 days after stopping medication.[91]

F. Mechanism of Action

Disulfiram and its metabolites have many varied effects on metabolism and probably have many mechanisms of action. Early reports suggested that disulfiram may produce histotoxic anoxia in the brain.[123] More recently, disulfiram has been found to be a potent inhibitor of dopamine-β-hydroxylase, which is the final step in the conversion of dopamine to noradrenalin.[124,125] Clinical investigations in man support the theory that psychiatric illness may be due to disulfiram interference in catecholamine levels in the brain.[126]

Interaction between disulfiram and central and peripheral axons is less clear. Kane,[127] Cavanagh,[128] and Rainey[100] have pointed out the similarities between disulfiram and CS_2-induced neurotoxicities. Since disulfiram is metabolized to two neurotoxins, diethyldithiocarbamate and CS_2, it may be that the mechanism of action of disulfiram is the same as those of its metabolites.

G. Other Effects

Like its metabolites, disulfiram has many effects on oxidative drug metabolism including inhibition of hepatic and plasma carboxyl esterase and cholinesterase activities. This may lead to increased sensitivity to organophosphate inhibition of cholinesterase.[129]

In rats given weekly injections of the carcinogen, acetoxymethylmethylnitrosamine, and also disulfiram by gavage, the rate of heart and lung tumor formation was reduced but neurogenic tumors which were not seen with the carcinogen alone were seen with the combination.[130] Disulfiram increases the toxicity of ethylene dibromide[131] and increases benzo(a)-pyrene monooxygenase activity.[132]

Although disulfiram was not teratogenic when suspended in carboxymethyl cellulose, the combination of disulfiram and dimethyl sulfoxide was more teratogenic than dimethyl sulfoxide was alone.[133]

IV. TETRAMETHYLTHIURAMDISULFIDE

A. Synonyms Thiram®
Bis(dimethylthiocarbamoyl)disulfide
Tetramethylthioperoxydicarbonic acid diamide
TMTD

B. Introduction

Tetramethylthiuram disulfide $[(CH_3)_2\text{-N-C(=S)-S-S-C(=S)-N-}(CH_3)_2]$ is an analogue of disulfiram and similarly may interfere with ethanol metabolism resulting in the acetaldehyde reaction. TMTD is more acutely toxic than disulfiram; therefore, it is not used for its antimetabolic effects.[91] It is a common commercial chemical used as a rubber accelerator, rubber vulcanizer, fungicide, seed disinfectant, insecticide, and bacteriostat in soaps and lotions. The greatest exposure to people is probably in handling and mixing powdered TMTD during formulation.

C. Neurotoxicity

A threshold limit value of 5 mg/m³ has been established for dust inhalation of TMTD.[134] No human neurotoxicity has been reported with this material possibly because exposure to TMTD followed by exposure to ethanol, either through alcohol-containing beverages or products such as cologne, may result in an acetaldehyde reaction. The reaction should not only inhibit the use of alcohol, it would be expected to reinforce good hygiene practices to reduce repeated exposure to TMTD.

Lee et al.[135] fed diets containing 0%, 0.01%, or 0.1% TMTD to male and female rats. Eight of twenty-four female rats fed the 0.1% diet consumed 66.9 mg/kg and developed ataxia and paralysis of the hindlimbs between 5 and 19 months of feeding. Poor body weight gain and alopecia were also observed in these animals. In a second experiment, 4 of 24 female rats consuming 65.8 mg/kg were also paralyzed while 9 other animals showed abnormal hindlimb clasping. Female rats consuming 25.5 or 66.9 mg/kg of TMTD which were not ataxic had significant differences in hindlimb stride width and stepping angle. Impaired jumping and climbing ability in females receiving 66.9 mg/kg and increased activity in these females and in males receiving 20.4 and 52.0 mg/kg were related to TMTD exposure.

D. Neuropathology

In rats exposed to TMTD, axonal degeneration and secondary demyelination were located in the sciatic nerve. Degenerative changes in the spinal cord included chromatolysis, pyknosis, and satellitosis of neurons.[135]

E. Mechanism of Action
TMTD appears to be readily absorbed through the intestinal tract and the lungs and is quickly and widely distributed throughout the body. Inhibition of hepatic dehydrogenases results in an acetaldehyde reaction on exposure to ethanol. Other effects may result from the known reactions of dithiocarbamates with metals, sulfhydryl-containing enzymes, or metabolism to reactive metabolites including CS_2.

F. Other Effects
TMTD feeding may also result in alopecia, squamous metaplasia of the thyroid, fatty infiltration of the pancreas, testicular germinal cell atrophy, and skeletal malformation.[135]

V. OTHER DITHIOCARBAMATE COMPOUNDS

Other neurotoxic dithiocarbamates are shown on Table 1. These materials have not been studied in a systematic manner for neurotoxicity, but the available evidence indicates that they produce axonal damage which may be similar to TMTD. Ferric dimethyldithiocarbamate, like TMTD, also produces testicular germinal cell damage[135,139] and is metabolized to the dimethyldithiocarbamate moiety, CS_2,[140] and dimethylamine,[140] suggesting the neurotoxicity of dithiocarbamates are metabolically related to each other and to CS_2.

VI. THIOPHENE

A. Synonyms
Thiaphene Thiotetrole
Thiophen Thiacyclopentadiene
Thiofuran Divinylenesulfide
Thiofuram Huile H5O
Thiofurfuran Huile HSO
Thiole

B. Introduction
Thiophene is a clear, colorless, volatile, aromatic liquid first recognized as a separate chemical by Meyer in 1882.[144] It occurs naturally in peat, coal shale, some crude oils, coal tar, coal gas, and technical-grade benzene. The crude benzene fraction of coal tar distillates contains about 0.5% by weight of thiophene. The purity of commercial-grade thiophene is 99%. The solvent properties of thiophene are similar to benzene but it has uses at higher and lower temperatures. Other uses include the manufacture of resins with phenol or formaldehyde and as an intermediate or starting material in the synthesis of dyes and pharmaceuticals.

C. Human Neurotoxicity
Human neurotoxicity due to thiophene has not been reported, probably because ventilation is used to control its odor and other toxic effects. Theoretically, thiophene might reach neurotoxic levels if a spill occurred in an enclosed area. Workers engaged in the synthesis of sulfur-containing compounds including thiophenes have been reported to show changes in the nervous system, the cardiovascular system, and particularly the liver.[145] Thiophene concentrations below the threshold odor perception (2.1 mg/m^3) level down to 1 mg/m^3 had effects on the light sensitivity of human eyes.[146,147] 0.8 mg/m^3 had no effect on light sensitivity but did produce changes in cerebral bioelectric activity.[146,147] 0.6 mg/m^3 was judged to be a no-effect level.[146,147]

Table 1
OTHER NEUROTOXIC DITHIOCARBAMATES

Chemical	Structure	Use	Comment	Ref.
Sodium diethyldithiocarbamate or dithiocarb	$(C_2H_5)_2\text{—NC—SNa}$ with S double-bonded	Medicinal chelator	Central-peripheral distal axonal damage and vacuolated dorsal root ganglia described	137, 138, 141, 142
Ferric dimethyldithiocarbamate or ferbam	$[(CH_3)_2\text{NC—S}]_3 Fe^{+3}$ with S double-bonded	Fungicide	Diets containing 0.1% may result in abnormal hindlimb clasping, ataxia, and/or paralysis after several months; dogs may show seizures	135, 139
Zinc dimethyldithiocarbamate or ziram	$(CH_3)_2\text{N—C(=S)—S—Zn—S—C(=S)—N(CH}_3)_2$	Fungicide; Rubber accelerator	Abnormal hindlimb clasping has been described in rats	3, 6
Tecoram	Oxidation product of disodium ethylenebisdithiocarbamate and sodium diethyldithiocarbamate with ammonium persulfate		Single dose of 0.01, 0.1, 1, or 10 mg/chicken egg results in peripheral distal axonal damage, vacuolated dorsal root ganglia, and muscle atrophy	143

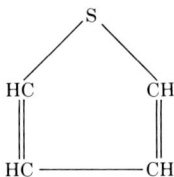

FIGURE 1. Thiophene.

D. Experimental Neurotoxicity

Inhalation of high concentrations of thiophene may cause central nervous system depression, narcosis, and death. Repeated administration of oral or s.c. routes may lead to degeneration of cerebellar granule cells in dogs and rats.[148-154,157] Repeated daily injections of 2 g of thiophene in dogs produced severe ataxia and paralysis.[148-153] Fourteen of twenty-five rats receiving thiophene (0.4 mℓ/day for 5 or 9 days) developed selective destruction of cerebellar granule cells.[153] Groups of five male rats were given 50, 100, 500, or 1000 mg/kg/day of thiophene. At 1000 mg/kg, one rat died after a single dose and four were killed after two doses because of severe CNS toxicity including weakness, staggering, twitching, head tremors, and limb jerking.[154] All remaining animals received 13 doses of thiophene. Two rats receiving 500 mg/kg also showed evidence of CNS toxicity.

The neurotoxic hazard by inhalation appears to be low. Mice exposed to 2900 ppm were narcotized and in some cases died.[155] 8700 ppm of thiophene killed mice in 20 to 80 min.[155] Six-hour exposure of rats to 800 to 1600 ppm of thiophene was without effect, whereas 3200 and 6400 ppm produced drowsiness and generalized muscle fasciculations.[154] All four exposed rats died at 6400 ppm. The LC_{50} for thiophene was calculated to be 4525 ppm.[154] Twelve repeated 6-hr exposure of rats to 3200 ppm thiophene produced tremors in all animals but tremors were not seen after the fourth exposure. 1600 ppm of thiophene produced tremors in one of five rats during the third day of the study but subsided in 24 hr. No lasting signs of CNS toxicity were noted and no neuropathological sequelae were identified.

E. Neuropathology

Repeated exposure of laboratory animals to thiophene produces a highly selective destruction of cerebellar granule cells. In less affected areas individual granule cells become pyknotic and undergo karyorrhexis and lysis. The pattern of granule cell necrosis is patchy resulting in complete destruction of granule cells in limited areas.[153,154,156] All phylogenetic areas of the cerebellum may be affected.[153]

Prior to condensation of granule cell nucleoplasm and pyknosis, no changes in cytoplasmic oxidative enzymes or acid phosphatases are observed.[153]

Purkinje cells which receive granule cell synaptic input show a temporary fall in NADH-tetrazolium reductase activity following necrosis of granule cells.[153] Loss of granule cells induces an increase in the size of Purkinje cell mitochondria followed by a compensatory rise in mitochondrial density and a reduction in the Purkinje cell dendritic tree.[156-158] The Purkinje cell response may be a form of trans-synaptic degeneration following the loss of

F. Metabolism

The toxicity of thiophene may require the formation of an activated metabolite by microsomal cytochrome P-450 oxygenases.

McMurtry and Mitchell[159] have shown that hepatic and renal toxicity of thiophene can be increased by pretreatment with phenobarbital, a P-450 inducer, and can be reduced by treatment with colbatous chloride and piperonyl butoxide, P-450 inhibitors. Presumably, neurotoxicity could also be altered by exposure to these agents.

Thiophene is rapidly absorbed, distributed widely in the body, and apparently excreted as a sulfur-containing metabolite. Subcutaneous injection of radiolabeled-S thiophene in rats produces a peak blood concentration in 30 min and peak organ levels (liver, skin, brain, and intestines) in 3 hr. Urinary excretion accounts for 61.1% of the labeled sulfur within 34 hr and 74.4% within 3 days.[160] Labeled sulfur is incorporated into the skin very quickly and eventually is found in hair.[161] Respiratory excretion accounts for only 5% of the dose (33 mg/kg of ^{14}C-labeled thiophene) when thiophene is given orally to mice.[162] Dogs and rabbits apparently excrete a neutral sulfur metabolite of thiophene.[149,163] Thiophene is metabolized to premercapturic acid and 2-thienyl-mercapturic acid.[164] Theoretically, an epoxide could be formed.[164]

G. Other Effects
The liver and kidneys are more severely affected by thiophene than is the central nervous system.[154,159] Circulating red and white blood cells and the thymus may be damaged by high doses (500 to 1000 mg/kg) of thiophene.[154]

VII. PYRIDINETHIONE DERIVATIVES

A. Introduction
Derivatives of 1-hydroxy-2-(1H)-pyridinethione are useful because of their antibacterial and antifungal properties. Some of the metallic salts or complexes have been sold under the name Omadine®. Iron, zinc, manganese, and copper complexes have been used to treat fungal plant diseases. The sodium salt has been used in the determination of copper and iron levels and as an adhesive preservative. The zinc salt (zinc pyridinethione, ZPT) and magnesium sulfate adduct of 2,2'-dithio-bis-pyridine-1-oxide (Omadine MDS, MDS) are used or have been proposed for use in antiseborrheic shampoos and hair care products. MDS has been proposed as a preservative in cosmetics. Aluminum, bismuth, cadmium, cobalt, nickel, and tin complexes have been studied for a variety of uses including antiperspirant, rodent repellants, tree wound dressings, and bactericides in plastics, adhesives, and textiles.

B. Neurotoxicity
Sodium pyridinethione (sodium pyridine-2-thiol-1-oxide, SPT), zinc pyridinethione (zinc pyridine-2-thiol-1-oxide, ZPT), and the magnesium sulfate adduct of 2,2'-dithio-bis-pyridine-1-oxide (MDS) experimentally produce neuropathy in laboratory rats.[165-170] No convincing data exist to demonstrate a similar effect in man although there is a single case report of neuropathy which developed in a man using a commercial shampoo containing 2% ZPT.[171] In this case, the neuropathy was not typical of the clinical pattern seen with other toxic neuropathies and did not appear similar to the ZPT-induced condition in laboratory animals.

Adams and colleagues[174] found that ZPT, SPT, and MDS produced cholinergic effects when administered i.v. to pigs. 50 mg/kg SPT, 5 mg/kg ZPT, or 4 mg/kg MDS resulted in salivation, vasodilation, urination, defecation, emesis, muscle fasciculation, increased motor activity, ataxia, and weakness which lasted for 30 to 60 min. Winek and Buehler[175] also reported acute cholinergic effects following ZPT administration to monkeys, dogs, and rabbits. Apart from the acute cholinergic effects of these materials, SPT, ZPT, and MDS produce neuropathies when repeatedly administered to laboratory species.

ZPT has been the most widely studied of these materials. Rats and rabbits receiving ZPT by gavage or in their diet develop weakness followed by paralysis as a result of central-peripheral distal axonal damage.[166-169] This occurs in females fed diets at or above 25 ppm, male rats fed at 50 ppm or above, and in rabbits fed 17.5 mg/kg.[166-173]

Rats consuming 166 mg/kg of ZPT in their diet for 4 to 7 days develop weakness in their hindlimbs resulting in abnormal supraspinal reflexes and paralysis in 14 to 17 days. Proprioception is affected little as sensory pathways appear relatively normal and forelimb weakness is mild. Removal of ZPT from the diet results in recovery in about 1 week.[168-172] Rats fed diets of 250 ppm ZPT for 2 weeks also show spinal kyphosis with muscle atrophy and penile prolapse, the later indicating effects on unmyelinated axons of the sympathetic nervous system.[169] Primates, dogs, and cats do not show similar results although the studies may not have been capable of detecting subtle neuropathic effects.[166,176]

Application of ZPT in dimethylsulfoxide or 300 mg/kg of a 10% aqueous solution of SPT paralyzed rabbits within 7 to 13 days, whereas 400 mg/kg of ZPT in a methyl cellulose suspension did not.[165]

MDS (1, 3, or 7.5 mg/kg) was given to male rats for 60 days, to females from 14 days prior to mating through lactation day 21, or to females from gestation day 15 through lactation day 20.[170] The 7.5-mg/kg dose was toxic, resulting in decreased body weight gain and mortality. Severe motor impairment evidenced by ataxia and hindlimb paralysis also occurred.

C. Neuropathology

Although the neuropathy produced by ZPT was recognized as early as 1958, it was not until relatively recently that the underlying axonal pathology was identified.[168,169]

Rats fed a diet containing 166 ppm of ZPT developed a progressive hindlimb weakness and paralysis with morphological lesions in the distal regions of the peripheral nerves and the lumbo-sacral spinal cord and cerebellum.[168] Motor axons were more severely damaged than sensory axons. This differential susceptibility correlates well with clinical findings of preservation of proprioception and pain in the face of severe motor impairment and limited effects on electrophysiological changes in the sural nerve,[169] a sensory nerve, with evidence of denervation in the musculature.[172]

Pathological changes occur most severely in larger myelinated axons to the hindlimb calf musculature than the more distal foot musculature. Characteristic lesions appear as swollen axons containing excessive tubulovesicular profiles, dense granular deposits, and progressive degeneration of the preterminal portion of the neuromuscular junction.[168,169] Degenerative axon changes in the peripheral nerves occur earlier, after about 1 week of exposure to 166 ppm, than central axons lesions which occur after 4 weeks of exposure.[168]

D. Metabolism

The metabolism of SPT, ZPT, and MDS are linked by conversion to common metabolites 2(pyridyl-N-oxide)sulfonic acid and 2-mercaptopyride.[174] Both SPT, ZPT, and MDS are absorbed through the skin and by the gastrointestinal tract.[177] When given i.v., SPT, ZPT, and MDS plasma decay curves follow a biphasic exponential function suggesting a two-compartment model for parent compound and metabolite distribution.[174] Twenty-four-hour urinary excretion of C^{14}-labeled SPT (95% elimination) was greater than for ZPT (59% elimination) or MDS (56% elimination).[174]

Radiochromatographic analysis of plasma and urine samples from pigs administered pyridinethiones suggested that SPT is rapidly metabolized to 2,2'-(pyridyl-N-oxide)disulfide within 30 min of dosing, that 2,2'-(pyridyl-N-oxide)disulfide is a primary urinary metabolite of SPT, and that it is hydrolyzed to a sulfonic acid (Figure 2). Reductive metabolism of SPT produces 2-mercaptopyridine. Further work in this area following i.v. dosing with pyridinethiones showed that the major metabolite is the glucuronide of 2-mercaptopyridine-N-oxide and a minor metabolite is the glucuronide of 2-mercaptopyridine.[173]

Following dermal application of pyridinethiones, the major metabolites were 2,2'-pyridyldisulfide (77% of extracted radiolabel) and 2-(pyridyl-N-oxide)sulfonic acid.[173]

In rats, SPT was excreted as 2-(pyridyl-N-oxide)sulfonic acid in the urine with only trace

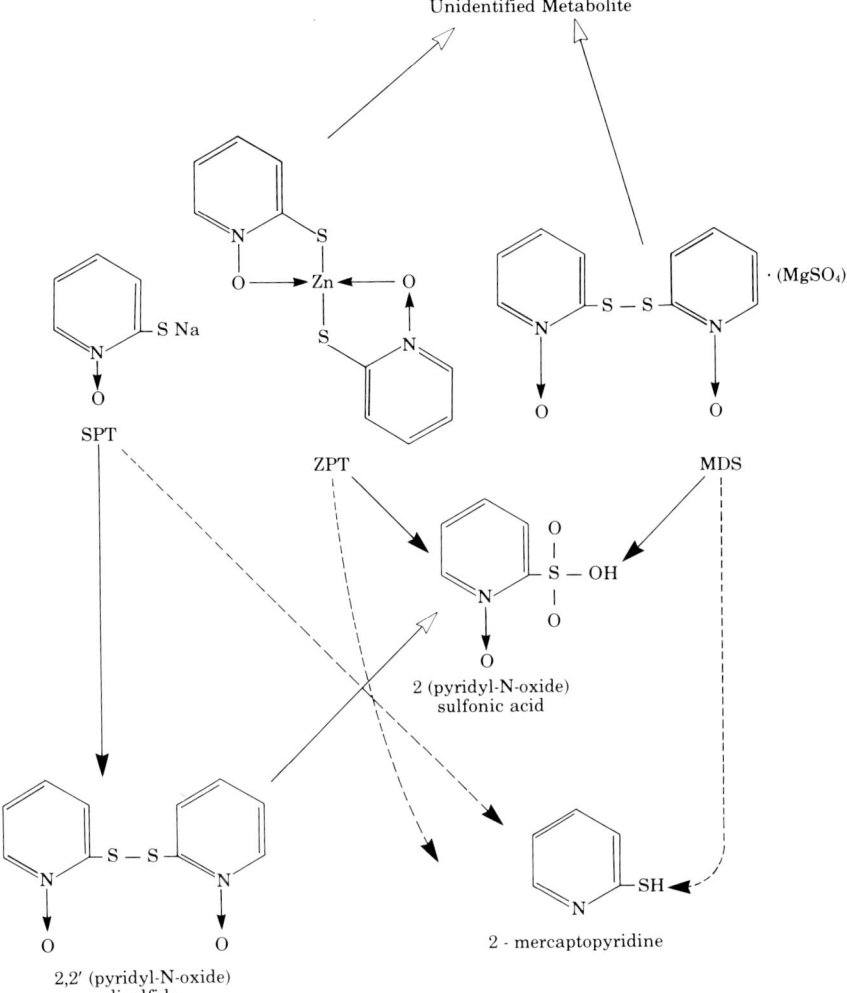

FIGURE 2. Summary of postulated metabolic pathways for ^{14}C-labeled SPT, ZPT, and MDS in swine following intravenous administration; based on thin-layer radiochromatographic analysis of urine samples. Dark arrows (→) indicate primary pathways, light arrows (⇒) represent secondary pathways, and dashed arrows (--→) refer to minor pathways. (From Adams, M. D., Wedig, J. H., Jordan, R. L., Smith, L. W., Henderson, R., and Borzelleca, J. F., *Toxicol. Appl. Pharmacol.*, 36, 523, 1976. With permission.)

amounts of 2,2'-(pyridyl-*N*-oxide)disulfide.[178] In rabbits, exposed topically, the sulfonic acid was also the main urinary metabolite.[179] In rabbits, rats, dogs, and monkeys, glucuronide conjugates are significant metabolites of SPT and ZPT.[172,180]

Differences in metabolism may account for the greater sensitivity of female rats to ZPT than male rats.[173] Absorption from the gastrointestinal tract decreased more rapidly in males than female rats as the dose increased. Also, the retention of the ^{14}C label from ZPT was higher in male livers, possibly due to greater enterohepatic bile circulation. This correlates with the longer period of time to maximum blood levels. Renal clearance of ^{14}C label was double in males compared to female rats (0.129 vs. 0.065 mℓ/min), possibly due to the larger amount bound to female red blood cells. Pyridinethione effects on blood zinc homeostasis may partially explain differences in blood levels.

E. Mechanism of Action

The ultimate neurotoxic agent responsible for SPT, ZPT, and MDS neurotoxicity has not been identified but probably is the pyridinethione moiety itself or a metabolite of it since it is a common link among all three materials. The neurotoxicity of these materials may derive from their ability to inactivate enzyme systems by chelating metal enzyme cofactors although no data exist to support such a hypothesis.

Sahenk and Mendell have accumulated evidence to show that ZPT-induced accumulation of tubulovesicular membranes within axons is related to a deficit in the axoplasmic transport resulting in a delay in the "turn-around" of fast transported axoplasmic materials.[182]

F. Other Effects of Pyridinethiones

Animals such as the dog and cat, having a tapetum lucidum as part of their retina, undergo retinal degeneration as a result of pyridinethione exposure, while animals such as monkeys, possessing nontapetal retinas, do not.[183-185] The effect is thought to be related to chelation of zinc which is highly concentrated in the tapetum lucidum.[183-185]

REFERENCES

1. **Timmerman, R. W.,** Carbon disulfide, in *Kirk-Othmer Encyclopedia of Chemical Technology,* Vol. 4, 3rd ed., Bushey, G. J., Campbell, L., Klingsberg, A., and van Nes, L., Eds., John Wiley-Interscience, New York, 1978, 742.
2. **Simpson, J. Y.,** Notes on the anesthetic effects of chloride of hydrocarbon, nitrate of ethyle, bezin, aldehyde, and bisulphuret of carbon, *Mon. J. Med. Sci.,* 8, 740, 1848.
3. **Delpech, A.,** Incidence which develop in rubber workers: inhalation of carbon disulfide vapors, *Union Med.,* 10, 265, 1856.
4. National Institute of Occupational Safety and Health, Occupational Exposure to Carbon Disulfide, Department of Health, Education, and Welfare, Publ. No. 77-156, Washington, D.C., 1977.
5. Survey of Carbon Disulfide and Hydrogen Sulfide Hazards in the Viscose Rayon Industry, Commonwealth of Pennsylvania, Harrisburg, Department of Labor and Industry, Bulletin No. 46, 1938.
6. **Coppock, R. W. and Buck, W. B.,** Toxicology of carbon disulfide: a review, *Vet. Hum. Toxicol.,* 23, 331, 1981.
7. **Cooper, P.,** Carbon disulphide toxicology: the present picture, *Food Cosmet. Toxicol.,* 14, 57, 1976.
8. **Davidson, M. and Feinleib, M.,** Carbon disulfide poisoning: a review, *Am. Heart J.,* 83, 100, 1972.
9. **Seppäläinen, A. M. and Haltia, M.,** Carbon disulfide, in *Experimental and Clinical Neurotoxicology,* Spencer, P. S. and Schaumburg, H. H., Eds., Williams & Wilkins, Baltimore, 1980, 356
10. **Tolonen, M.,** Vascular effects of carbon disulfide: a review, *Scand. J. Work Environ. Health,* 1, 63, 1975.
11. **Brieger, H. and Teisinger, J.,** *Toxicology of Carbon Disulfide,* Excerpta Medica, Amsterdam, 1967.
12. **Djerassi, L. S. and Lumbroso, R.,** Carbon disulphide poisoning with increased ethereal sulphate, *Br. J. Ind. Med.,* 25, 220, 1968.
13. **Besarabic, M.,** Accidental poisoning with CS_2 in production of viscose fibers, *Arh. Hig. Rada Toksikol.,* 28, 181, 1977.
14. **Yamada, J.,** A case of acute carbon disulfide poisoning by accidental ingestion, *Sangyo Igaku,* 19, 140, 1977.
15. **Kruse, A., Borch-Johnsen, K., and Pedersen, L. M.,** Cerebral damage following a single high exposure to carbon disulphide, *J. Soc. Occup. Med.,* 32, 44, 1982.
16. **Vigliani, E. C.,** Carbon disulphide poisoning in viscose rayon factories, *Br. J. Ind. Med.,* 11, 235, 1954.
17. **Manu, P., Lilis, R., Lancranjan, I., Ionescu, S., and Vasilescu, I.,** The value of electromyographic changes in the early diagnosis of carbon disulphide peripheral neuropathy, *Med. Lav.,* 61, 102, 1970.
18. **Hänninen, H.,** Psychological picture of manifest and latent carbon disulphide poisoning, *Br. J. Ind. Med.,* 28, 374, 1971.
19. **Knave, B., Kolmodin-Hedman, B., Persson, H. E., and Goldberg, J. M.,** Chronic exposure to carbon disulfide. Effects on occupationally exposed workers with special reference to the nervous system, *Work Environ. Health,* 11, 49, 1974.

20. **Seppäläinen, A. M. and Tolonen, M.**, Neurotoxicity of long-term exposure to carbon disulfide in the viscose rayon industry: a neurophysiological study, *Work Environ. Health*, 11, 145, 1974.
21. **Tolonen, M.**, Chronic subclinical carbon disulfide poisoning, *Work Environ. Health*, 11, 154, 1974.
22. **Tuttle, T. C., Wood, G. D., and Grether, C. B.**, Behavioral and Neurological Evaluation of Workers Exposed to Carbon Disulfide (CS_2), Department of Health, Education, and Welfare (NIOSH) Publ. No. 77-128, Washington, D.C., 1976.
23. **Styblova, V.**, Electroencephalography in diagnosis of early cerebral changes due to carbon disulfide, *Int. Arch. Occup. Environ. Health*, 38, 263, 1977.
24. **Gilioli, R., Bulgheroni, C., Bertazzi, P. A., Cirla, A. M., Tomasini, M., Cassitto, M. G., and Jacovone, M. T.**, Study of neurological and neurophysiological impairment in carbon disulphide workers, *Med. Lav.*, 69, 130, 1978.
25. **Vasilescu, C., Florescu, A., Alexianu, M., and Petrovici, A.**, Clinical value of early electrophysiological changes in sensory and motor fibers, *Rev. Roum. Med. Neurol. Psychiatr.*, 17, 159, 1979.
26. **Vasilescu, C. and Florescu, A.**, Clinical and electrophysiological studies of carbon disulphide polyneuropathy, *J. Neurol.*, 224, 59, 1980.
27. **Raitta, C., Teir, H., Tolonen, M., Nurminen, M., Helpio, E., and Malstrom, S.**, Impaired color discrimination among viscose rayon workers exposed to carbon disulfide, *J. Occup. Med.*, 23, 189, 1981.
28. **Misiakiewicz, Z., Szulinska, G., and Chyba, A.**, Effect of a carbon disulfide-hydrogen sulfide mixture in air on white rats during several months of continuous exposure, *Rocz. Panstw. Zakl. Hig.*, 23, 465, 1972.
29. **Bitterschl, G.**, Relations in action between carbon disulfide and hydrogen sulfide, *Med. Lav.*, 62, 554, 1971.
30. **Wakatsuki, T. and Higashikawa, H.**, Experimental studies on CS_2 and H_2S poisoning: the histological changes in hemotopoetic organs and other main internal organs, *Shikoku Igaku Zasshi*, 14, 549, 1959.
31. **Wakatsuki, T.**, Experimental study on the poisoning by carbon disulfide and hydrogen sulfide, *Shikoku Igaku Zasshi*, 15, 671, 1959.
32. Carbon disulfide, in Documentation of the Threshold Limit Values for Substances in Work Room Air, 4th ed., American Conference of Governmental Industrial Hygienists, Cincinnati, 1980, 70.
33. **Wiley, F. H., Hueper, W. C., and von Oettingen, W. F.**, On the toxic effects of low concentrations of carbon disulfide, *J. Ind. Hyg. Toxicol.*, 18, 733, 1936.
34. **Lewey, F. H.**, Experimental chronic carbon disulfide poisoning in dogs. A clinical biochemical and pathological study, *J. Ind. Hyg. Toxicol.*, 23, 415, 1941.
35. **Ferraro, A., Jervis, G. A., and Flicker, D. J.**, Neuropathological changes in experimental carbon disulfide poisoning in cats, *Arch. Pathol.*, 32, 723, 1941.
36. **Lukáš, E.**, Eight years experience with experimental CS_2 polyneuropathy in rats, *G. Ital. Med. Lav.*, 1, 7, 1979.
37. **Lukáš, E., Kotas, P., and Obrusnik, I.**, Copper and zinc levels in peripheral nerve tissues of rats with experimental carbon disulfide neuropathy, *Br. J. Ind. Med.*, 31, 288, 1974.
38. **Frantík, E.**, The development of motor disturbances in experimental chronic carbon disulphide intoxication, *Med. Lav.*, 61, 309, 1970.
39. **Knobloch, K., Stetkiewicz, J., and Wronska-Nofer, T.**, Conduction velocity in the peripheral nerves of rats with chronic carbon disulphide neuropathy, *Br. J. Ind. Med.*, 36, 148, 1979.
40. **Seppäläinen, A. M. and Linnoila, I.**, Electrophysiological findings in rats with experimental carbon disulphide neuropathy, *Neuropathol. Appl. Neurobiol.*, 2, 209, 1976.
41. **Maroni, M., Colombi, A., Rota, E., Antonini, C., Picchi, O., Foà, V., Caimi, L., Tettamanti, G., and Castano, P.**, Biochemical and morphological investigations on nervous tissue of rats inhaling carbon disulfide, *Med. Lav.*, 70, 443, 1979.
42. **Tilson, H. A., Cabe, P. A., Ellinwood, E. H., Jr., and Gonzalez, L. P.**, Effects of carbon disulfide on motor function and responsiveness to d-amphetamine in rats, *Neurobehav. Toxicol.*, 1, 57, 1979.
43. **Milivojevic, V.**, Study of pathomorphological changes in the liver and brain of rats chronically poisoned by carbon disulfide, *Acta Vet. (Belgrade)*, 31, 129, 1981.
44. **Alpers, B. J. and Lewey, F.**, Changes in the nervous system following carbon disulfide poisoning in animals and in man, *Arch. Neurol. Psychiat.*, 44, 725, 1940.
45. **Haltia, M. and Linnoila, I.**, Distal axonopathy induced by carbon disulfide (CS_2), *J. Neuropathol. Exp. Neurol.*, 37, 621, 1978.
46. **Szendzikowski, S., Stetkiewicz, J., Wrońska-Nofer, T., and Zdrajkowska, I.**, Structural aspects of experimental carbon disulfide neuropathy. I. Development of neurohistological changes in chronically intoxicated rats, *Int. Arch. Arbeitsmed.*, 31, 135, 1973.
47. **Szendzikowski, S., Karesek, M., Stetkiewicz, J., and Wrońska-Nofer, T.**, Ultrastructure of the peripheral nerve in rats chronically exposed to carbon disulfide, preliminary report, *Folia Histochem. Cytochem.*, 11, 353, 1973.

48. **Szendzikowski, S., Stetkiewicz, J., Wrońska-Nofer, T., and Karasek, M.**, Pathomorphology of the experimental lesion of the peripheral nervous system in white rats chronically exposed to carbon disulfide, in *Structure and Function of Normal and Diseased Muscle and Peripheral Nerve*, Haumanowa-Petrusewicz, I. and Jedrzejowska, H., Eds., Polish Medical Publishers, Warsaw, 1974, 319.
49. **Colombi, A., Maroni, M., Picchi, O., Rota, E., Castano, P., and Foà, V.**, Carbon disulfide neuropathy in rats. A morphological and ultrastructural study of degeneration and regeneration, *Clin. Toxicol.*, 18, 1463, 1981.
50. **Juntunen, J., Linnoila, I., and Haltia, M.**, Histochemical and electron microscopic observations on the myoneural junctions of rats with carbon disulfide induced polyneuropathy, *Scand. J. Work Environ. Health*, 3, 36, 1977.
51. **Dutkiewicz, T. and Baranowska, B.**, The significance of absorption of carbon disulphide through the skin in the evaluation of exposure, in *Toxicology of Carbon Disulphide*, Brieger, H. and Teisinger, J., Eds., Excerpta Medica, Amsterdam, 1967, 50.
52. **Cohen, A. E., Paulus, H. J., Keenan, R. G., and Scheel, L. D.**, Skin absorption of carbon disulfide vapor in rabbits. I. Associated changes in blood protein and zinc, *Arch. Ind. Health*, 17, 164, 1958.
53. **Hueper, W. C.**, Etiologic studies on the formation of skin blisters in viscose workers, *J. Ind. Hyg. Toxicol.*, 18, 432, 1936.
54. **Brieger, H.**, Carbon disulphide in the living organism: retention, biotransformation, and pathophysiologic effects, in *Toxicology of Carbon Disulphide*, Brieger, H. and Teisinger, J., Eds., Excerpta Medica, Amsterdam, 1967, 27.
55. **Teisinger, J. and Soucek, B.**, Absorption and elimination of carbon disulfide in man, *J. Ind. Hyg. Toxicol.*, 31, 67, 1949.
56. **McKee, R., Kiper, C., Fountain, J., Riskin, A., and Drinker, P.**, A solvent vapor: carbon disulfide — absorption, elimination, metabolism and mode of action, *JAMA*, 122, 217, 1943.
57. **Lieben, J.**, Third international symposium on toxicology of carbon disulfide, *Arch. Environ. Health*, 29, 173, 1974.
58. **McKenna, M. J. and DiStefano, V.**, Carbon disulfide. I. The metabolism of inhaled carbon disulfide in the rat, *J. Pharmacol. Exp. Ther.*, 202, 245, 1977.
59. **Vasak, V. and Kopecky, J.**, On the role of pyridoxamine in the mechanism of the toxic action of carbon disulphide, in *Toxicology of Carbon Disulphide*, Brieger, H. and Teisinger, J., Eds., Excerpta Medica, Amsterdam, 1967, 35.
60. **Scheel, L. D.**, Experimental carbon disulphide poisoning in rabbits — its mechanism and similarities with human case reports, in *Toxicology of Carbon Disulphide*, Breiger, H. and Teisinger, J., Eds., Excerpta Medica, Amsterdam, 1967, 107.
61. **Lukáš, E.**, Important factors influencing the etiology and development of experimental carbon disulfide neuropathy in rats, *Cesk. Neurol.*, 36, 169, 1973.
62. **Kotas, P., Obrusnik, I., Lukáš, E., and Krivanek, M.**, Determination of zinc and copper in the peripheral nerves of rats with carbon disulfide-induced neuropathy, *J. Radioanal. Chem.*, 19, 263, 1974.
63. **McKenna, M. J. and DiStefano, V.**, Carbon disulfide. II. A proposed mechanism for the action of carbon disulfide on dopamine β-hydroxylase, *J. Pharmacol. Exp. Ther.*, 202, 253, 1977.
64. **Freundt, K. J., Schnapp, E., and Dreher, W.**, Pharmacokinetics of inhaled carbon disulphide in rats in relation to its inhibitory effect on the side-chain oxidation of hexobarbital, *Int. Arch. Occup. Environ. Health*, 35, 173, 1975.
65. **Kromer, W. and Freundt, K. J.**, In vitro inhibition of oxidative N-demethylation by means of carbon disulfide, *Arzneim.-Forsch*, 26, 189, 1976.
66. **Freundt, K. J., Lieberwirth, K., Netz, H., and Poehlmann, E.**, Blood acetaldehyde in alcoholized rats and humans during inhalation of carbon disulfide vapor, *Int. Arch. Occup. Environ. Health*, 37, 35, 1976.
67. **Teisinger, J.**, Some mechanisms of chronic carbon disulfide poisoning, *Prac. Lek.*, 23, 306, 1971.
68. **DeMeio, R. and Brieger, H.**, Effects of CS_2 on tissue respiration and glycolysis, *J. Ind. Hyg. Toxicol.*, 31, 93, 1949.
69. **Spencer, P. S., Sabri, M. I., Schaumburg, H. H., and Moore, C. L.**, Does a defect of energy metabolism in the nerve fiber underlie axonal degeneration in polyneuropathies?, *Ann. Neurol.*, 5, 501, 1979.
70. **Tarkowski, S., Kolakowski, J., Gorny, R., and Opacka, J.**, Content of high energy phosphates and ultrastructure of mitochondria in the brain of rats exposed to carbon disulfide, *Toxicol. Lett.*, 5, 207, 1980.
71. **Savolainen, H. and Vainio, H.**, High binding of carbon disulfide sulfur in spinal cord axonal fraction, *Acta Neuropathol.*, 36, 251, 1976.
72. **Savolainen, H., Jarvisalo, J., Elovaara, E., and Vainio, H.**, The binding of carbon disulfide in central nervous system of control and phenobarbitone pre-treated rats, *Toxicology*, 7, 207, 1977.
73. **Sugimoto, K., Goto, S., Taniguchi, H., Baba, T., Raitta, Ch., Tolonen, M., and Hernberg, S.**, Ocular fundus photography of workers exposed to carbon disulfide — a comparative epidemiological study between Japan and Finland, *Int. Arch. Occup. Environ. Health*, 39, 97, 1977.

74. **Tomasini, M., Villa, A., and Cirla, A.,** Changes in coronary arteries due to chronic carbon disulfide poisoning, *Med. Lav.,* 63, 34, 1972.
75. **Artamonova, V. G. and Klishova, Z. N.,** A clinical experimental study of the pathogenesis of chronic carbon disulfide poisoning, *Gig. Tr. Prof. Zabol.,* 16, 22, 1972.
76. **Gondzik, M.,** Histology and histochemistry of rat testicles as affected by carbon disulfide, *Pol. Med. J.,* 10, 133, 1971.
77. **Lancranjan, I.,** Alterations of spermatic liquid in patients chronically poisoned by carbon disulphide, *Med. Lav.,* 63, 29, 1972.
78. **Lancranjan, I., Popescu, H. I., and Klepsch, I.,** Changes of the gonadic function in chronic carbon disulfide poisoning, *Med. Lav.,* 60, 566, 1969.
79. **Calabrese, E. J.,** Does use of oral contraceptives enhance the toxicity of carbon disulfide through interactions with pyridoxine and tryptophan metabolism?, *Med. Hypotheses,* 6, 21, 1980.
80. **Tabacova, S. and Balabaeva, L.,** Subtle consequences of prenatal exposure to low carbon disulphide levels, *Arch. Toxicol.,* Suppl. 4, 252, 1980.
81. **Tabacova, S., Hinkova, L., and Balabaeva, L.,** Carbon disulfide teratogenicity and postnatal effects in rat, *Toxicol. Lett.,* 2, 129, 1978.
82. **Lancranjan, I., Florea, O., and Petrescu, M.,** Adrenocortical insufficiency as a feature of carbon disulfide poisoning, *Med. Lav.,* 62, 557, 1971.
83. Anonymous, Dithiobiuret, *Cyanamid New Products Bulletin,* Collective Vol. 1, American Cyanamid Co., Princeton, N.J., 1952.
84. **Astwood, E. B., Hughes, A. M., Lubin, M., VanderLaan, W. P., and Adams, R. D.,** Reversible paralysis of motor function in rats from the chronic administration of dithiobiuret, *Science,* 102, 196, 1945.
85. **Seifter, S., Harkness, D. M., Muntwyler, E., and Seifter, J.,** The effect of dithiobiuret (DTB) on the electrolyte and water content of skeletal muscle and on carbohydrate metabolism, *J. Pharmacol. Exp. Ther.,* 93, 93, 1948.
86. **Peterson, R. E. and Sheth, N. K.,** Acute toxicity and effect of dithiobiuret and acrylamide on rotored performance in rats and mice, *Toxicol. Appl. Pharmacol.,* 33, 142, 1975.
87. **Atchison, W. D. and Peterson, R. E.,** Potential neuromuscular toxicity of 2,4-dithiobiuret in the rat, *Toxicol. Appl. Pharmacol.,* 57, 63, 1981.
88. **Atchison, W. D., Yang, K. H., and Peterson, R. E.,** Dithiobiuret toxicity in the rat: evidence of latency and cumulative dose thresholds, *Toxicol. Appl. Pharmacol.,* 61, 166, 1981.
89. **Atchison, W. D. and Peterson, R. E.,** Alteration of dithiobiuret induced rotarod failure by sulfur containing compounds, *Fed. Proc. Fed. Am. Soc. Exp. Biol.,* 36, 405, 1977.
90. **Atchison, W. D., Lalley, P. M., Cassens, R. G., and Peterson, R. E.,** Depression of neuromuscular function in the rat by chronic 2,4-dithiobiuret treatment, *Neurotoxicology,* 2, 329, 1981.
91. **Ritchie, J. M.,** The aliphatic alcohols, in *The Pharmacological Basis of Therapeutics,* 5th ed., Goodman, L. S. and Gilman, A., Eds., Macmillan, New York, 1975, 137.
92. **Haley, T. J.,** Disulfiram, tetraethylthioperoxydicarbonic diamide, a reappraisal of its toxicity and therapeutic application, *Drug Metab. Rev.,* 9, 319, 1979.
93. **Bouldin, T. W., Hall, C. D., and Krigman, M. R.,** Pathology of disulfiram neuropathy, *Neuropathol. Appl. Neurobiol.,* 6, 155, 1980.
94. **Graveleau, J., Ecoffet, M., and Villard, A.,** Disulfiram-induced peripheral neuropathies, *Nouv. Presse Med.,* 9, 2905, 1980.
95. **Mizon, J. P., Froissart, M., Gentit, F., and Morcamp, D.,** Peripheral neurological signs connected with prolonged use of disulfiram, *Lille Med.,* 22, 790, 1977.
96. **Moddel, G., Bilbao, J. M., Payne, D., and Ashby, P.,** Disulfiram neuropathy, *Arch. Neurol.,* 35, 658, 1978.
97. **Mokri, B., Ohnishi, A., and Dyck, P. J.,** Disulfiram neuropathy, *Neurology,* 31, 730, 1981.
98. **Olney, R. K. and Miller, R. G.,** Peripheral neuropathy associated with disulfiram administration, *Muscle Nerve,* 3, 172, 1980.
99. **Nukada, H. and Pollock, M.,** Disulfiram neuropathy. A morphometric study of sural nerve, *J. Neurol. Sci.,* 51, 51, 1981.
100. **Rainey, J. M.,** Disulfiram toxicity and carbon disulfide poisoning, *Am. J. Psychiatry,* 134, 371, 1977.
101. **Watson, C. P., Ashby, P., and Bilbao, J. M.,** Disulfiram neuropathy, *Calcif. Tissue Int.,* 32, 123, 1980.
102. **Dandelot, J. B. and Dupuis, M.,** Disulfiram: angel or devil?, *Concours Med.,* 101, 7666, 1979.
103. **Hotson, J. R. and Lungston, J. W.,** Disulfiram-induced encephalopathy, *Arch. Neurol.,* 33, 141, 1976.
104. **Liddon, S. C. and Satran, R.,** Disulfiram (antabuse) psychosis, *Am. J. Psychiatry,* 123, 1284, 1967.
105. **Lidy, C., Priollet, P., and Pepin, B.,** Pyramidal and extrapyramidal syndrome developing during acute disulfiram poisoning, *Nouv. Presse Med.,* 8, 3561, 1979.
106. **Sans, P., Deneux, A., Magne, C., and Besancon, G.,** Neurologic complications due to disulfiram. Neuropathies, encephalopathies, *Concours Med.,* 97, 3779, 1976.

107. **Weddington, W. W., Marks, R. C., and Verghese, J. P.,** Disulfiram encephalopathy as a cause of the catatonia syndrome, *Am. J. Psychiatry,* 137, 1217, 1980.
108. **Anon.,** Drug can cause optic neuritis in alcoholics. *JAMA,* 218, 808, 1971.
109. **Corydon, L.,** Optic neuritis and polyneuropathy and disulfiram (antabuse) therapy, *Ugeskr. Laeg.,* 135, 1470, 1973.
110. **Hansen, P. E. and Corydon, L.,** Optic neuritis and polyneuropathy during treatment with disulfiram, (Antabuse®), *Ugeskr. Laeg.,* 141, 3045, 1979.
111. **Norton, A. L. and Walsh, F. B.,** Disulfiram-induced optic neuritis, *Trans. Am. Acad. Ophthalmol. Otolaryngol.,* 76, 1263, 1972.
112. **Saraux, H. and Biais, B.,** Optic neuritis caused by disulfiram, *Ann. Ocul.,* 203, 769, 1970.
113. **Althoff, H.,** Multiple hemorrhage necrosis of the brain as a complication of disulfiram (Antabuse) therapy, *Ger. Med. Mon.,* 13, 180, 1968.
114. **Ueno, T., Miyagishi, T., Takahata, N., and Fujieda, T.,** Electron microscopic studies on the cerebral lesions of rats in experimental chronic disulfiram poisoning, *Acta Neuropathol.,* 38, 221, 1977.
115. **Anzil, A. P., Blinzinger, K., and Herrlinger, H.,** Preliminary ultra-structural studies of the central and peripheral nervous system following experimental tetraethyl thiouram disulfide intoxication, *Acta Neurol.,* 31, 15, 1976.
116. **Zuccarello, M. and Anzil, A. P.,** A localized model of experimental neuropathy by topical application of disulfiram, *Exp. Neurol.,* 64, 699, 1979.
117. **Child, G. P. and Crump, M.,** The toxicity of tetraethylthiuram disulfide (Antabuse) to mouse, rat, rabbit, and dog, *Acta Pharmacol. Toxicol.,* 8, 305, 1952.
118. **Ansbacher, L. E., Bosch, E. P., and Cancilla, P. A.,** Disulfiram neuropathy: a neurofilamentous distal axonopathy, *Neurology,* 32, 424, 1982.
119. **Faiman, M. D., Dodd, D. E., and Hanzlik, R. E.,** Distribution of S^{35} disulfiram and metabolites in mice, and metabolism of S^{35} disulfiram in the dog, *Res. Commun. Chem. Pathol. Pharmacol.,* 21, 34, 1978.
120. **Philips, M.,** Excretion of carbon disulfide in breath following subcutaneous injection of disulfiram, *Clin. Res.,* 28, 623A, 1980.
121. **Eldjarn, L.,** Metabolism of tetraethyl thiuram disulphide (antabuse, aversan) in the rat, investigated by means of radioactive sulphur, *Scand. J. Clin. Lab. Invest.,* 2, 198, 1950.
122. **Eldjarn, L.,** Metabolism of tetraethyl thiuram disulphide in man, investigated by means of radioactive sulphur, *Scand. J. Clin. Lab. Invest.,* 2, 202, 1950.
123. **Busse, E. N., Barnes, R. H., and Ebaugh, F. G.,** The effect of antabuse on the electroencephalogram, *Am. J. Med. Sci.,* 223, 126, 1952.
124. **Goldstein, M., Anagnoste, B., Lauber, E., and McKereghan, M. R.,** Inhibition of dopamine-β-hydroxylase by disulfiram, *Life Sci.,* 3, 763, 1964.
125. **Moore, K. E.,** Effects of disulfiram and diethyl dithiocarbamate on spontaneous locomotor activity and brain catecholamine levels in mice, *Biochem. Pharmacol.,* 18, 1627, 1969.
126. **Major, L. F., Murphy, D. L., Gershon, E. W., and Brown, G. L.,** The role of plasma amine oxidase, platelet monoamine oxidase, and red cell catechol-*O*-methyl transferase in severe behavioral reactions to disulfiram, *Am. J. Psychiatry,* 136, 679, 1979.
127. **Kane, F. J.,** Carbon disulfide intoxication from overdosage of disulfiram, *Am. J. Psychiatry,* 127, 690, 1970.
128. **Cavanagh, J. B.,** Peripheral neuropathy caused by chemical agents, *Crit. Rev. Toxicol.,* 3, 365, 1973.
129. **Zematis, M. A. and Greene, F. E.,** Impairment of hepatic microsomal and plasma esterases of the rat by disulfiram and diethyldithiocarbamate, *Biochem. Pharmacol.,* 25, 453, 1976.
130. **Habs, M., Schmahl, D., and Kretzer, H.,** Effect of disulfiram on acetoxymethyl-methyl-nitrosamine induced tumors in rats, *Oncology,* 38, 18, 1981.
131. **Millar, J. D.,** Ethylene Dibromide and Disulfiram Toxic Interaction, NIOSH Current Intelligence Bulletin No. 23, Department of Health, Education and Welfare, Publ. No. 79-146, Washington, D.C., 1978.
132. **Grafström, R. and Holmberg, B.,** The effect of long term treatment with disulfiram on content of cytochrome P-450 and on benzo(a)pyrene monooxygenase activity in microsomes isolated from the rat small intestinal mucosa, *Toxicol. Lett.,* 7, 79, 1980.
133. **Robens, J. F.,** Teratologic studies of carbaryl, diazinon, norea, disulfiram and thiram in small laboratory animals, *Toxicol. Appl. Pharmacol.,* 15, 152, 1969.
134. Thiram®, in *Documentation of the Threshold Limit Values,* 4th ed., American Conference of Governmental Industrial Hygienists, Cincinnati, 1981, 397.
135. **Lee, C.-C., Russell, J. Q., and Minor, J. L.,** Oral toxicity of ferric dimethyl dithiocarbamates (Ferbam) and tetramethylthiuram disulfide (Thiram) in rodents, *J. Toxicol. Environ. Health,* 4, 93, 1978.
136. **Rasul, A. R. and Howell, J. McC.,** The toxicity of some dithiocarbamate compounds in young and adult domestic fowl, *Toxicol. Appl. Pharmacol.,* 30, 63, 1974.

137. **Howell, J. M. and Edington, N.**, The neurotoxicity of sodium diethyldithiocarbamate in the hen, *J. Neuropathol. Exp. Neurol.*, 27, 464, 1968.
138. **Edington, H. and Howell, J. M.**, The neurotoxicity of sodium diethyldithiocarbamate in the rabbit, *Acta Neuropathol.*, 12, 339, 1969.
139. **Hodge, H. C., Maynard, E. A., Downs, W. L., Coye, R. D., Jr., and Steadman, L. T.**, Chronic oral toxicity of ferric dimethyldithiocarbamate (Ferbam) and zinc dimethyldithiocarbamate (Ziram), *J. Pharmacol. Exp. Ther.*, 118, 174, 1956.
140. **Hodgson, J. R., Hoch, J. C., Castles, T. R., Helton, D. O., and Lee, C.-C.**, Metabolism and disposition of ferbam in the rat, *Toxicol. Appl. Pharmacol.*, 33, 505, 1975.
141. **Rasul, A. R. and Howell, J. McC.**, A comparison of the effect of sodium diethyldithiocarbamate on the central nervous system of young and adult domestic fowl, *Acta Neuropathol.*, 24, 68, 1973.
142. **Rasul, A. R. and Howell, J. McC.**, Further observations on the response of the peripheral and central nervous system of the rabbit to sodium diethyldithiocarbamate, *Acta Neuropathol.*, 24, 161, 1973.
143. **Van Steenis, G. and van Logten, M. J.**, Neurotoxic effect of the dithiocarbamate tecoram on the chick embryo, *Toxicol. Appl. Pharmacol.*, 19, 675, 1971.
144. **Meth-Cohn, O.**, Thiophene, in *Kirk-Othmer Encyclopedia of Chemical Technology*, Vol. 20, 2nd ed., Mark, H. F., McKetla, J. J., Jr., and Othmer, D. F., Eds., John Wiley-Interscience, New York, 1969, 219.
145. **Mukhametova, G. M., Murtazina, L. F., Mustaeva, N. A., and Braginskaya, L. L.**, Some data on the effect of organic sulfur compounds on the human body, *Tezisy Dokl. Nauchn. Sess. Khim. Tekhnol. Org. Soedin. Sery Sernistykh Neftei*, 13, 87, 1974.
146. **Knikmatullaeva, S. S.**, Maximum permissible concentration of thiophene in the atmosphere, *Gig. Sanit.*, 32, 319, 1967.
147. **Knikmatullayeva, S. S.**, Study of the reflex and resorptive effects of thiophene, in *American Institute of Crop Ecology Survey of USSR Air Pollution Literature*, Vol. 15, Nuttonson, M. Y., Ed., American Institute of Crop Ecology, Silver Spring, Md., 1972, 46.
148. **Christomanos, A.**, Experimental production of cerebellar symptoms by thiophene, *Klin. Wochenschr.*, 9, 2354, 1930.
149. **Christomanos, A.**, Action of organic sulfur compounds on the dog organism: action and fate of thiophene in the metabolism of the dog, *Biochem. Z.*, 229, 248, 1930.
150. **Upners, T.**, Experimental studies concerning the local action of thiophene on the central nervous system, *Z. Ges. Neurol. Psychiat.*, 166, 623, 1939.
151. **Christomanos, A. and Scholz, W.**, Klinische beobachtungen und pathologisch anatomisches befunde am zentral nerven system mit thiophen vergifteter hande, *Z. Ges. Neurol. Psychiat.*, 144, 1, 1933.
152. **Anon.**, Toxicological review of thiophene and derivatives, *API Toxicological Reviews*, American Petroleum Institute, Washington, D.C., 1948, 2.
153. **Albrechtsen, R. and Jensen, H.**, Histochemical investigation of thiophen necrosis in the cerebellum of rats, *Acta Neuropathol.*, 26, 217, 1973.
154. Eastman Kodak Company, unpublished data, Rochester, N.Y., 1983.
155. **Flury, F. and Zernik, F.**, Toxicity of thiophene, *Chem. Ztg.*, 56, 149, 1932.
156. **Bradley, P. and Berry, M.**, Effects of thiophene on the Purkinje cell dendritic tree: a quantitative golgi study, *Neuropathol. Appl. Neurobiol.*, 5, 9, 1979.
157. **Albrechtsen, R., Diemer, N. H., and Nielsen, M. H.**, Size and density of the mitochondria in Purkinje cells of rats after thiophene intoxication as measured by image analyzing system, *Acta Pathol. Microbiol. Scand. Sect. A*, 82, 791, 1974.
158. **Herndon, R. M. and Oster-Granite, M. L.**, Effect of granule cell destruction on development and maintenance of the Purkinje cell dendrite, in *Advances in Neurology*, Vol. 12, Kreutzberg, G. W., Ed., Raven Press, New York, 1975, 361.
159. **McMurtry, R. J. and Mitchell, J. R.**, Renal and hepatic necrosis after metabolic activation of 2-substituted furans and thiophenes, including furosemide and cephaloridine, *Toxicol. Appl. Pharmacol.*, 42, 285, 1977.
160. **Bikbulatov, N. T. and Nigmatullina, G. N.**, Absorption, distribution, and elimination of thiophene from the body, *Sb. Nauchn. Tr. Bashk. Gos. Med. Inst.*, 19, 114, 1976.
161. **Bikbulatova, L. I., Bikbulatov, N. T., and Vazilo, V. E.**, Mechanism of action of thiophene on regenerative processes in the skin, *Vopr. Reakt. Org. Norme. Patol., Mater. Nauchn. Konf.*, 1974, 23.
162. **Chanal, J. L., Calmette, M. T., Bonnaud, B., and Cousse, H.**, Bioavailability of thiophene 2,5-14C after oral and rectal administration in mice, *Eur. J. Med. Chem.-Chim. Ther.*, 9, 641, 1974.
163. **Williams, R. T.**, *Detoxification Mechanisms*, John Wiley-Interscience, New York, 1959, 553.
164. **Hathway, D. E.**, *Foreign Compound Metabolism in Mammals*, Vol. 2, National Bureau of Economic Research, New York, 1972, 386.
165. **Collum, W. D. and Winek, C. L.**, Percutaneous toxicity of pyridinethiones in a dimethylsulfoxide vehicle, *J. Pharmaceut. Sci.*, 56, 1673, 1967.

166. **Snyder, F. H., Buehler, E. V., and Winek, C. L.**, Safety evaluation of zinc 2-pyridinethiol-1-oxide in a shampoo formulation, *Toxicol. Appl. Pharmacol.*, 7, 425, 1965.
167. **Snyder, D. R., Gralla, E. J., Coleman, G. L., and Wedig, J. H.**, Preliminary neurological evaluation of generalized weakness in zinc pyrithione-treated rats, *Food Cosmet. Toxicol.*, 15, 43, 1977.
168. **Sahenk, Z. and Mendell, J. R.**, Ultrastructural study of zinc pyridinethione-induced peripheral neuropathy, *J. Neuropathol. Exp. Neurol.*, 38, 532, 1979.
169. **Snyder, D. R., deJesus, C. P. V., Towfighi, J., Jacoby, R. O., and Wedig, J. H.**, Neurological, microscopic and enzyme-histochemical assessment of zinc pyrithione toxicity, *Food Cosmet. Toxicol.*, 17, 651, 1979.
170. **Johnson, D. E., Schardein, J. L., Goldenthal, E. I., Wazeter, F. Y., and Wedig, J. H.**, Reproductive toxicology of Omadine MDS, *Toxicologist*, 2, 75, 1982.
171. **Beck, J. E.**, Zinc pyridinethione and peripheral neuritis, *Lancet*, 1, 444, 1978.
172. **Sahenk, Z. and Mendell, J. R.**, Zinc pyridinethione, in *Experimental and Clinical Neurotoxicology*, Spencer, P. S. and Schaumburg, H. H., Eds., Williams & Wilkins, Baltimore, 1980, 578.
173. **Wedig, J. H.**, Disposition of zinc pyrithione in the rat, *Food Cosmet. Toxicol.*, 16, 553, 1978.
174. **Adams, M. D., Wedig, J. H., Jordan, R. L., Smith, L. W., Henderson, R., and Borzelleca, J. R.**, Urinary excretion and metabolism of salts of 2-pyridinethiol-1-oxide following intravenous administration to female Yorkshire pigs, *Toxicol. Appl. Pharmacol.*, 36, 523, 1976.
175. **Winek, C. L. and Buehler, E. V.**, Intravenous toxicity of zinc pyridinethiones and several zinc salts, *Toxicol. Appl. Pharmacol*, 9, 269, 1966.
176. **Cloyd, G. G., Wyman, M., Shadduck, J. A., Winrow, M. J., and Johnson, G. R.**, Ocular toxicity studies with zinc pyridinethione, *Toxicol. Appl. Pharmacol.*, 45, 771, 1978.
177. **Wedig, H., Feldman, R. J., and Maibach, H. I.**, Percutaneous penetration of the magnesium sulfate adduct of dipyrithione in man, *Toxicol. Appl. Pharmacol.*, 41, 1, 1977.
178. **Min, B. H., Parekh, C., Golberg, L., and McChesney, W. E.**, Experimental studies of sodium pyridinethione. II. Urinary excretion following topical application to rats and monkeys, *Food Cosmet. Toxicol.*, 8, 161, 1970.
179. **Howlett, H. C. S. and van Abbe, N. J.**, The action and fate of sodium pyridinethione when applied topically to the rabbit, *J. Soc. Cosmet. Chem.*, 26, 3, 1975.
180. **Kabacott, B. L., Fairchild, C. M., and Burnett, C.**, Pyridinethione glucuoride as a metabolite of sodium pyridinethione, *Food Cosmet. Toxicol.*, 9, 519, 1971.
181. **Spiker, R. C. and Ciuchta, H. P.**, Effects of pyrithiones and surfactants on zinc and enzyme levels in rabbits, *Am. Ind. Hyg. Assoc. J.*, 41, 248, 1980.
182. **Sahenk, Z. and Mendell, J. R.**, Evidence for the distal axon as the site of axoplasmic transport abnormality in ZPT-induced neuropathy, *Neurology*, 29A, 590, 1979.
183. **Moe, R. A., Kirpan, K., and Linegar, C. R.**, Toxicology of hydroxypyridinethione, *Toxicol. Appl. Pharmacol.*, 2, 156, 1960.
184. **Delahunt, C. S., Stebbins, R. B., Anderson, J., and Bailey, J.**, The cause of blindness in dogs given hydroxypyridinethione, *Toxicol. Appl. Pharmacol.*, 4, 286, 1962.
185. **Cloyd, G. G., Wyman, M., Shadduck, J. A., Winrow, M. J., and Johnson, G. R.**, Ocular toxicity studies with zinc pyridinethione, *Toxicol. Appl. Pharmacol.*, 45, 771, 1978.

Chapter 4

ALKANES, ALCOHOLS, KETONES, AND ETHYLENE OXIDE

John L. O'Donoghue

TABLE OF CONTENTS

I.	n-Hexane and Methyl n-Butyl Ketone	62
	A. Introduction	62
	B. Human Neurotoxicity	62
	C. Experimental Neurotoxicity	64
	D. Neuropathology	64
	E. Toxicokinetics	64
	F. Neurotoxicity of n-Hexane/MnBK Metabolites	76
	G. Role of 2,5-Hexanedione in n-Hexane/MnBK Neuropathies	76
	H. Diketone Neurotoxicity	77
	I. Mechanism(s) of Action	79
	J. Chemicals Tested for γ-Diketone Neurotoxicity	80
II.	Methanol	80
	A. Synonyms	80
	B. Introduction	80
	C. Human Neurotoxicity	82
	D. Experimental Neurotoxicity	84
	E. Metabolism and Mechanism of Action	84
III.	Ethylene Glycol	85
	A. Introduction	85
	B. Neurotoxicity	85
IV.	Ethylene Glycol Monomethyl Ether	86
	A. Introduction	86
	B. Neurotoxicity	86
V.	Ethylene Oxide	86
	A. Introduction	86
	B. Human Neurotoxicity	86
	C. Experimental Neurotoxicity	87
	D. Other Effects	87
	E. Related Substances	87
References		89

I. n-HEXANE AND METHYL n-BUTYL KETONE

A. Introduction

Aliphatic alkanes such as n-hexane and ketones such a methyl n-butyl ketone (MnBK or 2-hexanone) belong to large classes of substances with large scale commercial uses and production volumes in the billions of pounds per year. Aliphatic alkanes and ketones are used as solvents, extractants, fuels, chemical intermediates, and flavoring and fragrance agents. Exposures to high concentrations of alkanes or ketones result in narcosis or anesthesia but are generally not associated with lasting neurotoxic effects.

n-Hexane and methyl n-butyl ketone are exceptions and although there are other members of these classes which are neurotoxic, only n-hexane and MnBK have been associated with clinical neurotoxicity in humans. Commerical uses of n-hexane are very common while those of MnBK have largely been replaced by other ketones. The neurotoxicity of both these materials is linked by their metabolism which results in common neurotoxic metabolites which are more neurotoxic than the parent compounds.

In 1964, the earliest cases linking n-hexane to peripheral neuropathy were reported in Japanese printing plant[1] and pharmaceutical plant[2] workers. Yamamura[3] later reported 93 cases involving sandal makers employed in a cottage industry where industrial hygiene was often poor and inhalation exposures to n-hexane were high (500 to 2500 ppm) and prolonged over several months. In 1971, Herskowitz et al.[4] reported a similar neuropathy involving n-hexane in an American furniture finishing plant. Little attention was focused on these reports as they appeared to be isolated instances associated with poor hygenic conditions. In 1973, 86 workers in a plastics coating and printing plant developed clinical and subclinical peripheral neuropathy after working with a mixture of MnBK and methyl ethyl ketone (MEK).[5,6] The neuropathy in this instance was attributed to MnBK but it is likely that MEK potentiated the toxicity of MnBK.[7,8] Following this outbreak, few other cases of MnBK-related neuropathy have been reported[7,9-11] but awareness of the possibiity of neurotoxicity from similar materials has greatly expanded.[12-17]

n-Hexane has been recognized as the probable cause of peripheral neuropathy in about 400[18] cases involving Italian artisan shoe workers[18-27] and in cases from France,[28,29] Spain,[30] Morocco,[28] Portugal,[31] Japan,[32-36] and the U.S.A.[37,38,39]

Abuse of commercial products, particularly adhesives, lacquer thinners, paints, and solvents containing n-hexane, has been associated with peripheral neuropathies in several countries.[39-47] Although toluene has been cited as a cause of some of these neuropathies, there is no experimental evidence to substantiate these claims.[14,46,47] The exposures under abuse conditions are to very high air concentrations approaching or exceeding the anesthetic levels of the solvents and are repetitive in order to maintain a euphoric state for long periods of time.

B. Human Neurotoxicity

Minimum levels of n-hexane and MnBK which are neurotoxic to man have not been established. Many reports do not include atmospheric concentrations in their descriptions. When exposure conditions are indicated, it is often difficult to know how representative they are of actual exposures. Exposures are commonly to multicomponent solvent systems which may result in toxic interactions.

Exposures in Japanese and American industries associated with neuropathy were in the range of 500 to 1500 ppm of n-hexane for a month or more and included exposures to other hexanes. In the cases of Herskowitz et al.,[4] levels of n-hexane of 650 ppm with excursions to 1300 ppm resulted in clinical signs in 2 to 4 months. Sobue et al.[32] reported that 93 of 1662 shoe manufacturing workers exposed to a solvent which was >70% n-hexane developed polyneuropathy after exposure to 500 to 2500 ppm n-hexane. The maximum intensity of neuropathy was reached after 1 to 4 months of exposure in severe cases. A few reports refer to levels below 500 ppm. Ruff et al.[38] described the typical clinical appearance of n-hexane neuropathy in a janitor who had cleaned adhesive spills with n-hexane for several years. Limited environmental samples indicated exposures of 325 ppm which lasted for about 10 hr per week. Mutti et al.[27] found subjective neurologic complaints and changes in motor action potentials and nerve conduction velocities in a group of workers exposed to levels of 243 to 474 ppm of n-hexane and other solvents for from 1 to 25 years. Sanagi et al.[33] concluded that n-hexane vapor levels of less than 100 ppm for 8 hr/day were not likely to produce a clinical neuropathy, but mild subclinical changes in muscle strength and nerve conduction velocity may occur. Low levels of hydrocarbon mixtures including n-hexane are reported to affect peripheral nerve conduction velocity.[23] Levels below 100 ppm may be neuropathic if extensive skin exposures also occur.[35,36]

MnBK levels associated with human neurotoxicity have not been accurately determined. In the cases reported by Allen et al.[5] and Billmaier et al.,[6] a partial simulation of exposure conditions suggested MnBK levels of 9 to 36 ppm and MEK levels of 331 to 516 ppm but skin exposures included washing hands, equipment, and clothing with the solvents. The majority of cases occurred after 6 to 7 months of exposure with one case occurring with 5 weeks of exposure. The clinical courses of both n-hexane[3,24,25,32,38,48] and MnBK[5-7,9-11] neuropathy following occupational exposures are very similar. Prolonged exposures over a period of several weeks to months leads to a slowly developing insidious bilaterally, symmetrical, sensorimotor, peripheral neuropathy. Headache, nausea, anorexia, tiredness, and calf cramps occur in certain cases. Distal sensory changes involve numbness in the fingers and toes in a glove and stocking distribution. In more severe cases, distal weakness involves the hands and feet. Grasping objects such as coffee cups becomes difficult. Loss of touch, pressure, vibration, pinprick, temperature sensation, and Achilles tendon reflexes are common. Body weight loss which is not attributable to muscle atrophy may be prominent. In more severe cases, weakness involves proximal muscles and intrinsic muscles of the hand and feet atrophy. Progression of clinical signs may continue for 4 to 8 weeks following removal from exposure. In the most severe occupational cases, cranial nerve involvement may lead to blurring of vision, dysarthria, and facial numbness.[24,32]

The severity of clinical cases is dependent on the severity of exposure. The most severe cases have occurred in glue sniffers[39,40,43,45,46,49] and solvent huffers.[42] Clinical presentation in these cases may be similar to the Landry-Guillain-Barré syndrome with a progressive, symmetrical ascending polyneuropathy accompanied by severe muscle atrophy.[42,43,46] Sensory modalities, in contrast to the situation with occupational exposures, are affected little. In very rare cases, the respiratory muscles may atrophy or be paralyzed.[42,43]

Although changes in the visual system after n-hexane exposure have generally been infrequent, Raitta et al.[50] and Seppäläinen et al.[51] have described subclinical color discrimination defects and macular pigmentary changes in a group of workers exposed to n-hexane for ≥5 years.

Recovery from the neuropathic effects of n-hexane and MnBK is generally good but depends on the severity of the clinical cases. Mild to moderate cases may take several months to a year for complete recovery.[14] When significant muscle atrophy is apparent, complete recovery may not occur leaving the patient with residual sensory symptoms and weakness. Residual spasticity in two cases of n-hexane neuropathy indicates the potential for corticospinal tract damage.[14]

In many case studies, electrodiagnostic techniques including electromyography (EMG) and nerve conduction velocity have been used to follow the progress of the clinical condition. EMG effects appear early but slowing of nerve conduction velocity may not be useful in the individual cases because decreases in conduction velocity are mild initially.[14] However, studies involving large groups of workers may demonstrate subclinical effects if proper controls are used.[33]

C. Experimental Neurotoxicity

Peripheral neuropathy comparable to human cases has been reproduced using rats, cats, monkeys, hens, and pigeons exposed to n-hexane, practical grade hexanes which contain n-hexane, and benzine or gasoline containing n-hexane (Table 1). Exposures to commercial hexanes free of n-hexane and to purified components of commercial hexanes other than n-hexane have not caused peripheral neuropathy.[52]

Exposures of rats, cats, dogs, monkeys, hens, and guinea pigs to MnBK or MnBK mixtures have also resulted in peripheral neuropathies (Table 1). In experimental animals, the clinical course often appears to have a greater effect on the motor system, as seen in solvent abusers, rather than on the sensory system as occurs in occupational exposures. This is probably because the exposures are frequently to high concentrations or carried out for long periods of time, sometimes continuously. Weakness of the hindquarters with incomplete extension, base-wide stance, waddling gait, and drooping of the tail are commonly observed effects. Bowel and bladder functions are typically normal. Sensory changes such as loss of tactile placing occur but are not usually seen prior to the onset of weakness. Patellar reflex loss in dogs given 300 mg/kg of MnBK was an early sign of toxicity.[53] Pupillary reflex impairment is reported in guinea pigs given 0.1% or 0.25% MnBK in their drinking water for 24 weeks.[54] Damage to visual pathways and brain stem nuclei associated with vision has been reported in cats and rats given 0.5% 2,5-hexanedione, a metabolite of n-hexane and MnBK.[55,56]

D. Neuropathology

Both n-hexane and MnBK produce a central-peripheral distal axonopathy in experimental animals.[7,8,57-65] The term is used to indicate that pathology occurs earliest in the distal ends of axons in both the central and peripheral nervous system.[57-59] In rats exposed to n-hexane or MnBK, PNS axonal damage occurs first in the tibial nerve branches innervating the calf musculature.[59] At the same time, swollen axonal terminals appear in the gracile nucleus of the medulla oblongata indicating that both CNS and PNS axons are equally vulnerable in the rat.

The characteristic lesion associated with n-hexane and MnBK axonopathy is focal or "giant" axonal swelling occurring in the preterminal axon. Swelling is present in nerve terminals and axons in the gracile nucleus, cuneate nucleus, the medullary reticular formation, spinal grey matter and at neuromuscular junctions. Swollen axons are common within the tibial nerves, spinal cord, spinocerebellar tracts, cerebellar peduncles, and cerebellar folia. Swollen axons are less common in the sciatic nerve and spinal roots.

Axonal swellings are most common adjacent to nodes of Ranvier but also occur in internodal segments. Swellings contain masses of poorly aligned neurofilaments, and aggregates of organelles including mitochondria and glycogen. Neurotubules are misaligned in swollen segments and are not distributed normally within the axon.[7,65] Large myelinated axons are involved. "Giant" or focally swollen axons are also seen in spontaneous neuropathies occurring in man[66,67] and dog[68] and with neuropathies induced by acrylamide (Chapter 9), carbon disulfide, (Chapter 3), and β,β'-iminodipropionitrile (Chapter 2).

Myelin damage has usually been regarded as a secondary effect following axonal damage. Paranodal myelin sheaths retract over axonal swellings leading to demyelinated segments and followed by remyelination. In human neuropathies, similar lesions with typical "giant" axons have been reported.[7,8-11,40,43,44,61,69]

E. Toxicokinetics

Aliphatic alkanes and ketones can be metabolized by similar pathways in organisms from bacteria to mammals (Figure 1).[70-75] In vertebrates, the cytochrome P-450 oxidase system in the liver and other cell types plays a major role in oxidative metabolism of these materials.[72,73,77-81] Oxidation of alkanes and ketones can occur at any carbon location and

Table 1
ALKANES, ALCOHOLS, AND KETONES PRODUCING γ-DIKETONE NEUROTOXICITY

Test substance	Species	Route	Exposure conditions	Duration of exposure (wk)	Effects[a]	Ref.
n-Hexane	Rat	Inhal.	24,000 or 48,000 ppm, 10-min exposures 6, 12, or 24 times/d, 5 d/wk	18—22	Slight behavioral effects at 6 or 12 times/d exposures, for 18 wk; exposure increased to 24 times/d for 4 more weeks increased severity of neurobehavioral effects; no axonopathy	138—140
	Rat	Inhal.	10,000 ppm, 8 hr/d, 5 d/wk	15—19	Weight loss, paralysis, axonopathy	8
n-Hexane/MEK	Rat	Inhal.	9000 ppm n-hexane/1000 ppm MEK, 8 hr/d, 5 d/wk	15—19	Weight loss, paralysis, axonopathy	8
n-Hexane	Rat	Inhal.	5000 ppm, 9 hr/d, 5 d/wk	14	Dec. weight gain, axonopathy	130
			3000 ppm, 12 hr/d	16	Clin. neuropathy, axonopathy, nerve conduction effects	129
	Pigeon	Inhal.	3000 ppm, 5 hr/d, 5 d/wk	17	No neuropathy or axonopathy	141
	Rat	Inhal.	2500 ppm, 10 hr/d, 6 d/wk	30	Axonopathy	130
			2000 ppm 5 hr/d, 5 d/wk	20—24	Nerve conduction altered; myelin damage	142
			1500 ppm, 9 hr/d, 5 d/wk	14	No clinical neuropathy; no axonopathy	130
			1000 ppm, 24 hr/d, 5 d/wk	11	Auditory evoked response latency inc., nerve conduction effects	140
	Preg. Rat	Inhal.	1000 ppm, 6 hr/d	0.5	Minimal effect on perinatal development	143
	Rat	Inhal.	1000 ppm, 12 hr/d, 7 d/wk	16	Considerable nerve function impairment	144
			1000 ppm, 24 hr/d, 5 d/wk	11	Limb strength, motor activity, and avoidance, response reduced	138, 139
n-Hexane/Toluene	Rat	Inhal.	1000 ppm n-hexane/1000 ppm toluene 12 hr/d, 7 d/wk	16	Toluene reduced the neurotoxicity of n-hexane	144

Table 1 (continued)
ALKANES, ALCOHOLS, AND KETONES PRODUCING γ-DIKETONE NEUROTOXICITY

Test substance	Species	Route	Exposure conditions	Duration of exposure (wk)	Effects[a]	Ref.
n-Hexane	Rat	Inhal.	500 ppm, 12 hr/d, 7 d/wk	24	Nerve function impaired, axonopathy	145
			500 ppm, 9 hr/d, 5 d/wk	30	Dec. body weight, no neuropathy	130
			400—600 ppm, continuous	23	Clin. neuropathy after 45 days, axonopathy	57
			200 ppm, 12 hr/d, 7 d/wk	24	Slight nerve function impairment; axonopathy	145
		Gavage	4000 mg/kg/d, 5 d/wk	17	Dec. wt. gain, paralysis, axonopathy	51
n-Hexane/Benzine	Rat	Inhal.	125 ppm, 21 hr/d, 7 d/wk	24	No axonopathy	14
		Inhal.	Benzine with n-hexane levels of 200 or 500 ppm	24	Nerve impairment less than that observed with n-hexane alone	145
Practical hexane	Rat	Gavage	4000 mg/kg/d, 5 d/wk	17	Dec. wt. gain, no clinical neuropathy, axonopathy	51
n-Hexane	Rat	Gavage	1140 mg/kg/d, 5 d/wk	13	Dec. wt. gain, no neuropathy	51
			0.4—1.2 ml/kg/d, 7 d/wk	8	Dec. conduction velocity	145
			570 mg/kg/d, 5 d/wk	13	Dec. wt. gain, no neuropathy	51
	Hen	Gavage	100 mg/kg/d, 7 d/wk	13	Weakness with rapid recovery, no axonopathy	64
	Rat	s.c.	325 mg/kg/d, 5 d/wk	Several	Amplitude of sensory nerve action potential dec.	147
		i.p.	0.5 mg/kg/d, 5 d/wk	20	Excitability of sweat glands dec.	146
		i.p.	650—2000 mg/kg/d, 5 d/wk	35	No neuropathy; axonopathy	57
	Hen	i.p.	200 mg/kg/d, 7 d/wk	13	Weakness with rapid recovery; no axonopathy	64
		i.p.	100 mg/kg/d, 7 d/wk	13	Weakness with rapid recovery; no axonopathy	64
MnBK	Rat	Inhal.	1300 ppm, 6 hr/d, 5 d/wk	7	Weight loss; clin. neuropathy: axonopathy	40
			1050—1460 ppm, 6 hr/d, 5 d/wk	15	Neuropathy in 11 wk; axonopathy	128

Compound	Species	Route	Dose/Exposure	Duration	Effects	Ref.
	Monkey	Inhal.	1000 ppm, 6 hr/d, 5 d/wk	25	Hindlimbs dragged; nerve conduction velocity dec.; behavioral response rate dec.	136, 148
		Inhal.	1000 ppm, 6 hr/d, 5 d/wk	25	Hindlimbs dragged; nerve conduction velocity dec.; visual evoked potential latency inc., wt. loss	136, 148
	Rat	Inhal.	700 ppm, 72 hr/wk	11	Clinical neuropathy after 2 wk; axonopathy	149
		Inhal.	600—400 ppm, continuous		Hindlimbs dragged in 4—5 wk; axonopathy	150
	Cat	Inhal.	600—400 ppm, continuous	13	Clinical neuropathy; axonopathy	7, 150
MnBK/MiBK	Hen	Inhal.	400 ppm of a 70/30 MnBK/MiBK mixture, continuous		Wt. loss, paralysis, axonopathy	151
MnBK	Rat	Inhal.	400 ppm, continuous	6	Paralysis after 42 days, axonopathy	7
		Inhal.	330 ppm, 6 hr/d, 5 d/wk	104	Dec. wt. gain, minimal axonopathy	17
	Cat	Inhal.	330 ppm, 6 hr/d, 5 d/wk	104	No clinical neuropathy; axonopathy	17
	Rat	Inhal.	225 ppm, continuous	7	Paralysis after 66 days; axonopathy	7
			225 ppm continuous with or without concurrent exposure to 0.1 g% phenobarbital in their drinking water	11	Phenobarbital protected against induction of neuropathy	77, 79
MnBK/MEK	Rat	Inhal.	225 ppm MnBK/1125 ppm MEK, continuous	7	Paralysis after 25 days; axonopathy	7
			225 ppm MnBK/750 ppm MEK, continuous	3	Lethal exposure; severe neuropathy	77, 79
MnBK	Rat	Inhal.	200 ppm, 8 hr/d, 5 d/wk	6	Axonopathy	152
	Hen	Inhal.	200—100 ppm, continuous	5	Inability to stand; axonopathy	150
MnBK/MiBK	Rat	Inhal.	200 ppm of a 70/30 MnBK/MiBK mixture, continuous	13	Wt. loss, paralysis, axonopathy	151
MnBK/MEK	Rat	Inhal.	200 ppm MnBK/2000 ppm MEK, 8 hr/d, 5 d/wk	6	Axonopathy	152

Table 1 (continued)
ALKANES, ALCOHOLS, AND KETONES PRODUCING γ-DIKETONE NEUROTOXICITY

Test substance	Species	Route	Exposure conditions	Duration of exposure (wk)	Effects[a]	Ref.
M*n*BK	Rat	Inhal.	100 ppm, 22 hr/d, 7 d/wk	24	Axonopathy	52
	Monkey	Inhal.	100 ppm, 6 hr/d, 5 d/wk	29	Nerve conduction velocity dec.	136, 148
	Rat	Inhal.	100 ppm, 6 hr/d, 5 d/wk	41	Nerve conduction velocity dec.	136, 148
		Inhal.	100 ppm, 6 hr/d, 5 d/wk	104	Dec. wt. gain, no neuropathy or axonopathy	17
	Cat	Inhal.	100 ppm, 6 hr/d, 5 d/wk	104	No adverse effect	17
M*n*BK/MiBK	Hen	Inhal.	100 ppm of a 70/30 M*n*BK/MiBK mixture, continuous	13	Ataxia; axonopathy	151
M*n*BK	Rat	Inhal.	50 ppm, 8 hr/d, 5 d/wk	24	Nerve conduction velocity dec.; minimal axonopathy	153
M*n*BK/MiBK	Hen	Inhal.	50 ppm of a 70/30 M*n*BK/MiBK mixture, continuous	13	Ataxia; axonopathy	151
		Inhal.	10 ppm of a 70/30 M*n*BK/MiBK mixture, continuous	13	No adverse effect	151
M*n*BK	Rat	Water	1.0% in drinking water	17	Dec. wt. gain; clinical neuropathy	131
			1.0% in drinking water (C.560 mg/kg)	57	Clinical neuropathy in 6 wk; axonopathy; dec. wt. gain	17
			0.5% in drinking water (C.266 mg/kg)	57	Clinical neuropathy in 10 wk; axonopathy; dec. wt. gain	17
			0.25% in drinking water (C.143 mg/kg)	58	No clinical neuropathy; axonopathy; dec. wt. gain	17
			0.1% or 0.25% in drinking water	24	Impaired pupillary response; dec. motor activity	54
		Gavage	660 mg/kg/d, 5 d/wk	13	Clinical neuropathy; paralysis in 56 days, dec. wt. gain, axonopathy	92
			400 mg/kg, 7 d/wk	40	Transient weakness	154
	Hen	Gavage	100 mg/kg/d, 7 d/wk	13	Ataxia, axonopathy	64

Compound	Species	Route	Dose	Effect	Ref.
	Rat	s.c.	415 mg/kg/d, 5 d/wk	Several Nerve conduction velocity and action potential dec.	147
	Cat	s.c.	300 mg/kg/d, 5 d/wk	8.5 Clinical neuropathy after 8—10 wk; paralysis after 16 wk; axonopathy	60
	Dog	s.c.	300 mg/kg/d, 5 d/wk	18—44 Subclinical neuropathy to paralysis; axonopathy	17
MnBK/MEK	Cat	s.c.	300 mg/kg/d, 5 d/wk of a 1/9 MnBK/MEK mixture	8.5 Segmental remyelination in the tibial nerve	60
	Dog	s.c.	300 mg/kg/d, 5 d/wk of a 1/9 MnBK/MEK mixture	11 No adverse effect	17
MnBK	Rat	s.c.	0.5 mg/kg/d, 5 d/wk	16 Dec. excitability of sweat glands	146
	Hen	i.p.	200 mg/kg/d, 7 d/wk	13 Ataxia with near paralysis; axonopathy	64
	Rat	i.p.	100 mg/kg/d, 7 d/wk	13 Ataxia; paralysis; axonopathy	64
	Rat	i.p.	10—20, 30—60, or 100—200 mg/kg/d, 5 d/wk	35 No clinical neuropathy; no axonopathy	17
MnBK/MEK	Rat	i.p.	10—20, 30—60, or 100—200 mg/kg/d, 5 d/wk	35 No clinical neuropathy; no axonopathy	17
MnBK	G.P.	Skin	2 mℓ with or without DMSO, 5 d/wk	31 No clinical neuropathy; no axonopathy	17
MnBK/MEK	G.P.	Skin	2 mℓ of a 1/9 MnBK/MEK mixture, 5 d/wk	31 No clinical neuropathy; no axonopathy	17
2,5-Hexanedione	Mouse	Water	0.5% or 2% in drinking water	15 No clinical neuropathy; no wt. loss; prolonged resolution of a demyelinating lesion	155
	Rat	Water	1% in drinking water (C.1600 mg/kg)	18 Transient hyperalgesia; no clinical neuropathy	156
	Rat	Water	1% in drinking water (C.750 mg/kg)	7 Lethal level; clinical neuropathy in 43 days; paralysis in 46 days; axonopathy; wt. loss	157
		Water	1% in drinking water	6 Testes atrophied; clin. neuropathy	107, 108, 110
			1—0.5% in drinking water	13—23 Clinical neuropathy; axonopathy	158
	Mouse	Water	1% in drinking water	13—23 Clinical neuropathy; axonopathy	158
	Rat	Water	0.5% in drinking water	12 Clinical neuropathy in 28 days; paralysis in 10—12 wk	104

Table 1 (continued)
ALKANES, ALCOHOLS, AND KETONES PRODUCING γ-DIKETONE NEUROTOXICITY

Test substance	Species	Route	Exposure conditions	Duration of exposure (wk)	Effects[a]	Ref.
			0.5% in drinking water	3—9	Axonopathy in optic nerve tracts, glucose utilization dec. in superior colliculus, clin. neuropathy	56
			0.5% in drinking water (C.450 mg/kg)	16	Clinical neuropathy in 52 days; paralysis; axonopathy; dec. wt. gain; testes atrophied	157
			0.5% in drinking water	12	Clinical neuropathy; axonopathy; dec. wt. gain	96, 97
	Mouse	Water	0.5% in drinking water	16	Slight decline in rotarod performance; slight inc. in hindlimb splay; slight dec. wt.	159
	Cat	Water	0.5% in drinking water	19	Clinical neuropathy in 66—75 days; quadriparesis; axonal degeneration in the mamillary bodies and visual nuclei	55
	G.P.	Water	0.5% in drinking water	8	Lethal level; decreased motor activity	54
	Rat	Water	0.1 and 0.5% in drinking water	8	Clinical neuropathy in 8 wk and dec. wt. gain at 0.5%; biophysical changes in nerve lipids at both doses	111
			0.25% in drinking water (C.240 mg/kg)	24	Clinical neuropathy in 101 days; paralysis in 134 days; axonopathy; dec. wt. gain; testes atrophied	157
	G.P.	Water	0.1% in drinking water	24	Pupillary response impaired	54

Compound	Species	Route	Dose	Duration	Effects	Ref.
	Crayfish	Water	1.48 g% in an aquarium	3	Nerve conduction velocity amplitude, and resting potential decreased	160
	Rat	Gavage	755 mg/kg/d, 5 d/wk	13	Paralysis in 17 days; wt. loss; testes atrophied	92
			400 mg/kg/d, 7 d/wk	5.5	Clinical neuropathy in 3 wk; paresis in 5 wk; dec. wt.	154
	Hen	Gavage	200 mg/kg/d, 5 d/wk	5	Clinical neuropathy; axonopathy; wt. loss; pyrrole adducts in CNS and PNS tissue	100, 114
	Dog	Oral	110 mg/kg/d, 7 d/wk	20	Clinical neuropathy in 2 mos., paresis in 4—6 mos.; axonopathy	161
	Hen	Gavage	100 mg/kg/d, 7 d/wk	13	Ataxia; axonopathy	64
	Rat	s.c.	485 mg/kg/d, 5 d/wk	Several	Nerve conduction velocity and action potential dec.	147
			200 or 400 mg/kg/d, 5 d/wk	4—14	Nerve conduction altered	162
			340 mg/kg/d, 7 d/wk	19—23	Clinical neuropathy; axonopathy	162
			0.5 mg/kg, 5 d/wk	2	Sweat gland excitability dec.	146
		i.p.	456 mg/kg/d, 5 d/wk	5	Paralysis after 5 wk; axonopathy	98
			450—300 mg/kg/d, 7 d/wk	5	Clinical neuropathy in 13 days; axonopathy; change in endplate activity	164
			400 mg/kg/d, 7 d/wk	1—4	Clinical neuropathy; axonopathy; ganglionic transmission impaired	165
			200 mg/kg/d, 5 d/wk	6—30	No clinical neuropathy; axonopathy; chronic peritonitis	17
	Hen	i.p.	100 or 200 mg/kg/d, 7 d/wk	13	Ataxia; paralysis; axonopathy	64
	Rat	i.p.	35 mg/kg/d, 5 d/wk	8	No clinical neuropathy	98
	Rat	i.p.	35.5 mg/kg/d, 5 d/wk	2.8	Paralysis in 2.8 wk; proximal axonopathy; more neurotoxic than 2,5-hexanedione	98
3,4-Dimethyl-2,5-hexanedione						
3,3-Dimethyl-2,5-hexanedione	Rat	Gavage	1000 mg/kg/d, 5 d/wk	13	Clinical neuropathy, axonopathy, less neurotoxic than 2,5-hexanedione	168
2,5-Hexanediol	Rat	Water	0.5% in drinking water	68	Demyelination and remyelination secondary to axonopathy	166
			0.5% in drinking water	14—24	Clinical neuropathy; axonopathy; dec. wt. gain	96

Table 1 (continued)
ALKANES, ALCOHOLS, AND KETONES PRODUCING γ-DIKETONE NEUROTOXICITY

Test substance	Species	Route	Exposure conditions	Duration of exposure (wk)	Effects[a]	Ref.
		Gavage	780 mg/kg/d, 5 d/wk	13	Paralysis in 29 days; testes atrophied	92
			400 mg/kg/d, 7 d/wk	10.5	Clinical neuropathy in 5—6 wk; paresis in 10 wk.; dec. wt. gain	154
	Hen	Gavage	100 mg/kg/d, 7 d/wk	13	Ataxia; near paralysis; axonopathy	64
		i.p.	100 or 200 mg/kg/d, 7 d/wk	13	Ataxia; paralysis, axonopathy	64
	Rat	Water	2%—1% in drinking water	7	Clinical neuropathy; axonopathy	167
2-Hexanol	G.P.	Water	0.5% in drinking water	24	Pupillary response impaired; motor activity decreased	54
			0.1% in drinking water	24	Pupillary response impaired	54
	Rat	Gavage	675 mg/kg/d, 5 d/wk	13	Paralysis in 84 days; testes atrophied	92
		i.p.	101—152 mg/kg/d, 6 d/wk	34	Conduction velocity dec.; distal latency prolonged; axonopathy	133
		s.c.	400 mg/kg/d, 5 d/wk	21	Clinical neuropathy; nerve conduction velocity reduced; dec. wt. gain	137
5-Hydroxy-2-hexanone	Rat	Gavage	765 mg/kg/d, 5 d/wk	13	Paralysis in 22 days; axonopathy; testes atrophied	92
		Water	0.5% in drinking water	5—14	Clinical neuropathy in 28 days; paralysis; axonopathy; dec. wt. gain	17
			0.25% in drinking water	11—26	Clinical neuropathy in 42 days; paralysis; axonopathy; dec. wt. gain	17
Ethyl n-butyl ketone	Rat	Inhal.	700 ppm, 72 hr/wk	24	No clinical neuropathy; no axonopathy	149
		Water	1% drinking water	17	No clinical neuropathy; no axonopathy	131

Compound	Species	Route	Dose	Duration	Result	Ref.
		Gavage	2000 mg/kg/d, 5 d/wk	14	Clinical neuropathy, seizures; axonopathy, dec. wt. gain	118
			250, 500, or 1000 mg/kg/d, 5 d/wk	14	No clinical neuropathy or axonopathy	118
Ethyl n-butyl ketone/MEK	Rat	Gavage	1000 mg/kg EBK/1500 mg/kg MEK, 5 d/wk	14	Clinical neuropathy; axonopathy; dec. wt. gain	118
			250 or 500 mg/kg EBK/750 or 1500 mg/kg 5M2O, 5 d/wk	14	No clinical neuropathy or axonopathy	118
Ethyl n-butyl ketone/5-methyl-2-octanone	Rat	Gavage	250, 500 or 1000 mg/kg EBK/1500 mg/kg 5M2O, 5 d/wk	14	No clinical neuropathy or axonopathy	118
5-Nonanone	Rat	Gavage	150—1500 mg/kg/d, 7 d/wk	14	Paralysis; axonopathy	121
			2000 mg/kg/d, 5 d/wk	13	Clinical neuropathy in 11 days; paralysis in 3—4 wk; axonopathy; wt. loss	119
			1000 mg/kg/d, 5 d/wk	13	Clinical neuropathy in 28 days; axonopathy; dec. wt. gain	119
			233 mg/kg/d, 5 d/wk	13	No clinical neuropathy; minimal axonal damage	109
5-Nonanone/MEK	Rat	Gavage	233 mg/kg/d 5-nonanone plus 742 mg/kg/d MEK, 5 d/wk	13	No clinical neuropathy; minimal axonal damage	109
Commercial grade 5-methyl-2-octanone (methyl heptyl ketone) containing 12% 5-nonanone	Rat	Gavage	2000 mg/kg/d, 5 d/wk	13	Clinical neuropathy in 59 days; axonopathy; dec. wt. gain	109
5-Nonanone/5-methyl-2-octanone	Rat	Gavage	233 mg/kg/d 5-nonanone plus 1540 mg/kg 5-methyl-2-octanone, 5 d/wk	13	Clinical neuropathy in 38 days; axonopathy; dec. wt. gain	109

[a] Clinical neuropathy means evidence of peripheral nerve damage on physical examination. Axonopathy is used to indicate histological or ultrastructural verification of axon damage usually axonal swelling.

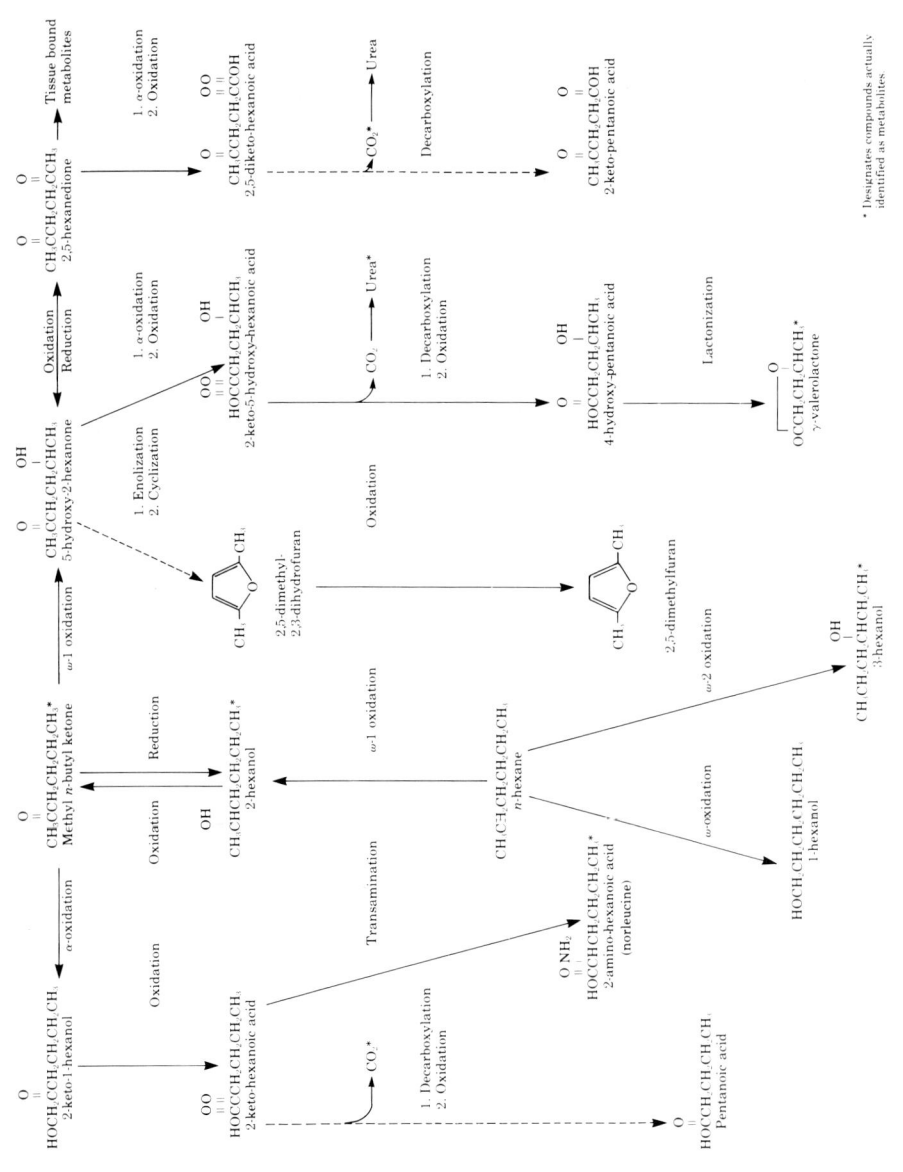

FIGURE 1. Proposed metabolic scheme for methyl *n*-butyl ketone, *n*-hexane, and 2,5-hexanedione.

there appear to be different oxidases with different affinities for each location.[72,73,80] In preparations of rabbit liver microsomal cytochrome P-450LM$_{2-4}$, P-450LM$_2$ was the most active in metabolizing n-hexane to 2-hexanol.[80] Oxidation of n-hexane at ω, ω-1, or ω-2 locations results in the formation of 1, 2, or 3-hexanol, respectively.[72,73] Oxidation occurs most readily at the weakest carbon bond; therefore, it is most likely to occur at tertiary bonds, followed by secondary bonds and is least likely to occur at primary bonds.[72] 1-Hexanol then is the least likely oxidation product of n-hexane; 2-hexanol and 3-hexanol should have equal chances of being formed.[72] Possibly due to steric hinderance at the ω-2 position, the ω-1 position is preferred and 2-hexanol is the predominant oxidation product of n-hexane.[72] Further oxidation of 2-hexanol results in the production of MnBK or 2,5-hexanediol followed by the formation of 5-hydroxy-2-hexanone, 2-5,hexanedione (2,5-HD) and other metabolites.[16,74,78,79]

Baker and Rickert[82] described n-hexane uptake, distribution, and elimination in Fischer rats exposed to air levels of 500 to 10,000 ppm. n-Hexane was rapidly absorbed and steady-state tissue levels were reached within 2 hr. Respiratory uptake of n-hexane in man varies widely and is at least partially dependent on body fat.[83] Nomiyama and Nomiyama[84] reported respiratory uptake of n-hexane in man was 28%. Tissue elimination in the rat was rapid with half-lives of 1 to 2 hr except for the kidneys in which the half-life was 6 hr.[82] n-Hexane was rapidly metabolized to MnBK but MnBK levels were not directly related to n-hexane exposure levels.[82] Peak blood concentrations of 2,5-HD were 6.1 μg/mℓ after exposure to 1000 ppm n-hexane and did not increase with increasing n-hexane concentration.[74] 3000 ppm and 10,000 ppm n-hexane produced only 4.1 μg/mℓ and 3.8 μg/mℓ of 2,5-HD, repectively. In the first 12-hr postexposure, 50 to 80% of metabolites were excreted in the urine. 5-Hydroxy-2-hexanone, 2-hexanol, 2,5-dimethylfuran, MnBK, and 2,5-HD were detected in the urine.[82] Glucuronides of 2-hexanol and 2,5-dimethylfuran were present in the urine.[82] Acid hydrolysis increased the urine detection of these materials and 2,5-hexanedione.[82]

Perbellini et al.[85] reported a species difference in urinary excretion of n-hexane metabolites. Rats excreted 2-hexanol, 3-hexanol, MnBK, γ-valerolactone, and 2,5-dimethylfuran. The principal urinary metabolite was 2,5-dimethylfuran in rats and 2-hexanol in rabbits and monkeys. Postexposure excretion of 2,5-hexanedione continued for 60 hr indicating a residence time much longer than the parent compound. Humans exposed to 10 to 140 ppm of n-hexane excreted 0.4 to 21.7 mg/ℓ of 2,5-HD.[85] In man, urinary metabolites include 2,5-HD, 2-hexanol, 2,5-dimethylfuran, and γ-valerolactone. 2,5-HD was the principal urinary metabolite excreted by man.[86,87]

Modulation of cytochrome P-450 oxidases alters the metabolism and neurotoxicity of n-hexane. Exposure to toluene decreases the neurotoxic potency of n-hexane[88] and decreases urinary excretion of all n-hexane metabolites.[89] Phenobarbital increases the excretion of n-hexane urinary metabolites[78,89] and stimulates hydroxylation of n-hexane to 2- and 3-hexanol.[73] Certain phthalate esters increase urinary output of 2- and 3-hexanol.[90] Exposure to n-hexane at 900 ppm does not appear to alter P-450-mediated reactions and did not induce total liver microsomal cytochrome P-450 or its subfractions.[81] Excess n-hexane may inhibit metabolism of 2-hexanol to 2,5-hexanediol.[73]

MnBK is also metabolized by hepatic cytochrome P-450 oxidases resulting in 5-hydroxy-2-hexanone and 2,5-hexanedione.[74,75,78,79] Hepatic cytosol fractions are required to reduce MnBK to 2-hexanol.[77] 2,5-hexanediol may be formed either by further oxidation of 2-hexanol or reduction of 5-hydroxy-2-hexanone. In rats given 20 or 200 mg/kg of MnBK by gavage, approximately 44% of the dose was eliminated in the breath as CO_2 (38%) or MnBK (6%), 40% of the dose was excreted in the urine, 1.4% in the feces, and 14% was retained in the body 48 hr after dosing and 8% remained after 8 days.[76] Pretreatment was SKF 525A markedly increased excretion of CO_2 and decreased urinary metabolite excretion.[76] Phenobarbital pretreatment increased breath elimination of CO_2, decreased MnBK breath excretion, and increased urinary metabolite excretion.[76] Pretreatment with MnBK slightly increased excretion of CO_2 and decreased urinary metabolite excretion.[76]

Urinary metabolites of MnBK in the rat include 2-hexanol, 2,5-HD, 5-hydroxy-2-hexanone, γ-valerolactone, norleucine, urea, and 2,5-dimethylfuran. 2-Hexanol, 5-hydroxy-2-hexanone, and 2-5-dimethylfuran are excreted as glucuronides.[76]

Human volunteers ingesting 0.1 mg/kg of [1-^{14}C] MnBK excreted 40% of the dose in their breath and 26% in urine.[75] 66% of the dose was recovered over an 8-day period. The excretion of MnBK in man was slower than that of the rat (92% in 6 days), suggesting an increased potential for accumulation of metabolites by humans.[75,76]

Chemicals which affect cytochrome P-450 activity may affect MnBK metabolism and neurotoxicity. Combined exposures to MnBK/MEK resulted in neurotoxicity sooner than with MnBK alone.[7,8,77,78] Total urinary metabolites, urinary excretion of 2-hexanol and 2,5-HD, and hepatic microsomal enzyme activity were increased by MnBK/MEK exposure compared to MnBK exposure.[78] Phenobarbital pretreatment enhanced in vitro conversion of MnBK to 2-hexanol by hepatic cytosol fractions and enhanced the production of 2,5-HD by liver microsome fractions.[70] Phenobarbital treatment of rats and guinea pigs increased blood levels of 2-hexanol and 2,5-HD, enhanced urinary excretion of 2,5-HD in the rat, and enhanced MnBK, 2-hexanol, and 2,5-HD excretion, particularly glucuronides, of 2-hexanol and MnBK in the guinea pig.[79] Increased excretion of glucuronide-conjugated metabolites may explain the protective effect of phenobarbital in MnBK exposed rats.[79]

Although MnBK and n-hexane share common metabolites, there is a major difference in their relative neurotoxic potency. Krasavage et al.[92] have shown that n-hexane is <0.08 times as neurotoxic as MnBK given by gavage to rats and Abou-Donia et al.[64] found it about 0.3 times as neurotoxic as MnBK given orally or i.p. to hens. Inhalation exposures may result in additional differences because of the wide differences in uptake of n-hexane (28%)[84] and MnBK (98%)[75] in man. The higher water solubility of MnBK is thought to increase absorption and distribution in tissue. The primary reason for differences in neurotoxicity is presumed to be the lower, approximately fivefold difference in production of 2,5-HD from metabolism of n-hexane.[92]

F. Neurotoxicity of n-Hexane/MnBK Metabolites

The metabolites of n-hexane and MnBK, including 2-hexanol, 2,5-hexanediol, 5-hydroxy-2-hexanone, and 2,5-HD, have not been reported to cause neuotoxicity in man but all produce a central-peripheral distal axonopathy (Table 1) in laboratory species which is morphologically identical to that produced by the parent compounds. These metabolites have very little commercial use. 2-Hexanol and 2,5-hexanediol have been produced for use in synthetic chemistry. 2,5-HD has had limited used as a lacquer solvent and in synthetic chemistry. 5-Hydroxy-2-hexanone has not been produced commercially.

The relative neurotoxicity of n-hexane, MnBK, and their metabolites have been studied in rats[92] and hens[64] with similar results. The neurotoxicity of the parent compounds and metabolites increases with oxidation and is directly proportional to the amount of 2,5-HD resulting from metabolism of each material.[92]

G. Role of 2,5-Hexanedione in n-Hexane/MnBK Neuropathies

2,5-Hexanedione is 3.3 times more neurotoxic than MnBK and 38 times more neurotoxic than n-hexane in rats.[92] Using different criteria, Abou-Donia et al.[64] reported results suggesting 2,5-HD was 1 to 2 times more neurotoxic than MnBK and 4 to 7 times more neurotoxic than n-hexane in hens. The high neurotoxic potency of 2,5-HD is responsible for the neurotoxic properties of n-hexane and MnBK. Another is its prolonged residence time and slow excretion from the body.[74,75] Angelo[93] has found that 2,5-HD is extensively metabolized to CO_2 (up to 45%) but that incorporation into macromolecular and cellular elements of tissues is responsible for the persistence of 2,5-HD in tissues. Brain and muscle have the longest 2,5-HD half-lives, 33 and 32 days, respectively, after 21 days of i.p. exposures. In tissues, the majority of radioactivity recovered from subcellular tissue fractions from rats given ^{14}C-2,5-HD was in the residual fraction suggesting extensive incorporation of 2,5-HD. Radioactivity recovered from the brains of animals given ^{14}C-2,5-hexanedione appears to partition into two regions; one corresponds to 2,5-HD and the other to polar metabolites. These data further support the hypothesis that 2,5-HD is the ultimate neurotoxin in these neuropathies.

H. Diketone Neurotoxicity

Realization of the importance of 2,5-HD in the production of neurotoxicity led to studies on molecular requirements necessary to induce axonopathy with 2,5-HD.[95-97] Several diones including 2,4-pentanedione, 2,3-hexanedione, 2,4-hexanedione, 2,5-hexanedione, 2,5-heptanedione, 3,3-dimethyl-2,5-hexanedione, 3,4-dimethyl-2,5-hexanedione, 2,5-heptanedione, 2,6-heptanedione, 3,5-heptanedione, and 3,6-octanedione have been studied for neurotoxic effects (Table 2).[17,60,95-97] These compounds provide a series of compounds of differing chain length, dione spacing, and symmetry. Only those with γ-spacing of carbonyl groups produced axonal pathology similar to n-hexane and MnBK. Increasing the chain length progressively from hexanedione to octanedione resulted in a decrease in activity.[95] Similarly, addition of two methyl groups to the 3-position on 2,5-hexanedione dramatically lessens neurotoxicity[60] while addition of single methyl groups to the 3 and 4 positions on 2,5-hexanedione dramatically increases neurotoxicity.[98] Changes in activity following addition of methyl groups between 2,5-HD carbonyls are likely the result of enhancing enolization in the case of 3,4-dimethyl-2,5-HD[88] and reducing enolization of carbonyls by 3,3-dimethyl-2,5-HD.

2,3-Hexanedione, 2,4-hexanedione, 2,6-heptanedione, and 3,5-heptanedione were not neurotoxic.[95-97] 2,4-pentanedione produced an anatomically, morphologically, and clinically different neurotoxic effect in rats than did γ-diketones.[17] Rats given 125 to 150 mg/kg of 2,4-pentanedione by gavage repeatedly suddenly develop weakness, ataxia, tremors, head tilt, and alternating rolling movements of the head. Microscopic changes in the brain included perivascular edema and hemorrhage and endothelial cell swelling in the brainstem and cerebellum in animals examined shortly after onset of signs.[17] Chronic lesions were bilateral, symmetrical malacic foci and gliosis in the cerebellar peduncles, olivary nuclei, and lower brainstem.[17] Rabbits did not develop similar lesions.

The importance of the γ-spacing for neurotoxicity suggests two possible interactions between 2,5-HD and possible sites important for neurotoxicity. These are imine formation resulting in inter- or intramolecular cross-linking by 2,5-hexanedione or a more stable reaction with amino groups resulting in pyrrole formation.[95,99,100] Graham[99,101,102] studied the reaction of 2,5-HD with bovine serum albumin, chicken ovalbumin, polylysine, and glyceraldehyde-3-phosphate dehydrogenase and reported the production of an orange chromophore which was interpreted as due to formation of a conjugated imine.

2,5-HD has been used in synthetic chemical preparations for the production of 2,5-substituted pyrroles from amino acids. DeCaprio et al.[100] reacted 2,5-HD with primary amines resulting in the formation of substituted pyrroles. 2,5-HD and other γ-diketones also reacted with the ε-amino groups(s) of lysine and polylysine. γ-Diketones but not acetone, 2,3-hexanedione, or 2,4-hexanedione formed pyrroles with ε-aminocaproic acid. Pyrrole adducts can also be demonstrated in serum proteins from hens intoxicated with 2,5-HD and from protein hydrolysates exposed to 2,5-HD in vitro.[100] Orange chromophores developing in these γ-diketone/amine mixtures were identified as oxidation products of pyrrole reaction products.[100]

Table 2
STRUCTURE OF DIKETONES TESTED FOR CENTRAL-PERIPHERAL DISTAL AXONOPATHY WITH GIANT AXONAL SWELLING

Diketone	Structure	Diketone spacing	Giant axonal neuropathy
2,4-Pentanedione	$CH_3CCH_2CCH_3$ (O, O)	β	No[a]
2,3-Hexanedione	$CH_3CCCH_2CH_2CH_3$ (O, O)	α	No
2,4-Hexanedione	$CH_3CCH_2CCH_2CH_3$ (O, O)	β	No
2,5-Hexanedione	$CH_3CCH_2CH_2CCH_3$ (O, O)	γ	Yes
2,5-Heptanedione	$CH_3CCH_2CH_2CCH_2CH_3$ (O, O)	γ	Yes
2,6-Heptanedione	$CH_3CCH_2CH_2CH_2CCH_3$ (O, O)	δ	No
3,5-Heptanedione	$CH_3CH_2CCH_2CCH_2CH_3$ (O, O)	β	No
3,6-Octanedione	$CH_3CH_2CCH_2CH_2CCH_2CH_3$ (O, O)	γ	Yes
3,3-Dimethyl-2,5-hexanedione	$CH_3CCHCH_2CCH_3$ with OCH_3, O and CH_3	γ	Yes
3,4-Dimethyl-2,5-hexanedione	$CH_3CCHCHCCH_3$ with CH_3CH_3 (O, O)	γ	Yes

[a] 2,4-Pentanedione produces multifocal malacia in the brainstem of rats but does not produce nerve damage similar to 2,5-hexanedione.

I. Mechanism(s) of Action

Although several possible mechanisms of action have been identified and evidence gathered in support of some, no theory has received general acceptance. Spencer, Sabri, and colleagues[14,103,104] have postulated that inhibition of glycolytic pathways by 2,5-HD, M*n*BK, and other neurotoxic materials may be the underlying cause of central-peripheral distal axonopathies. Neurologic functions, particularly axonal transport functions, are highly dependent on glycolysis. Mendell et al.[65] have shown that fast axoplasmic flow is inhibited by M*n*BK. Sabri et al.[103,104] found that 2,5-HD and M*n*BK inhibited the important glycolytic enzyme glyceraldehyde-3-phosphate dehydrogenase (GAPDH) from endogenous neural sources and in crystalline form. Graham and Abou-Donia[101] reported that 2,5-HD irreversibly bound to GAPDH probably by reacting with amino groups in the enzyme rather than sulfhydryl groups and seriously questioned the role of GAPDH inhibition in *n*-hexane neuropathy. Sabri et al.[104] also reported inhibition of another important glycolytic enzyme, fructose-6-phosphate kinase (PFK), by 2,5-HD and m*n*BK in vitro and in brain homogenates from rats intoxicated with 2,5-HD. Protection of sulfhydryl groups on GAPDH and PFK by dithiothreitol prevented inhibition of GAPD and PFK although it did not restore inhibited enzyme activity. Howland et al.[105] confirmed the in vitro inhibition of GAPDH and PFK and also reported inhibition of a third glycolytic enzyme, enolase. Lactic dehydrogenase[104,105] was not inhibited by 2,5-HD and 2,4-hexaneione did not inhibit glycolytic enzymes suggesting a specificity for γ-diketones by glycolytic enzymes. Glucose utilization in the superior colliculus of 2,5-HD-intoxicated rats was reduced prior to the onset of axonal damage in optic pathways.[56] Inactivation of thiamin pyrophosphate, an important vitamin involved in energy production, has been proposed as a possible site of reaction for 2,5-HD.[106]

Gillies et al.[107-110] found little effect of 2,5-HD on glycolysis as measured by incorporation into free fatty acids, triacyl glycerols, and phospholipids. They did find significant in vivo and in vitro inhibition of incorporation of radiolabeled precursors into sterols indicating that sterologenesis was impaired in the testes and nerves of 2,5-HD intoxicated rats but not in the liver.[110] Sterologenesis inhibition may play a role in the findings of Couri and Nachtman[111] who reported that myelin from rats intoxicated with 2,5-HD was more viscous than myelin from control rats and that loss of phase transition in viscosity-temperature plots preceded neuropathy.

Interference with axonal cytoskeletal components such as neurofilaments and neurotubules may account for the axonal damage resulting from γ-diketone related substances. Alteration of neurotubular function could account for slowing of fast axoplasmic flow in rat sciatic nerve following M*n*BK exposure.[65] Accumulations of neurofilaments at nodes of Ranvier and abnormal distribution of neurotubules support the concept of cytoskeletal interaction with 2,5-HD.[7,65] On the other hand, morphologic axonal changes may not be a primary effect but a secondary effect.[65,92] Krasavage et al.[92] reported that the number of axonal swellings induced by *n*-hexane/M*n*BK metabolites correlated with the length of the clinical course and not with neurotoxic potency. Those materials which produce neurotoxicity most rapidly had the fewest number of swellings, implying that axonal swelling is a secondary effect and not a requirement for axonal dysfunction. There is also no evidence that 2,5-HD affects neurotubular function. Couri and Nachtman[111] reported that ^{3}H-colchicine binding by tubulin was not affected by 2,5-HD. Polymerization of sheep[112] and bovine[113] brain tubulin was not altered by incubation with 2,5-HD but αβ-dicarbonyls did inhibit polymerization.[112]

Effects on neurofilaments have not yet been demonstrated which would account for neurotoxicity, although 2,5-HD induction of pyrroles with neurofilaments has been suggested as a possible mechanism of action.[95,99-102] Selkoe et al.[113] found that 2,5-HD, *n*-hexane, and M*n*BK inhibited cellular proliferation and extension or maintenance of neurite processes. Inhibited cells did not show a proliferation of 10 nm cytoplasmic filaments implying that neurofilamentous hyperplasia is not a prerequisite for toxicity.[113] Julien et al.[68] have shown that alteration of neurofilament polypeptide composition is not necessary for the production of neurofilamentous axonal swellings in spontaneous canine giant axonal neuropathy.

The best evidence for interaction of 2,5-HD with neuronal protein is the work of De-Caprio,[100,114] who reported that 2,5-HD exposure of hens resulted in 2,5-dimethylpyrrole adducts in proteins and that 2,5-HD formed ϵ-N-(2,5-methylpyrrolyl) adducts with lysine in vivo. Axonal and axolemmal proteins, particularly myelin basic protein, bound 2,5-HD.[114]

Other mechanisms associated with n-hexane, MnBK, or 2,5-HD toxicity are inhibition of mitochondrial function,[115] calcium chelation,[116] acetylcholinesterase inhibition,[117] or iontophoretic effects,[14] but there is little supporting evidence for a cause-effect relationship.

J. Chemicals Tested for γ-Diketone Neurotoxicity

A large number of solvents and related chemicals have been tested for neurotoxicity. Table 3 lists those which have been shown not to have lasting effects on the nervous system. Table 1 lists those which have produced central-peripheral distal axonopathy with giant axonal swellings in experimental species. All of the compounds on Table 1 either are γ-diketones or can be metabolized to γ-diketones.

Ethyl n-butyl ketone (EnBK) or 3-heptanone and 5-nonanone or dibutyl ketone are two materials with limited commercial use which demonstrate that materials other than six carbon alkanes and ketones may produce γ-diketone neurotoxicity.[118-121] EnBK is metabolized (Figure 2) by ω-1 oxidation to its corresponding alcohol and then to 2,5-heptanedione, a neurotoxic γ-diketone.[118] 2,5-heptanedione is at least partially metabolized to 2,5-hexanedione. Of the total diketones excreted by rats given EnBK, 20 to 30% was 2,5-hexanedione and the remainder was 2,5-heptanedione.

5-Nonanone has been reported to occur in commercial samples of 5-methyl-2-octanone (methyl heptyl ketone) and is used in synthetic organic chemistry.[119] It is also metabolized (Figure 2) by ω-1 oxidation yielding 2-hydroxy-5-nonanone and 2,5-nonanedione, a γ-diketone.[120] MnBK has been detected in the serum and urine of rats given 5-nonanone.[119,120] Other urinary metabolites of 5-nonanone included 2,5-HD, 5-hydroxy-2-hexanone, 2,5-hexanediol, and γ-valerolactone. These studies emphasize the importance of ω-1 oxidation in alkane, alcohol, and ketone metabolism and the central role of 2,5-hexanedione in γ-diketone neuropathies.

Case reports of neuropathy due to exposure to methyl isobutyl ketone,[122,124] methyl ethyl ketone,[125] and n-heptane[126] are not supported by data from experimental exposure to these materials (Table 3).

II. METHANOL

A. Synonyms

Methylalcohol	Methylhydroxide
Woodalcohol	Wood spirits
Methylol	Colonialspirits
Carbinol	Columbianspirits

B. Introduction

A wide variety of products contain methanol (CH_3OH) as a solvent or ingredient. Examples include cosmetics, dyes, lacquers, wood stains, paint strippers, antifreeze, windshield washing solutions, and many other cleaning, painting, and automotive products. It is also used as a chemical intermediate, industrial solvent, and ethanol denaturant. Alcoholic beverages may contain low levels of methanol.[169,170] Interest in toxicity has increased because of the possible increased use of methanol in synthetic fuels.[171]

Table 3
ALKANES, ALCOHOLS, KETONES, AND ALDEHYDES REPORTED NOT TO PRODUCE NEUROTOXICITY AFTER CHRONIC AND SUBCHRONIC EXPOSURES

Test substance	Species	Route	Exposure conditions	Duration of exposure (wk)	Ref.
Acetone	Rat	Water	0.5 and 1.0% in drinking water	4—8	96, 97
Methyl ethyl ketone	Rat	Inhal.	6000 ppm, 8 h/d, 7 d/wk	15	8, 91
			1250, 2500, and 5000 ppm, 6 hr/d, 5 d/wk	13	127
			4740 ppm, repeated	4	128
			2150 ppm, repeated	6	128
			1125 ppm, continuous	20	7
			500 ppm, 22 hr/d, 7 d/wk	24	52
	Cat	Inhal.	300 mg/kg/d, 5 d/wk	34	60
	Dog	Inhal.	300 mg/kg/d, 5 d/wk	48	17
	Rat	i.p.	100—200 mg/kg/d, 5 d/wk	35	17
1,4-Butanediol	Rat	Water	0.5% in drinking water	10	96, 97
n-Pentane	Rat	Inhal.	3000 ppm, 12 hr/d, 5 d/wk	16	129
			3000 ppm, 9 hr/d, 5 d/wk	30	130
Methyl-n-propyl ketone (2-pentanone)	Rat	Water	0.25, 0.5, and 1.0% in drinking water	43—56	17
Diethyl ketone	Rat	Inhal.	1503 ppm, 72 hr/wk	17	17
		Water	Exposure condition unstated		131
Glutaraldehyde	Rat	Water	0.25, 0.5, and 1.0% in drinking water	11	96, 97
Cyclohexane	Rat	Inhal.	1500 ppm, 9 hr/d, 5 d/wk	30	130
			2000 ppm, 10 hr/d, 6 d/wk	30	130
Cyclohexanol	Rat	i.p.	400 mg/kg, 5 d/wk	6	132
Cyclohexanone	Rat	i.p.	400 mg/kg, 5 d/wk	13	132
Commercial hexanes free of n-hexane	Rat	Inhal.	500 ppm, 22 hr/d, 7 d/wk	24	52
2-Methyl pentane	Rat	Inhal.	1500 ppm, 9 hr/d, 5 d/wk	14	130
3-Methyl pentane	Rat	Inhal.	1500 ppm, 9 hr/d, 5 d/wk	14	130
1-Hexanol	Rat	Water	0.25 and 0.5% in drinking water	27	17
		i.p.	102 mg/kg/d, 6 d/wk	30	133
6-Amino-1-hexanol	Rat	Water	0.25 and 0.5% in drinking water	27	17
Methyl isobutyl ketone	Rat	Inhal.	1500 ppm, 6 hr/d, 5 d/wk	20	62
	Cat	s.c.	300 mg/kg/d, 5 d/wk	34	96, 97
	Dog	s.c.	300 mg/kg/d, 5 d/wk	48	17
	Rat	i.p.	100—200 mg/kg/d, 5 d/wk	35	17
1,6-Hexanediol	Rat	Water	0.5% in drinking water	12	96, 97
2,3-Hexanedione	Rat	Water	0.5% in drinking water	12	96, 97
			0.25 and 0.5% in drinking water	30	17
2,4-Hexanedione	Rat	Water	0.5% in drinking water	7	96, 97
			0.25 and 0.5% in drinking water	30	17
3,5-Hexanedione	Rat	Water	0.5% in drinking water	12	96, 97
n-Heptane	Rat	Inhal.	3000 ppm, 12 hr/d, 7 d/wk	16	129
			1500 ppm, 9 hr/d, 5 d/wk	30	130
		Gavage	4000 mg/kg/d, 5 d/wk	13	17
Methyl n-amyl ketone (2-heptanone)	Rat	Inhal.	131 and 1025 ppm, 6 hr/d, 5 d/wk	40	134, 135, 136

Table 3 (continued)
ALKANES, ALCOHOLS, KETONES, AND ALDEHYDES REPORTED NOT TO PRODUCE NEUROTOXICITY AFTER CHRONIC AND SUBCHRONIC EXPOSURES

Test substance	Species	Route	Exposure conditions	Duration of exposure (wk)	Ref.
	Monkey	Inhal.	131 and 1025 ppm, 6 hr/d, 5 d/wk	40	134, 135, 136
	Rat	Water	0.5% in drinking water	12	96, 97
Di-n-propyl ketone (4-heptanone)	Rat	Gavage	1000 mg/kg/d, 5 d/wk	13	17
2,6-Heptanedione	Rat	Gavage	500—1000 mg/kg/d, 5 d/wk	13	17
Methyl isoamyl ketone (5-methyl-2-hexanone)	Rat	Gavage	2000 mg/kg/d, 5 d/wk	13	17
2-Octanone	Rat	s.c.	400 mg/kg/d, 5 d/wk	21	137
Diisobutyl ketone	Rat	Gavage	2000 mg/kg/d, 5 d/wk	13	17
Methyl n-heptyl ketone (2-nonanone)	Rat	Gavage	2000 mg/kg/d 5 d/wk	13	17
5-Methyl-2-octanone[a] (methyl heptyl ketone)	Rat	Gavage	1540 mg/kg/d, 5 d/wk	13	17
Diisoamyl ketone	Rat	Gavage	2000 mg/kg/d, 5 d/wk	13	17

[a] Commercial samples of 5-methyl-2-octanone may contain 5-nonanone which is neurotoxic. 5-nonanone neurotoxicity is potentiated by 5-methyl-2-octanone.

C. Human Neurotoxicity

Methanol toxicity in man has a long history extending back to 1855,[172] but it was not until 1904, following a report by Wood and Buller,[173] that methanol, and not contaminants in crude commercial products, was identified as a cause of blindness and death. Temporary or permanent blindness is now a well-recognized hazard of acute methanol poisoning.

Toxic exposures may be from ingestion, inhalation, or skin application. In the older literature, inhalation or percutaneous absorption in work environments is reported to result in blindness or death.[174-177] Toxic effects are the same no matter the route of exposure.[178]

The usual route of exposure is by ingestion either as a substitute for ethanol, in attempts to commit suicide, or in adulterated or mislabeled alcoholic beverages. Poisonings have involved many people. For example, 6% of all blindness in American forces during World War II was estimated to be due to methanol ingestion.[177] Benton and Calhoun[179] described ocular effects occurring in 320 people who had consumed bootleg whiskey containing 35% methanol. Scrimgeour[180] reported a recent (1978) mass poisoning involving 372 men who had consumed a solution of 82% methanol and 18% isopropanol.

Methanol is not as inebriating as ethanol. Poisoning produces moderate central nervous system depression which is followed by a characteristic asymptomatic delay period of 12 to 24 hr.[169,171,173,177-179] Following the delay, patients exhibit weakness, abdominal pains, visual disturbances, metabolic acidosis, and hypernea.[169,171,173,177-179] In later stages, permanent blindness, coma, or death follow.

Visual disturbances include blurring, photophobia, feelings of a covering over the eyes, seeing spots or flashes, and partial or total blindness.[169,171,173,177-179] Pupils may be dilated and pupillary reflexes diminished or absent.[169,179] The optic disc and retina may show edema and hyperemia or appear relatively normal.

The seriousness of the clinical case is often not correlated with the amount of methanol ingested.[169,177-180] In the outbreak described by Scrimgeour,[180] blindness and death occurred in a man who had consumed 100 mℓ of a methanol solution, while two men who had

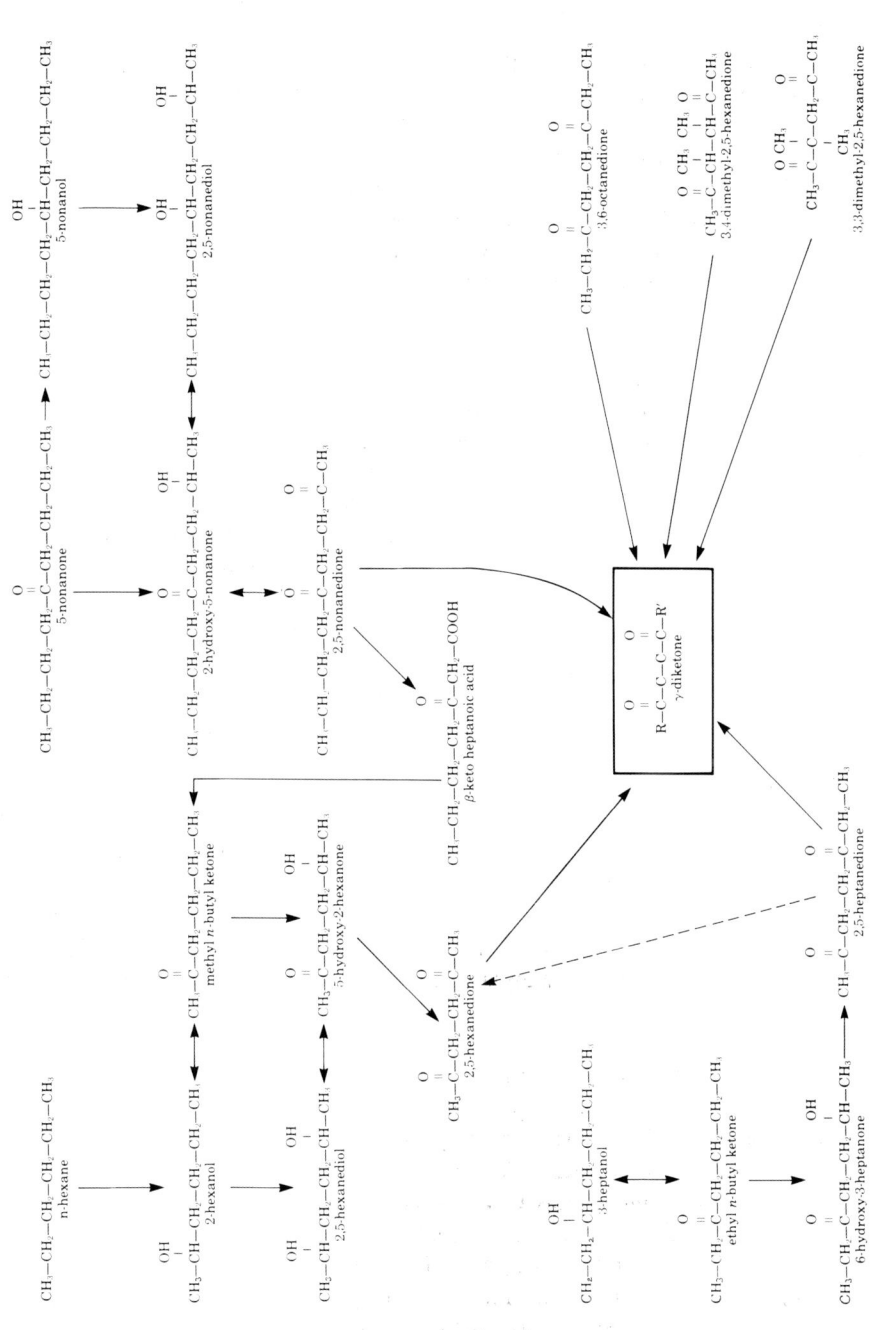

FIGURE 2. Interrelationships among neurotoxic alkanes, ketones, and alcohols.

consumed 500 mℓ showed no apparent disability. The reasons for differences in individual susceptibility are unknown but ingestion of ethyl alcohol is known to affect methanol metabolism,[169,171] the amount of food in the stomach may affect absorption,[169,171] and prior exposure to methanol on a regular basis may affect metabolism.[169]

Effects on the nervous system other than the retina and optic nerve are unusual.[181,186] Riegel and Wolf[182] described a case with optic atrophy and delayed onset of focal cranial nerve deficits and a Parkinson-like extrapyramidal syndrome. Permanent pyramidal and extrapyramidal damage was seen in a 13-year-old girl, 4 weeks after ingestion of a 60% methanol solution.[183] Similar clinical findings were reported by McLean et al.[186] Necrosis of the putamen may be associated with the presence of extrapyramidal symptoms.[181,185,186] Rarely associated with severe methanol poisoning are dementia,[129,186] pseudobulbar palsy,[180] and peripheral neuropathy.[169,186] A single case of bilateral lateral geniculate body necrosis resulting in sudden blindness occurred in a woman who reportedly drank methanol.[184] In this case, typical visual system changes were not observed and there was no confirmation of methanol exposure.

D. Experimental Neurotoxicity

Studies of methanol poisoning, until recently, have been hindered by the absence of a reproducible model stimulating the human condition. The more common laboratory species (rat, mouse, rabbit, and dog) do not develop visual system damage consistently present in man.[177,178,187-192,194-197] Toxicity in these species is characterized by progressive depression of neurologic function.[187,197] Severe metabolic acidosis, so common in man, is not a feature of methanol toxicity in these species. The apparent reason for these differences lies in species differences in methanol metabolism and reduced rates of methanol-derived formate metabolism to carbon dioxide in primates as compared to rodents.[177,178,187-192,194-196,198]

Only nonhuman primate models show close similarity to man with regard to metabolic acidosis and visual deficits.[188-192,194,195] Large single doses of methanol approximating 4 g/kg bodyweight,[188] smaller doses (2 g/kg) followed by repetitive infusion of methanol,[189-195] or use of folate-deficient monkeys[198] produces a syndrome including initially depressed neurologic function followed by a latency period of 8 to 12 hr. Subsequently, anorexia, vomiting, severe metabolic acidosis, ocular toxicity, coma, and death may result.[188-192,194-198] Optic disc edema routinely occurs in intoxicated primates.[192] The optic nerve head and intraorbital optic nerve appear to be the primary sites of ocular damage and not the retinal ganglion cells as previously thought.[192] Edema is seen as intraaxonal swelling and not enlargement of the extracellular space. Demyelination has been reported in a similar location in four human cases.[193] Interruption of axoplasmic flow has been proposed as the basis for axonal swelling.[192]

E. Metabolism and Mechanism of Action

Methanol toxicity as it occurs in man is generally believed to be due to metabolic products and not the parent compound. This is suggested by the consistency of the delay period prior to onset of serious metabolic and ocular effects.

Experimental observations support this view. In monkeys, inhibitors of alcohol dehydrogenase, an important enzyme for oxidative metabolism of methanol in man and monkeys, delay or prevent metabolic acidosis and ocular toxicity.[188,189]

Important metabolites of methanol include formaldehyde which is oxidized to formic acid.[178] The roles played by these metabolites and severe acidosis have received considerable attention. Unfortunately, much of this work has been confusing or contradictory since the importance of species differences in metabolism has only recently been apparent.[177,188,189,196]

Metabolism of methanol to formic acid appears to play the major role in its toxicity.[188,190,198] Metabolic acidosis develops as formic acid accumulates in the blood and plasma bicarbonate levels decrease.[20,26] Metabolic inhibitors such as 4-methyl-pyrazole prevent formation of formic acid and the appearance of acidosis.[188,189] The general opinion has been that production of formic acid was insufficient to account for the degree of acidosis observed although recent reports suggest that formic acid production is a major factor in producing acidosis.[188]

The importance of the formate ion for ocular damage was emphasized by the production of optic disc edema, papillary changes, and intraaxonal swelling in sodium formate/formic acid infused monkeys.[194]

Formaldehyde is also a potentially toxic metabolite of methanol and in earlier reports was thought responsible for ocular toxicity.[169] More recent studies indicate that formaldehyde does not play an important role,[195] although it has not been possible to eliminate the possibility that formaldehyde has intracellular effects.

III. ETHYLENE GLYCOL

A. Introduction

Ethylene glycol (1,2-ethanediol, $OHCH_2CH_2OH$) is widely used in paints, lacquers, automotive antifreeze, and other heat transfer fluids, hydraulic fluids, waxes, polishes, cosmetics, and many other products. It is not considered a particularly hazardous substance in industry or during normal use of commercial products containing it. Toxic hazards associated with ethylene glycol stem from intentional ingestion for suicide and accidental ingestion particularly when it has been transferred to unlabeled or mislabeled containers. It is a leading cause of poisoning in the U.S. Haggerty[199] estimated that between 40 and 60 cases of ethylene glycol poisoning occur in the U.S. annually.

B. Neurotoxicity

The toxicity of ethylene glycol is well-recognized and has been reviewed by Parry and Wallach,[200] and Rowe and Wolf.[201] Three stages of poisoning are recognized. The first is an acute central nervous system stage starting shortly after ingestion and lasting several hours. This stage is presumed to be due to the direct toxicity of ethylene glycol which initially causes CNS stimulation and then depression of higher cortical function. People appear inebriated and visual difficulty may occur as well as vomiting, confusion, tremors, depressed reflex activity, coma, and convulsions. The second stage involves cardiorespiratory failure and is followed by renal failure. Berger and Ayyar[202] and Fellman[203] described two unusual cases with cranial nerve damage occurring 6 to 14 days after ethylene glycol ingestion. Recovery was seen after 1 month in one case and after 7 months in the second. These cases showed dysarthria, dysphagia, dysfunction of the fifth, seventh, and twelfth cranial nerves, visual disturbances, and cerebellar dysfunction. Berger and Ayyar[202] reviewed many of the signs and symptoms reported after ethylene glycol ingestion. They include: personality change, hallucinations, convulsions, coma, meningism, myoclonus, fixed pupils, decreased or loss of vision, loss of accommodation, papilledema, diplopia, nystagmus, strabismus, abnormal eye movements, optic nerve atrophy, cranial nerve palsies, ataxia, tremor, myositis, muscle twitching, tetany, hyperreflexia, and areflexia.

Troisi[204] reported the only case of ethylene glycol poisoning due to industrial exposure. Fourteen of thirty-eight women producing electrolytic condensers were exposed to ethylene glycol vapors when a mixture of ethylene glycol/boric acid/ammonia was heated to 105°C. Toxicity was seen as nystagmus or a sudden loss of consciousness for 5 to 10 min with nystagmus occurring 2 to 3 times weekly until exposures were eliminated.

IV. ETHYLENE GLYCOL MONOMETHYL ETHER

A. Introduction
This substance, also known as 2-methoxy ethanol ($CH_3OCH_2CH_2OH$) and by several trade names, is one of a number of derivatives of ethylene glycol. It is used in detergents, paints, lacquers, inks, resins, and similar products.

B. Neurotoxicity
Donley[205] described the first case of ethylene glycol monomethyl ether (EGME) poisoning in 1936 in a plant producing "fused" shirt collars. Parsons and Parsons,[206] and Greenburg et al.[207] described 19 cases in another "fused" collar plant. Seven other cases were reported by Zavon,[208] and Ohi and Wegman[209] in plastic and textile printing plants.

Clinical abnormalities are similar among the reported cases. These include prominent personality changes, mental retardation, drowsiness, tiredness, impaired memory, nervousness, speech difficulties, headache, staggering, ataxia, tremor, positive Romberg test, increased reflex activity, and ankle clonus. Persistent pupillary dilation and bed-wetting occurred in some cases.

Exposures leading to toxicity have occurred under poor hygienic conditions with absorption by both inhalation and dermal routes. In the cases reported by Ohi and Wegman,[209] the skin was probably the only route of exposure. Moderately severe anemia, leukopenia, and renal damage may complicate clinical cases.

Savolainen[210] reported paresis and a shuffling gait in rats exposed to 400 ppm EGME, 6 hr/day, 5 days/week for 2 weeks. Histochemical changes were found in brain glial cells from rats exposed to 50, 100, and 400 ppm EGME for 2 weeks.

V. ETHYLENE OXIDE

A. Introduction
Ethylene oxide (C_2H_4O), also referred to as oxirane, ethene oxide, 1,2-epoxyethane, dimethyl oxide, and dihydrooxirene, is a highly reactive, explosive, sweet smelling gas at room temperature. It is used in the production of ethylene glycol, glycol ethers, polyethylene glycol, ethanolamines, lubricants, as a fumigant for furs and certain foods, and as a sterilant of heat sensitive materials both in industry and in health care facilities.

B. Human Neurotoxicity
The odor of ethylene oxide does not serve as a warning in preventing exposure. Industrial and occupational exposure is generally the result of inhalation of ethylene oxide vapor released from leaking or faulty equipment, valves, or fittings. Acute exposures may result in irritation to the eyes, nose, throat, and lungs, dyspnea, pulmonary edema, skin burns, frostbite from contact with liquid ethylene oxide, nausea, vomiting, headache, drowsiness, weakness, flu-like symptoms, aphonia, convulsions, or death.[211-216]

Gross et al.[212] reported four cases of ethylene oxide toxicity among men operating a defective ethylene oxide sterilizer. The exposure was estimated to have lasted about 2 months at levels over 700 ppm, at least intermittently. One of the men had an acute clinical course with repeated major motor seizures. The other three showed more subacute to chronic courses, primarily with evidence of sensorimotor peripheral neuropathy typical of a toxic axonopathy. Abnormalities included increased fatigue, weakness in hand and foot muscles, ataxia, gait disturbances, loss or reduction in stretch reflexes, abnormalities of pinprick, position and vibratory sensations, and slowing of nerve conduction velocity. One man had central nervous system and cranial nerve changes including difficulty with memory and thinking, slurred speech, difficulty swallowing, and facial weakness. All men showed marked improvement within 2 weeks.

In a study of 223 workers involved in producing ethylene and ethylene oxide, Spasovski et al.[217] found evidence of "vasodystonic, vasospastic and vegetative neuropathy syndromes with neurosis-like manifestations."

Joyner[218] found no adverse health effects in a group of 37 chemical operators who had worked for a mean period of 10 years in an area with an ethylene oxide level of approximating 5 to 10 ppm (range of 0 to 55 ppm).

C. Experimental Neurotoxicity

Neurotoxicity has been reproduced in dogs,[219-221] cats,[220,221] monkeys,[222] rats,[219,222] rabbits,[220,222] and guinea pigs,[220] but not mice.[219,222] Most of the data available on the effects of ethylene oxide on laboratory species come from two studies[219,222] which have been summarized on Table 4.

At lethal or near lethal single doses of ethylene oxide, neural effects are limited to increased movement and preening in rats and convulsions in dogs.[219]

Ethylene oxide at 357 ppm, 35 hr/week for 12 weeks produced a sensorimotor neuropathy in rats, rabbits, and monkeys but not guinea pigs or mice.[222] Continued exposure resulted in paralysis and muscle atrophy of the hindlimbs.[222] At 204 ppm, 35 hr/wk for 32 weeks, rabbits and monkeys, but not guinea pigs, rats, or mice, developed a clinical neuropathy.[222] The monkeys had decreased tendon reflexes, loss of withdrawal from superficial pain over the hindquarters, partial paralysis, and muscle atrophy indicative of toxic axonopathy.[222] A positive Babinski reflex in these monkeys[222] indicated that upper motor neurons or their axons were also affected. Dogs showed occasional tremors, transient weakness, and atrophy and fatty replacement of skeletal muscle following ethylene oxide exposures of 292 ppm, 30 hr/week for 6 weeks.[219] Levels of about 100 ppm repeatedly were without neurotoxicity in rats,[219,222] rabbits,[222] guinea pigs,[222] mice,[219] dogs,[219] and monkeys.[222]

D. Other Effects

A variety of other effects has been reported occurring in test animals including skin, eye, and respiratory irritation,[219] hepatic, renal and adrenal gland lesions,[222] anemia,[219] splenic hemosiderosis,[219] and testicular atrophy.[219] Mutagenic and teratogenic studies have been reviewed.[211]

E. Related Substances

Baer and Griepewtrog[223] fed diets fumigated with ethylene oxide to rats for 2 years without finding toxic effects. Oser et al.[224] administered ethylene chlorhydrin or 2-chloroethanol, a reaction product of ethylene oxide and sodium chloride, to dogs, rats, and monkeys for 90 days without evidence of neurotoxicity.

Rats fed a diet containing 2.5% of a copolymer of propylene glycol and ethylene oxide for 90 days did not show clinical evidence of neurotoxicity.[225]

Propylene oxide has not been reported to result in neurotoxicity.[211,226]

Further studies on ethylene oxide and propylene oxide have been announced.[227]

Table 4
SOME EFFECTS OF INHALATION OF ETHYLENE OXIDE ON LABORATORY ANIMALS

Exposure	Species	Effect(s)	Ref.
882—2298 ppm, 4 hr	Rat	LC_{50} 1460 ppm ocular and respiratory irritation, diarrhea, inc. activity	219
533—1365 ppm, 4 hr	Mouse	LC_{50} 835 ppm, as above, no diarrhea	219
2800 ppm, 4 hr	Dog	Ocular and nasal irritation, labored breathing, convulsions, vomiting, diarrhea, 100% mortality	219
1400 ppm, 4 hr	Dog	Ocular and nasal irritation, vomiting, 100% mortality	219
327 or 710 ppm, 4 hr	Dog	No effect	219
841 ppm, 7 hr/d, max. 8 exp. in 10 days	Rat; mouse; rabbit; guinea pig; monkey	Respiratory irritation, hepatic, renal, and adrenal lesions in rats and guinea pigs, 100% mortality	222
400 ppm, 6 hr/d, 5 d/wk, 6 wk	Rat	Nasal discharge, labored breathing, diarrhea, splenic hemosiderosis, weight loss, deaths, hindlimbs dragged; recovery from paralysis occurred in several months	219
400 ppm, 6 hr/d, 5 d/wk, 6 wk	Mouse	80% mortality	219
357 ppm, 7 hr/d, 5 d/wk, 12 wk	Rat; mouse; rabbit; guinea pig; monkey	In rats, rabbits, and monkeys, there was sensory and motor impairment, paralysis, and muscle atrophy with recovery in 100—132 days; guinea pigs—testicular atrophy, female guinea pigs—adrenal cortical degeneration	222
292 ppm, 6 hr/d, 5 d/wk, 6 wk	Dog	Vomiting, anemia, tremors, transient hindlimb weakness, muscle atrophy	219
204 ppm, 7 hr/d, 5 d/wk, 32 wk	Rat; mouse; rabbit; guinea pig; monkey	Rat — dec. growth, slight mortality, and testicular atrophy; rabbit — limb paralysis after 196 days; monkey — abnormal tendon reflexes and sensation, paralysis, and muscle atrophy	222
113 ppm, 7 hr/d, 5 d/wk, 25—32 wk	Rat; rabbit; guinea pig; monkey	Rat — dec. growth and inc. lung wt.	222
100 ppm, 6 hr/d, 5 d/wk, 26 wk	Rat; mouse; dog	Rat — dec. growth; dog — slight anemia	219
49 ppm, 7 hr/d, 5 d/wk, 26 wk	Rat; rabbit; guinea pig; monkey (54 days)	No effect	222

REFERENCES

1. **Oishi, A., Mineno, K., Yamada, M., Chib, A. K., and Shibata, K.,** Polyneuropathy caused by an organic solvent (*n*-hexane), *Saigai Igaku,* 7, 218, 1964.
2. **Yamada, S.,** Intoxication polyneuritis in the workers exposed to *n*-hexane, *Jpn. J. Ind. Health,* 9, 651, 1967.
3. **Yamamura, Y.,** *n*-Hexane polyneuropathy, *Folia Psych. Neurol. Jpn.,* 23, 45, 1969.
4. **Herskowitz, A., Ishii, N., and Schaumburg, H.,** *n*-Hexane neuropathy — a syndrome occurring as a result of industrial exposure, *N. Eng. J. Med.,* 285, 82, 1971.
5. **Allen, N., Mendell, J. R., Billmaier, D. J., Fontaine, R. E., and O'Neill, J., Jr.,** Toxic polyneuropathy due to methyl *n*-butyl ketone, *Arch. Neurol.,* 32, 209, 1975.
6. **Billmaier, D., Yee, H. T., Allen, N., Craft, B., Williams, N., Epstein, S., and Fontaine, R.,** Peripheral neuropathy in a coated fabrics plant, *J. Occup. Med.,* 16, 665, 1974.
7. **Saida, K., Mendell, J. R., and Weiss, H. S.,** Peripheral nerve changes induced by methyl *n*-butyl ketone and potentiation by methyl ethyl ketone, *J. Neuropathol. Exp. Neurol.,* 35, 207, 1976.
8. **Altenkirch, H., Stoltenburg, G., and Wagner, H. M.,** Experimental studies on hydrocarbon neuropathies induced by methyl-ethyl-ketone (MEK), *J. Neurol.,* 219, 159, 1978.
9. **Wickersham, C. W., III and Fredericks, E. J.,** Toxic polyneuropathy secondary to methyl *n*-butyl ketone, *Conn. Med.,* 40, 311, 1976.
10. **Davenport, J. G., Farrell, D. F., and Sumi, S. M.,** "Giant axonal neuropathy" caused by industrial chemicals: neurofilamentous axonal masses in man, *Neurology,* 26, 919, 1976.
11. **Mallov, J. S.,** MBK neuropathy among spray painters, *JAMA,* 235, 1455, 1976.
12. **Lande, S. S., Durkin, P. R., Christopher, D. H., Howard, P. H., and Saxena, J.,** Investigation of Selected Potential Environmental Contaminants; Ketonic Solvents, U.S. Environmental Protection Agency Rep. No. 560/2-76-003, Washington, D.C., 1976.
13. **Walter, P., Craigmill, A., and Dukick, A.,** Methyl Ketone Toxicity, Consumer Product Safety Commission, FIRL 80G-C422502, Washington, D.C., 1975.
14. **Spencer, P. S., Schaumburg, H. H., Sabri, M. I., and Veronesi, B.,** The enlarging view of hexacarbon neurotoxicity, *Crit. Rev. Toxicol.,* 7, 279, 1980.
15. **Griffin, J. W.,** Hexacarbon neurotoxicity, *Neurobehav. Toxicol. Teratol.,* 3, 437, 1981.
16. **Couri, D. and Milks, M.,** Toxicity and metabolism of the neurotoxic hexacarbons *n*-hexane, 2-hexanone, and 2,5-hexanedione, *Annu. Rev. Pharmacol. Toxicol.,* 22, 145, 1982.
17. **Krasavage, W. J., O'Donoghue, J. L., and DiVincenzo, G. D.,** Ketones, in *Patty's Industrial Hygiene and Toxicology,* Vol. 2C, 3rd rev. ed., Clayton, G. D. and Clayton, F. E., Eds., John Wiley-Interscience, New York, chap. 56.
18. **Abbritti, G., Siracusa, A., Cianchetti, C., Coli, C. A., Curradi, F., Perticoni, G. F., and DeRosa, F.,** Shoe-makers' polyneuropathy in Italy: the aetiological problem, *Br. J. Ind. Med.,* 33, 92, 1976.
19. **Caruso, G. and Santoro, L.,** Polyneuropathy due to industrial adhesive: longterm clinical and electromyographic study, *Acta Neurol.,* 30, 416, 1975.
20. **Cianchetti, C., Abbritti, G., Perticoni, G., Siracusa, A., and Curradi, F.,** Toxic polyneuropathy of shoe industry workers: a study of 122 cases, *J. Neurol. Neurosurg. Psychiat.,* 39, 1151, 1976.
21. **Rizzuto, N., Terzian, H., and Galiazzo-Rizzuto, S.,** Toxic polyneuropathies in Italy due to leather cement poisoning in shoe industries: a light-and-electron-microscopic study, *J. Neurol. Sci.,* 31, 343, 1977.
22. **Buiatti, E., Cecchini, S., Ronchi, O., Dolora, P., and Bulgarelli, G.,** Relationship between clinical and electromyographic findings and exposure to solvents, in shoe and leather workers, *Br. J. Ind. Med.,* 35, 168, 1978.
23. **Aiello, I., Rosati, G., Serra, G., and Manca, M.,** Subclinical neuropathic disorders and precautionary measures in the shoe industry: an electrophysiological investigation, *Acta Neurol.,* 35, 285, 1980.
24. **Rizzutto, N., DeGrandis, D., DiTrapani, G., and Pasinato, E.,** *n*-Hexane polyneuropathy. An occupational disease of shoe makers, *Eur. Neurol.,* 19, 308, 1980.
25. **Scelsi, R., Poggi, P., Fera, L., and Gonella, G.,** Industrial neuropathy due to *n*-hexane. Clinical and morphological findings in three cases, *Clin. Toxicol.,* 18, 1387, 1981.
26. **Perbellini, L., Brugnone, F., and Faggionato, G.,** Urinary excretion of the metabolites of *n*-hexane and its isomers during occupational exposure, *Br. J. Ind. Med.,* 38, 20, 1981.
27. **Mutti, A., Cavatorta, A., Lommi, G., Lotta, S., and Franchini, I.,** Neurophysiological effects of long-term exposure to hydrocarbon mixtures, *Arch. Toxicol. Suppl.,* 5, 120, 1982.
28. **Assouly, M., Sioul, G., and Cavignaux, A.,** Polyneuritis par *n*-hexane, *Arch. Med. Prof.,* 33, 20, 1972.
29. **Gaultier, M., Rancorel, G., Piva, C., and Etthymiou, L.,** Polyneuritis and aliphatic hydrocarbons, *J. Eur. Toxicol.,* 6, 294, 1973.
30. **Palao, A. and Lajo, J. L.,** Toxic polyneuropathy as a consequence of the use of glue and chemical dissolvents in the shoe industry, *Arch. Neurobiol.,* 44, 35, 1981.

31. **Oliveira, C. Q., Lopes, J. M., Junqueira, M. A., and Polonia, J.**, IR, GC, and mass-spectroscopic analysis of the volatile fraction of industrial products with peripheric polyneuropathic effects, *Rev. Port. Farm.*, 31, 39, 1981.
32. **Sobue, I., Iida, M., Yamamura, Y., and Takayanagui, T.**, n-Hexane polyneuropathy, *Int. J. Neurol.*, 11, 317, 1978.
33. **Sanagi, S., Seki, Y., Sugimoto, K., and Hirata, M.**, Peripheral nervous system functions of workers exposed to n-hexane at a low level, *Int. Arch. Occup. Environ. Health*, 47, 69, 1980.
34. **Takahashi, M., Takeuchi, H., Kyo, S., Yorifugi, S., Sanagi, S., Seki, Y., and Hara, I.**, n-Hexane polyneuropathy. A case report with a review of the literature, *Med. J. Osaka Univ.*, 28, 77, 1977.
35. **Yoshida, T., Yanagisawa, H., Muneyuki, T., and Shigiya, R.**, Four cases of n-hexane polyneuropathy and its electrophysiological studies, *Rinsho Shinkeigaku*, 14, 454, 1974.
36. **Takahashi, M., Takeuchi, H., Kyo, S., Yorifuji, S., Sanagi, S., Seki, Y., and Hara, I.**, n-Hexane polyneuropathy. A case report with a review of the literature, *Med. J. Osaka Univ.*, 28, 77, 1977.
37. **Paulson, G. W. and Waylonis G. W.**, Polyneuropathy due to n-hexane, *Arch. Intern. Med.*, 136, 880, 1976.
38. **Ruff, R. L., Petito, C. K., and Acheson, L. S.**, Neuropathy associated with chronic low level exposure to n-hexane, *Clin. Toxicol.*, 18, 515, 1981.
39. **Korobkin, R., Asbury, A. K., Sumner, A. J., and Nielson, S. L.**, Glue-sniffing neuropathy, *Arch. Neurol.*, 32, 158, 1975.
40. **Towfighi, J., Gonatas, N. K., Pleasure, D., Cooper, H. S., and McCree, L.**, Glue sniffer's neuropathy, *Neurology*, 26, 238, 1976.
41. **Chauplannaz, G., Bady, B., Kopp, N., Levrat, R., and Trillet, M.**, Peripheral neuropathy due to n-hexane in a drug addict, *Rev. Neurol.*, 138, 249, 1982.
42. **Means, E. D., Prockop, L. D., and Hooper, G. S.**, Pathology of lacquer thinner induced neuropathy, *Ann. Clin. Lab. Sci.*, 6, 240, 1976.
43. **Altenkirch, H., Mager, J., Stoltenberg, G., and Helmbrecht, J.**, Toxic polyneuropathies after sniffing a glue thinner, *J. Neurol.*, 214, 137, 1977.
44. **Asbury, A. K., Nielsen, S. L., and Telfer, R.**, Glue sniffing neuropathy, *J. Neuropathol. Exp. Neurol.*, 33, 191, 1974.
45. **Goto, I., Matsumara, M., Inoue, N., Murai, Y., Shida, K., Santa, T., and Kuroiwa, Y.**, Toxic polyneuropathy due to glue sniffing, *J. Neurol. Neurosurg. Psychiatry*, 37, 848, 1974.
46. **Suzuki, T., Shimbo, S., Nishitani, H., Oga, T., Imamura, T., and Ikeda, M.**, Muscular atrophy due to glue sniffing, *Int. Arch. Arbeitsmed.*, 33, 115, 1974.
47. **Shirabe, T. T., Terau, A., and Araki, S.**, Toxic polyneuropathy due to glue-sniffing, *J. Neurol. Sci.*, 21, 101, 1974.
48. **Schaumburg, H. H.**, Chronic hexacarbon intoxication in man, in *Advances in Neurotoxicology*, Proc. Int. Congr., Manzo, L., Ed., Pergamon Press, Oxford, 1979, 187.
49. **Gonzalez, E. G. and Downey, J. A.**, Polyneuropathy in a glue sniffer, *Arch. Phys. Med. Rehabil.*, 53, 333, 1972.
50. **Raitta, C., Seppäläinen, A. M., and Huuskonen, M. S.**, n-Hexane maculopathy in industrial workers, *Albrecht V. Graefes Arch. Klin. Exp. Opthal.*, 209, 99, 1978.
51. **Seppäläinen, A. M., Raitta, C., and Huuskonen, M. S.**, n-Hexane-induced changes in visual evoked potentials and electroretinograms of industrial workers, *Electroencephalogr. Clin. Neurophysiol.*, 47, 492, 1979.
52. **Egan, G., Spencer, P., Schaumburg, H., Murray, K. J., Bischoff, M., and Scala, R.**, n-Hexane — "Free" hexane mixture fails to produce nervous system damage, *Neurotoxicology*, 1, 515, 1980.
53. **O'Donoghue, J. L. and Krasavage, W. J.**, Identification and characterization of methyl n-butyl ketone neurotoxicity in laboratory animals, in *Experimental and Clinical Neurotoxicology*, Spencer, P. S. and Schaumburg, H. H., Eds., Williams & Wilkins, Baltimore, 1980, chap. 59.
54. **Abdel-Rahman, M. S., Saladin, J. J., Bohman, C. E., and Couri, D.**, The effect of 2-hexanone and 2-hexanone metabolites on pupillomotor activity and growth, *Am. Ind. Hyg. Assoc. J.*, 39, 94, 1978.
55. **Schaumburg, H. H. and Spencer, P. S.**, Environmental hydrocarbons produce degeneration in cat hypothalamus and optic tract, *Science*, 199, 199, 1978.
56. **Griffiths, I. R., Kelly, P. A. T., Carmichael, S., McCulloch, M., and Waterston, M.**, The relationship of glucose utilization and morphological change in the visual system in hexacarbon neuropathy, *Brain Res.*, 222, 447, 1981.
57. **Schaumburg, H. H. and Spencer, P. S.**, Degeneration in central and peripheral nervous systems produced by pure n-hexane: an experimental study, *Brain*, 99, 183, 1976.
58. **Spencer, P. S. and Schaumburg, H. H.**, Ultrastructural studies of the dying-back process. III. The evolution of experimental peripheral giant axonal degeneration, *J. Neuropathol. Exp. Neurol.*, 36, 276, 1977.

59. **Spencer, P. S. and Schaumburg, H. H.,** Ultrastructural studies of the dying-back process. IV. Differential vulnerability of peripheral nervous system and central nervous system fibers in experimental central-peripheral distal axonopathies, *J. Neuropathol. Exp. Neurol.*, 36, 300, 1977.
60. **Spencer, P. S. and Schaumburg, H. H.,** Feline nervous system response to chronic intoxication with commercial grades of methyl n-butyl ketone, methyl isobutyl ketone, and methyl ethyl ketone, *Toxicol. Appl. Pharmacol.*, 37, 301, 1976.
61. **Schaumburg, H. H.,** Morphological studies of toxic distal axonopathy, *Neurobehav. Toxicol.*, 1(Suppl. 1), 187, 1979.
62. **Spencer, P. S., Schaumburg, H. H., Raleigh, R. L., and Terhaar, C. J.,** Nervous system degeneration produced by the industrial solvent methyl n-butyl ketone, *Arch. Neurol.*, 32, 219, 1975.
63. **Raleigh, R. L., Spencer, P. S., and Schaumburg, H. H.,** Toxicity of methyl butyl ketone, *Arch. Environ. Health*, 30, 317, 1975.
64. **Abou-Donia, M., Makkawy, H. A. M., and Graham, D. G.,** The relative neurotoxicities of n-hexane, methyl n-butyl ketone, 2,5-hexanediol, and 2,5-hexanedione following oral or intraperitoneal administration in hens, *Toxicol. Appl. Pharmacol.*, 62, 369, 1982.
65. **Mendell, J. R., Sahenk, Z., Saida, K., Weiss, H. S., Savage, R., and Couri, D.,** Alterations of fast axoplasmic transport in experimental methyl n-butyl ketone neuropathy, *Brain Res.*, 133, 107, 1977.
66. **Carpenter, S., Karpati, G., Andermann, F., and Gold, R.,** Giant axonal neuropathy, *Arch. Neurol.*, 31, 312, 1974.
67. **Asbury, A. K., Gale, M. K., Cox, S. C., Baringer, J. R., and Berg, B. O.,** Giant axonal neuropathy — a unique case with segmental neurofilamentous masses, *Acta Neuropathol.*, 20, 237, 1972.
68. **Julien, J.-P., Mushynski, W. E., Duncan, I. D., and Griffiths, I. R.,** Giant axonal neuropathy: neurofilaments isolated from diseased dogs have a normal polypeptide composition, *Exp. Neurol.*, 72, 619, 1981.
69. **Vallat, J. M., Leboutet, M. J., Loubet, A., Piva, C., and Dumas, M.,** n-Hexane and methylethylketone-induced polyneuropathy. Abnormal accumulation of glycogen in myelinated axons. Report of a case, *Acta Neuropathol.*, 55, 275, 1981.
70. **Lukins, H. B. and Foster, J. W.,** Methyl ketone metabolism in hydrocarbon utilizing mycobacteria, *J. Bacteriol.*, 85, 1074, 1963.
71. **Markovetz, A. J.,** Intermediates from the microbial oxidation of aliphatic hydrocarbons, *J. Am. Oil Chem. Soc.*, 55, 430, 1978.
72. **Frommer, U., Ullrich, V., Staudinger, H., and Orrenius, O.,** The monooxygenation of n-heptane by rat liver microsomes, *Biochim. Biophys. Acta*, 280, 487, 1972.
73. **Frommer, U. and Ullrich, V.,** Influence of inducers and inhibitors on the hydroxylation pattern of n-hexane in rat liver microsomes, *FEBS Lett.*, 41, 14, 1974.
74. **DiVincenzo, G. D., Kaplan, C. J., and Dedinas, J.,** Characterization of the metabolites of methyl n-butyl ketone, methyl isobutyl ketone, and methyl ethyl ketone in guinea pig serum and their clearance, *Toxicol. Appl. Pharmacol.*, 36, 511, 1976.
75. **DiVincenzo, G. D., Hamilton, M. L., Kaplan, C. J., Krasavage, W. J., and O'Donoghue, J. L.,** Studies on the respiratory uptake and excretion and the skin absorption of methyl n-butyl ketone in humans and dogs, *Toxicol. Appl. Pharmacol.*, 44, 593, 1978.
76. **DiVincenzo, G. D., Hamilton, M. L., Kaplan, C. S., and Dedinas, J.,** Metabolic fate and disposition of carbon-14 labeled methyl-n-butyl ketone in the rat, *Toxicol. Appl. Pharmacol.*, 41, 547, 1977.
77. **Couri, D., Hetland, L. B., Abdel-Rahman, M. S., and Weiss, H.,** The influence of inhaled ketone solvent vapors on hepatic microsomal biotransformation activities, *Toxicol. Appl. Pharmacol.*, 41, 285, 1977.
78. **Couri, D., Abdel-Rahman, M. S., and Hetland, L. B.,** Biotransformation of n-hexane and methyl n-butyl ketone in guinea-pigs and mice, *Am. Ind. Hyg. Assoc. J.*, 39, 295, 1978.
79. **Abdel-Rahman, M. S., Hetland, L. B., and Couri, D.,** Toxicity and metabolism of methyl n-butyl ketone, *Am. Ind. Hyg. Assoc. J.*, 37, 95, 1976.
80. **Nilsen, O. G., Toftgard, R., Eng, L., and Gustafsson, J. O.,** Regioselectivity of purified forms of rabbit liver microsomal cytochrome P-450 in the metabolism of benzo (a) pyrene, n-hexane and 7-ethoxy-resorufin, *Acta Pharmacol. Toxicol.*, 48, 369, 1981.
81. **Toftgard, R., Nilsen, O. G., and Gustafsson, J. O.,** Changes in rat liver microsomal cytochrome P-450 and enzymatic activities after the inhalation of n-hexane, xylene, methyl ethyl ketone and methyl chloroform for four weeks, *Scand. J. Work Environ. Health*, 7, 31, 1981.
82. **Baker, T. S. and Rickert, D. E.,** Dose-dependent uptake, distribution, and elimination of inhaled n-hexane in the Fischer-344 rat, *Toxicol. Appl. Pharmacol.*, 61, 414, 1981.
83. **Veulemans, H., Van Vlem, E., Janssens, H., Masschelein, R., and Leplat, A.,** Experimental human exposure to n-hexane. Study of the respiratory uptake and elimination, and of n-hexane concentrations in peripheral venous blood, *Int. Arch. Occup. Environ. Health*, 49, 251, 1982.

84. **Nomiyama, K. and Nomiyama, H.,** Respiratory retention, uptake and excretion or organic solvents in man. Benzene, toluene, *n*-hexane, trichloroethylene, acetone, ethyl acetate and ethyl alcohol, *Int. Arch. Arbeits Med.,* 32, 75, 1974.
85. **Perbellini, L., Amantini, M. C., Brugnone, F., and Frontali, N.,** Urinary excretion of *n*-hexane metabolites — a comparative study in rat, rabbit and monkey, *Arch. Toxicol.,* 50, 203, 1982.
86. **Perbellini, L., Brugnone, F., Pastorello, G., and Grigolini, L.,** Urinary excretion of *n*-hexane metabolites in rats and humans, *Int. Arch. Environ. Health,* 42, 349, 1979.
87. **Perbellini, L., Brugnone, F., and Pavan, I.,** Identification of the metabolites of *n*-hexane, cyclohexane, and their isomers in men's urine, *Toxicol. Appl. Pharmacol.,* 53, 220, 1980.
88. **Takeuchi, Y., Ono, Y., and Hisanaga, N.,** An experimental study on the combined effects of *n*-hexane and toluene on the peripheral nerve of the rat, *Br. J. Ind. Med.,* 38, 14, 1981.
89. **Perbellini, L., Leone, R., Fracasso, M. E., Brugnone, F., and Venturini, M. S.,** Metabolic interaction between *n*-hexane and toluene in the vivo and in vitro, *Int. Arch. Occup. Environ. Health,* 50, 351, 1982.
90. **Walseth, F., Toftgard, R., and Nilsen, O. G.,** Phthalate esters. I. Effects on cytochrome P450 mediated metabolism in rat liver and lung, serum, enzymatic activities and serum protein levels, *Arch. Toxicol.,* 50, 1, 1982.
91. **Altenkirch, H., Stoltenburg-Didinger, G., and Wagner, H. M.,** Experimental data on the neurotoxicity of methyl-ethyl-ketone (MEK), *Experientia,* 35, 503, 1979.
92. **Krasavage, W. J., O'Donoghue, J. L., and Terhaar, C. J.,** The relative neurotoxicity of methyl-*n*-butyl ketone and its metabolites, *Toxicol. Appl. Pharmacol.,* 45, 251, 1978.
93. **Angelo, M. J.,** The Pharmacokinetics of the Neurotoxin 2,5-Hexanedione: Distribution Elimination and Model Simulations, Ph.D. thesis, University of Delaware, Newark, 1981.
94. **Angelo, M. J. and Bischoff, K. B.,** A physiologically-based pharmacokinetic model for 2,5-hexanedione, Proc. Conf. Environ. Toxicol. 12th, Air Force Aerosp. Med. Res. Lab., Tech. Rep. 81-149, Wright-Patterson Air Force Base, Dayton, Ohio, 1981, 250.
95. **O'Donoghue, J. L. and Krasavage, W. J.,** Hexacarbon neuropathy: a gamma diketone neuropathy?, *J. Neuropathol. Exp. Neurol.,* 38, 333, 1979.
96. **Spencer, P. S., Bischoff, M. C., and Schaumburg, H. H.,** On the specific molecular configuration of neurotoxic aliphatic hexacarbon compounds causing central-peripheral distal axonopathy, *Toxicol. Appl. Pharmacol.,* 44, 17, 1978.
97. **Spencer, P. S. and Schaumburg, H. H.,** Neurotoxic properties of certain aliphatic hexacarbons, *Proc. R. Soc. Med.,* 70, 37, 1977.
98. **Anthony, D. C. and Graham, D. G.,** Evidence for pyrrole formation in the pathogenesis of hexane neuropathy, *Toxicologist,* 2, 139, 1982.
99. **Graham, D. G.,** Hexane neuropathy: a proposal for pathogenesis of a hazard of occupational exposure and inhalant abuse, *Chem.-Biol. Interact.,* 32, 339, 1980.
100. **DeCaprio, A. P., Olajos, E. J., and Weber, P.,** Covalent bindings of a neurotoxic *n*-hexane metabolite: conversion of primary amines to substituted pyrrole adducts by 2,5-hexanedione, *Toxicol. Appl. Pharmacol.,* 65, 440, 1982.
101. **Graham, D. G. and Abou-Donia, M. B.,** Studies of the molecular pathogenesis of hexane neuropathy. I. Evaluation of the inhibition of glyceraldehyde-3-phosphate dehydrogenase by 2,5-hexanedione, *J. Toxicol. Environ. Health,* 6, 621, 1980.
102. **Graham, G. D., Anthony, D. C., Boekelheide, K., Maschmann, N. A., Richards, R. G., Wolfram, J. W., and Shaw, B. R.,** Studies on the molecular pathogenesis of hexane neuropathy. II. Evidence that pyrrole derivation of lysyl residues leads to protein cross linking, *Toxicol. Appl. Pharmacol.,* 64, 415, 1982.
103. **Sabri, M. I., Moore, C. L., and Spencer, P. S.,** Studies on the biochemical basis of distal axonopathies. I. Inhibition of glycolysis by neurotoxic hexacarbon compounds, *J. Neurochem.,* 32, 683, 1979.
104. **Sabri, M. I., Ederle, K., Holdsworth, C. E., and Spencer, P. S.,** Studies on the biochemical basis of distal axonopathies. II. Specific inhibition of fructose-6-phosphate kinase by 2,5-hexanedione and methyl butyl ketone, *Neurotoxicology,* 1, 285, 1979.
105. **Howland, R. D., Vyas, I. L., Lowndes, H. E., and Argentieri, T. M.,** The etiology of toxic peripheral neuropathies: in vitro effects of acrylamide and 2,5-hexanedione on brain enolase and other glycolytic enzymes, *Brain Res.,* 202, 131, 1980.
106. **Schoental, R. and Cavanagh, J. B.,** Mechanisms involved in the "dying-back" process — an hypothesis implicating coenzymes, *Neuropathol. Appl. Neurobiol.,* 3, 145, 1977.
107. **Gillies, P. J., Norton, R. M., and Bus, J. S.,** Effect of 2,5-hexanedione on lipid biosynthesis in sciatic nerve and brain of the rat, *Toxicol. Appl. Pharmacol.,* 54, 210, 1980.
108. **Gillies, P. J., Norton, R. M., White, E. L., and Bus, J. S.,** Inhibition of sciatic nerve sterologenesis in hexacarbon-induced distal axonopathy in the rat, *Toxicol. Appl. Pharmacol.,* 54, 217, 1980.

109. **Gillies, P. J., Norton, R. M., and Bus, J. S.,** Inhibition of sterologenesis but not glycolysis in 2,5-hexanedione-induced distal axonopathy in the rat, *Toxicol. Appl. Pharmacol.,* 59, 287, 1981.
110. **Gillies, P. J., Norton, R. M., Baker, T. S., and Bus, J. S.,** Altered lipid metabolism in 2,5-hexanedione-induced testicular atrophy and peripheral neuropathy in the rat, *Toxicol. Appl. Pharmacol.,* 59, 293, 1981.
111. **Couri, D. and Nachtman, J. P.,** Biochemical and biophysical studies of 2,5-hexanedione neuropathy, *Neurotoxicology,* 1, 269, 1979.
112. **Agutter P. S. and Mack, A. P.,** The effects of the α β-dicarbonyl compounds on the polymerization of sheep brain tubulin in vitro, *Biochem. Soc. Trans.,* 7, 691, 1979.
113. **Selkoe, D. J., Luckenbill-Edds, L., and Shelanski, M. L.,** Effects of neurotoxic industrial solvents on cultured neuroblastoma cells; methyl *n*-butyl ketone, *n*-hexane and derivatives, *J. Neuropathol. Exp. Neurol.,* 37, 768, 1978.
114. **DeCaprio, A. P.,** Neurotoxicity and Amine Reactivity of 2,5-hexanedione and Related Diketones in the Hen, Ph.D. thesis, Albany Medical College of Union University, Albany, 1981.
115. **Borgatti, A. R., Trigari, G., Ventrella, V., and Pagliarani, A.,** Interactions of n-alkanes with respiration and oxidative phosphorylation by rabbit heart mitochondria: *n*-hexane, *Boll.-Soc. Ital. Biol. Sper.,* 57, 1569, 1981.
116. **Goldstein, M. N.,** Effect of Methyl-*n*-butyl Ketone and Other Ketone Homologues on Mammalian Cells in Culture, National Institute of Occupational Safety and Health Report No. 210-75-0073, Washington, D.C., 1981.
117. **Dafforn, A., Jewell, M., Anderson, M., Ash, D., Horvath, D., Kitson, R., Margiotta, S., and Rych, G.,** Aliphatic ketones are acetylcholine-sterase inhibitors but not transition state analogs, *Biochem. Biophys. Acta,* 569, 23, 1979.
118. **O'Donoghue, J. L., Krasavage, W. J., DiVincenzo, G. D., and Katz, G. V.,** Further studies on ketone neurotoxicity and interactions, *Toxicol. Appl. Pharmacol.,* 72, 201, 1984.
119. **O'Donoghue, J. L., Krasavage, W. J., DiVincenzo, G. D., and Ziegler, D. A.,** Commercial-grade methyl heptyl ketone (5-methyl-2-octanone) neurotoxicity: contribution of 5-nonanone, *Toxicol. Appl. Pharmacol.,* 62, 307, 1982.
120. **DiVincenzo, G. D., Ziegler, D. A., O'Donoghue, J. L., and Krasavage, W. J.,** Possible role of metabolism in 5-nonanone neurotoxicity, *Neurotoxicology,* 3, 55, 1982.
121. **Shifman, M. A., Graham, D. G., Priest, J. W., and Bouldin, T. W.,** The neurotoxicity of 5-nonanone: preliminary reports, *Toxicol. Lett.,* 8, 283, 1981.
122. **Oh, J. S. and Kim, J. M.,** Giant axonal swelling in "Huffer's" neuropathy, *Arch. Neurol.,* 33, 583, 1976.
123. **Aubuchon, J., Robins, H. I., and Viseskul, C.,** Ultrastructural evaluation of toxic polyneuropathies, *Lancet,* 2, 1012, 1979.
124. **AuBuchon, J., Robins, H. I., and Viseskul, C.,** Peripheral neuropathy after exposure to methyl-isobutyl ketone in spray paint, *Lancet,* 2, 363, 1979.
125. **Dyro, F. M.,** Methyl ethyl ketone polyneuropathy in shoe factory workers, *Clin. Toxicol.,* 13, 371, 1978.
126. **Crespi, V., DiCostanzo, M., Ferrario, F., and Tredici, G.,** Electrophysiological findings in workers exposed to *n*-heptane fumes, *J. Neurol.,* 222, 135, 1979.
127. **Gralla, E. J.,** A 90-day Toxicology Study in Fischer-344 Rats Exposed to Methyl Ethyl Ketone, Chemical Industry Institute of Toxicology, Research Triangle Park, N.C., 1981.
128. **DeJesus, P. V., Pleasure, D. F., Asbury, A. K., Brown, M. J., and Paradise, C. M.,** Effects of Methyl Butyl Ketone on Peripheral Nerves and its Mechanism of Action, Contract No. TENN CDC 99-76-16, New Haven, Conn., Yale University and West Haven, Conn. Veterans Administration Hospital, 1977.
129. **Takeuchi, Y., Ono, Y., Hisanaga, N., Kitch, J., and Sugiura, Y.,** A comparative study on the neurotoxicity of *n*-pentane, *n*-hexane, and *n*-heptane in the rat, *Br. J. Ind. Med.,* 37, 241, 1980.
130. **Frontali, N., Amantini, M. C., Spagnolo, A., Guarcini, A. M., and Saltari, M. C.,** Experimental neurotoxicity and urinary metabolites of the C_5-C_7 aliphatic hydrocarbons used as glue solvents in shoe manufacture, *Clin. Toxicol.,* 18, 1357, 1981.
131. **Homan, E. R. and Maronpot, R. R.,** Neurotoxic evaluation of some alphatic ketones, *Toxicol. Appl. Pharmacol.,* 45, 215A, 1978.
132. **Perbellini, L., DeGrandis, D., Semenzato, F., and Bongiovanni, L. G.,** On the neurotoxicity of cyclohexanol and cyclohexanenone, *Med. Lav.,* 72, 102, 1981.
133. **Perbellini, L., DeGrandis, D., Semenzato, F., Rizzuto, N., and Simonati, A.,** An experimental study on the neurotoxicity of *n*-hexane metabolites hexanol-1 and hexanol-2, *Toxicol. Appl. Pharmacol.,* 46, 421, 1978.
134. **Lynch, D. W., Lewis, T. R., Moorman, W. J., Plotnick, H. B., Schuler, R. L., Smallwood, A. W., and Kommineni, C.,** Inhalation toxicity of methyl *n*-amyl ketone (2-heptanone) in rats and monkeys, *Toxicol. Appl. Pharmacol.,* 58, 341, 1981.

135. **Johnson, B. L., Setzer, J. V., Lewis, T. R., and Hornung, R. W.,** An electrodiagnostic study of the neurotoxicity of methyl n-amyl ketone, *Am. Ind. Hyg. Assoc. J.*, 39, 866, 1978.
136. **Johnson, B. L., Anger, W. K., Setzer, J. V., Lynch, D. W., and Lewis, T. R.,** Neurobehavioral effects of methyl n-butyl ketone and methyl n-amyl ketone in rats and monkeys: a summary of NIOSH investigations, *J. Environ. Pathol. Toxicol.*, 2, 113, 1979.
137. **Misumi, J., Nagano, M., and Nomura, S.,** An experimental study on the neurotoxicity of 2-octanone and 2-hexanol, a metabolite of n-hexane, *Sangyo Igaku*, 24, 475, 1982.
138. **Pryor, G. T., Bingham, L. R., Dickerson, J., Rebert, C. S., and Howd, R. A.,** Importance of schedule of exposure to hexane in causing neurotoxicity, *Neurobehav. Toxicol. Teratol.*, 4, 71, 1982.
139. **Howd, R. A., Bingham, L. R., Steeger, T. M., Rebert, C. S., and Pryor, G. T.,** Relation between schedules of exposure to hexane and plasma levels of 2,5-hexanedione, *Neurobehav. Toxicol. Teratol.*, 4, 87, 1982.
140. **Rebert, C. S., Houghton, P. W., Howd, R. A., and Pryor, G. T.,** Effects of hexane on the brainstem auditory response and caudal nerve action potential, *Neurobehav. Toxicol. Teratol.*, 4, 79, 1982.
141. **Foa, B., Gilioli, R., Bulgheroni, C., Maroni, M., and Chiappino, G.,** Sulla e ziologia delle polineuriti da collanti: in dagine spermentale intorno alla neurotossicita del n-esano, *Med. Lav.*, 67, 136, 1976.
142. **Truhaut, R., Laget, P., Piat, G., Phu-Lich, N., Dutertre-Catella, H., and Huyen, V. N.,** Premiers résultats électrophysiologiques après intoxications expérimentales par l'hexane et par l'heptane techniques chez le rat blanc, *Arch. Mal. Prof.*, 34, 29, 1973.
143. **Bus, J. S., White, E. L., Tyl, R. W., and Barrow, C. S.,** Perinatal toxicity and metabolism of n-hexane in Fischer-344 rats after inhalation exposure during gestation, *Toxicol. Appl. Pharmacol.*, 51, 295, 1979.
144. **Takeuchi, Y., Ono, Y., and Hisanaga, N.,** An experimental study on the combined effects of n-hexane and toluene on the peripheral nerve of the rat, *Br. J. Ind. Med.*, 38, 14, 1981.
145. **Ono, Y., Yasuhiro, T., Hisanaga, N., Masamitsu, I., Kitoh, J., and Sugiura, Y.,** Neurotoxicity of petroleum benzine compared to n-hexane, *Int. Arch. Occup. Environ. Health*, 50, 219, 1982.
146. **Abe, K., Misumi, J., Kawakami, M., and Nomura, S.,** Effects of n-hexane, methyl n-butyl ketone, and 2,5-hexanedione and the excitability of sweat glands in rats to mecholyl, *Sangyo Igaku*, 22, 380, 1980.
147. **Misumi, J., Kawakami, M., Hitoshi, T., and Nomura, S.,** Effects of n-hexane, methyl n-butyl ketone and 2,5-hexanedione on the conduction velocity of motor and sensory nerve fibers in rats' tail, *Sangyo Igaku*, 21, 180, 1979.
148. **Johnson, B. L., Setzer, J. V., Lewis, T. R., and Anger, W. K.,** Effects of methyl n-butyl ketone on behavior and the nervous system, *Am. Ind. Hyg. Assoc. J.*, 38, 567, 1977.
149. **Katz, G. V., O'Donoghue, J. L., DiVincenzo, G. D., and Terhaar, C. J.,** Comparative neurotoxicity and metabolism of ethyl n-butyl ketone and methyl n-butyl ketone in rats, *Toxicol. Appl. Pharmacol.*, 52, 153, 1980.
150. **Mendell, J. R., Saida, K., Ganansia, M. F., Jackson, D. B., Weiss, H., Gardier, R. W., Chrisman, C., Allen, N., Couri, D., O'Neill, J., Jr., Marks, B., and Hetland, L.,** Toxic polyneuropathy produced by methyl n-butyl ketone, *Science*, 185, 787, 1974.
151. **Abdo, K. M., Graham, D. G., Timmons, P. R., and Abou-Donia, M. B.,** Neurotoxicity of continuous (90 days) inhalation of technical grade methyl butyl ketone in hens, *J. Toxicol. Environ. Health*, 9, 199, 1982.
152. **Duckett, S., Williams, N., and Francis, S.,** Peripheral neuropathy associated with inhalation of methyl-n-butyl ketone, *Experientia*, 30, 1283, 1974.
153. **Duckett, S., Streletz, L. J., Chambers, R. A., Auroux, M., and Galle, P.,** 50 ppm MnBK subclinical neuropathy in rats, *Experientia*, 35, 1365, 1979.
154. **Eben, A., Flucke, W., Mihail, F., Thyseen, J., and Kimmerle, G.,** Toxicological and metabolic studies of methyl n-butylketone, 2,5-hexanedione, and 2,5-hexanediol in male rats, *Ecotoxicol. Environ. Saf.*, 3, 204, 1979.
155. **Hall, S. M. and Gregson, N. A.,** The effects of 2,5-hexanedione on remyelination in the peripheral nervous system of the mouse, *J. Neuropathol. Exp. Neurol.*, 41, 642, 1982.
156. **Mennear, J. H.,** A short lived effect of 2,5-hexandeione on thermal perception in mice, *Toxicol. Appl. Pharmacol.*, 62, 205, 1982.
157. **O'Donoghue, J. L., Krasavage, W. J., and Terhaar, C. J.,** Toxic effects of 2,5-hexanedione, *Toxicol. Appl. Pharmacol.*, 45, 269, 1978.
158. **Powell, H. C., Koch, T., Garrett, R., and Lampert, P. W.,** Schwann cell abnormalities in 2,5-hexanedione neuropathy, *J. Neurocytol.*, 7, 517, 1978.
159. **Gilbert, S. G. and Maurissen, J. P. J.,** Assessment of the effects of acrylamide, methylmercury, and 2,5-hexanedione on motor functions in mice, *J. Toxicol. Environ. Health*, 10, 31, 1982.
160. **Ramsey, C. C.,** 2,5-Hexanedione Induced Axonopathy in the Crayfish, *Procambarous clarkii*, Medial Giant Axon, Ph.D. theis, Ohio State University, Columbus, 1980.

161. **Krinke, G., Schaumburg, H. H., Spencer, P. S., Thomann, P., and Hess, R.,** Clioquinol and 2,5-hexanedione induce different types of distal axonopathy in the dog, *Acta Neuropathol.*, 47, 213, 1979.
162. **Nagano, M., Misumi, J., Kaisaku, J., and Hitoshi, T.,** Experimental studies on peripheral neuropathy due to chemical substance: electrophysiological technique for detecting peripheral neuropathy and its application to 2,5-hexanedione-treated rats, *Kyushu Yakugakkai Kaiho*, 36, 91, 1982.
163. **Spencer, P. S. and Schaumburg, H. H.,** Experimental neuropathy produced by 2,5-hexanedione — a major metabolite of neurotoxic industrial solvent methyl *n*-butyl ketone, *J. Neurol. Neurosurg. Psychiat.*, 38, 771, 1975.
164. **Cangiano, A., Lutzemberger, L., Rizzuto, N., Simonati, A., Rossi, A., and Toschi, G.,** Neurotoxic effects of 2,5-hexanedione in rats: early morphological and functional changes in nerve fibers and neuromuscular functions, *Neurotoxicology*, 2, 25, 1981.
165. **Rossi, A., Simonati, A., Rizzuto, N., and Toschi, G.,** Neurotoxic action of 2,5-hexanedione on the autonomic nervous system: ultrastructural and functional alterations in the rat sympathetic superior cervical ganglion, *Brain Res.*, 243, 373, 1982.
166. **Griffin, J. W. and Price, D. L.,** Secondary demyelination in axonal neuropathies, *Neurology*, 29, 589, 1979.
167. **Cavanagh, J. B.,** The pattern of recovery of axons in the nervous system of rats following 2,5-hexanediol intoxication: a question of rheology?, *Neuropathol. Appl. Neurobiol.*, 8, 19, 1982.
168. Eastman Kodak Company, Unpublished data, Rochester, N.Y., 1983.
169. **Schneck, S. A.,** Methyl alcohol, in *Handbook of Clinical Neurology*, Vol. 36, Vinkin, P. J. and Bruyn, G. W., Eds., Elsevier/North-Holland, Amsterdam, 1979, chap. 14.
170. **Guillot, J. G., Schwartz, T. H., Nantel, A. J., and Savoie, J. Y.,** Methanol in alcoholic beverages evaluation of the federal norm, *Clin. Toxicol.*, 12, 622, 1978.
171. **Posner, H. S.,** Biohazards of methanol in proposed new uses, *J. Toxicol. Environ. Health.*, 1, 153, 1975.
172. **Mac Farland, J. F.,** The methylated spirit and some of its preparations, *Pharm. J. Trans.*, 15, 310, 1955.
173. **Wood, C. A. and Buller, F.,** Poisoning by wood alcohol. Cases of death and blindness from Columbian spirits and other methylated preparations, *JAMA*, 43, 972, 1904.
174. **Tyson, H. H. and Schoenberg, M. J.,** Experimental researches in methyl alcohol inhalation, *JAMA*, 63, 915, 1914.
175. **Eisenberg, A. A.,** Visceral changes in wood alcohol. Poisoning by inhalation, *Am. J. Public Health*, 7, 765, 1917.
176. **Scott, E., Helz, M. K., and McLord, C. P.,** The histopathology of methyl alcohol poisoning, *Am. J. Clin. Pathol.*, 3, 311, 1933.
177. **Cooper, J. R. and Kini, M. M.,** Biochemical aspects of methanol poisoning, *Biochem. Pharmacol.*, 11, 405, 1962.
178. **Röe, O.,** The metabolism and toxicity of methanol, *Pharmacol. Rev.*, 7, 399, 1955.
179. **Benton, C. D., Jr. and Calhoun, F. P., Jr.,** The ocular effects of methyl alcohol poisoning. Report of a catastrophe involving 320 persons, *Am. J. Opthalmol.*, 36, 1677, 1953.
180. **Scrimgeour, E. M.,** Outbreak of methanol and isopropanol poisoning in New Britain, Papua, New Guinea, *Med. J. Aust.*, 2, 36, 1980.
181. **Crook, J. E. and McLaughlin, J. S.,** Methyl alcohol poisoning, *J. Occup. Med.*, 6, 467, 1965.
182. **Riegel, J. and Wolf, G.,** Schwere neurologische ausfälle als folge einer methyl alkohol vergiftung, *Fortschr. Neurol. Psychiat.*, 34, 346, 1966.
183. **Guggenheim, M. A., Cough, J. R., and Weinberg, W. S.,** Motor dysfunction as a permanent complication of methanol ingestion, *Arch. Neurol.*, 24, 550, 1971.
184. **Merren, M. D.,** Bilateral lateral geniculate body necrosis as a cause of ambylopia, *Neurology*, 22, 263, 1972.
185. **Aquilonius, S. M., Bergstrom, K., and Enoksson, P.,** Cerebral computed tomography in methanol intoxication, *J. Comput. Assisted Tomogr.*, 4, 425, 1980.
186. **McLean, D. R., Jacobs, H., and Mielke, B. W.,** Methanol poisoning: a clinical and pathological study, *Ann. Neurol.*, 8, 161, 1980.
187. **Gilger, A. P. and Potts, A. M.,** Studies, on the visual toxicity of methanol. V. The role of acidosis in experimental methanol poisoning, *Am. J. Ophthalmol.*, 39, 63, 1955.
188. **Clay, K. L., Murphy, R. C., and Watkins, W. D.,** Experimental methanol toxicity in the primate: analysis of metabolic acidosis, *Toxicol. Appl. Pharmacol.*, 34, 49, 1975.
189. **McMartin, K., Makar, A. B., Martin, A., Palese, M., and Tephly, T. R.,** Methanol poisoning. I. The role of formic acid in the development of metabolic acidosis in monkey and the reversal by 4-methylpyrazole, *Biochem. Med.*, 13, 319, 1975.
190. **Martin-Amat, G., Tephyl, T. R., McMartin, K. E., Makar, A. B., Hayreh, M. S., Hayreh, S. J., Baumbach, G., and Cancilla, P.,** Methyl alcohol poisoning. II. Development of a model for ocular toxicity in methyl alcohol poisoning using the Rhesus monkey, *Arch. Ophthalmol.*, 95, 1847, 1977.

191. **Hayreh, M. S., Hayreh, S. S., Baumbach, G. L., Cancilla, P., Martin-Amat, G., Tephyl, T. R., McMartin, K. E., and Makar, A. B.,** Methyl alcohol poisoning. III. Ocular toxicity, *Arch. Ophthalmol.,* 95, 1851, 1977.
192. **Baumbach, G. L., Cancilla, P. A., Martin-Amat, G., Tephyl, T. R., McMartin, K. E., Makar, A. B., Hayreh, M. S., and Hayreh, S. S.,** Methyl alcohol poisoning. IV. Alterations of the morphological findings of the retina and optic nerve, *Arch. Ophthalmol.,* 95, 1859, 1977.
193. **Sharpe, J. A., Hostovsky, M., Bilbao, J. M., and Rewcastle, N. B.,** Methanol optic neuropathy: a histopathological study, *Neurology,* 32, 1093, 1982.
194. **Martin-Amat, G., McMartin, K. E., Hayreh, M. S., Hayreh, S. S., and Tephly, T. R.,** Methanol poisoning: ocular toxicity produced by formate, *Toxicol. Appl. Pharmacol.,* 45, 201, 1978.
195. **McMartin, K. E., Martin-Amat, G., Noker, P. E., and Tephly, T. R.,** Lack of a role of formaldehyde in methanol poisoning in the monkey, *Biochem. Pharmacol.,* 28, 645, 1979.
196. **Koivusalo, M.,** Studies on the metabolism of methanol and formaldehyde in the animal organisms, *Acta Physiol. Scand. Suppl.,* 131, 39, 1956.
197. **Rowe, V. K. and McCollister, S. B.,** Alcohols, in *Patty's Industrial Hygiene and Toxicology,* Vol. 2C, 3rd rev. ed., Clayton, G. D. and Clayton, F. E., Eds., John Wiley-Interscience, New York, 1982, 4528.
198. **McMartin, K. E., Martin-Amat, G., Makar, A. B., and Tephly, T. R.,** Methanol poisoning. V. Role of formate metabolism in the monkey, *J. Pharmacol. Exp. Ther.,* 201, 564, 1977.
199. **Haggerty, R. J.,** Toxic hazards — deaths from permanent antifreeze ingestion, *N. Engl. J. Med.,* 261, 1296, 1959.
200. **Parry, M. F. and Wallach, R.,** Ethylene Glycol poisoning, *Am. J. Med.,* 57, 143, 1974.
201. **Rowe, V. K. and Wolf, M. A.,** Glycols, in *Patty's Industrial Hygiene and Toxicology,* Vol. 2C, 3rd rev. ed., Clayton, G. D. and Clayton, F. E., Eds., John Wiley-Interscience, New York, 1982, 3817.
202. **Berger, J. R. and Ayyar, D. R.,** Neurological complications of ethylene glycol intoxication, *Arch. Neurol.,* 38, 724, 1981.
203. **Fellman, D. M.,** Facial diplegia following ethylene glycol ingestion, *Arch. Neurol.,* 39, 739, 1982.
204. **Troisi, F. M.,** Chronic poisoning by ethylene glycol vapor, *Br. J. Ind. Med.,* 7, 65, 1950.
205. **Donley, D. E.,** Toxic, encephalopathy and volatile solvents in industry. Report of a case, *J. Ind. Hyg. Toxicol.,* 18, 571, 1936.
206. **Parsons, C. E. and Parsons, M. E. M.,** Toxic encephalopathy and "granulopenic anemia" due to volatile solvents in industry: report of two cases, *J. Ind. Hyg. Toxicol.,* 20, 124, 1938.
207. **Greenburg, L., Mayers, M. R., Goldwater, L. J., Burke, W. J., and Moskowitz, S.,** Health hazards in the manufacture of "fused collars". I. Exposure to ethylene glycol monomethyl ether, *J. Ind. Hyg. Toxicol.,* 20, 134, 1938.
208. **Zavon, M. R.,** Methyl cellosolve intoxication, *Am. Ind. Hyg. Assoc. J.,* 24, 36, 1963.
209. **Ohi, G. and Wegman, D. H.,** Transcutaneous ethylene glycol monomethyl ether poisoning in the work setting, *J. Occup. Med.,* 20, 675, 1978.
210. **Savolainen, H.,** Glial cell toxicity of ethyleneglycol monomethylether vapor, *Environ. Res.,* 22, 423, 1980.
211. **Hine, C., Rowe, V. K., White, E. R., Darmer, K. I., Jr., and Youngblood, G. T.,** Epoxy compounds, in *Patty's Industrial Hygiene and Toxicology,* Vol. 2A., 3rd rev. ed., Clayton, G. and Clayton, F., Eds., John Wiley-Interscience, New York, 1981, 2141.
212. **Gross, J. A., Haas, M. L., and Swift, T. R.,** Ethylene oxide neurotoxicity: report of four cases and review of the literature, *Neurology,* 29, 978, 1979.
213. **Bryan, R. M. and Bland, L. A.,** Occupational exposure to ethylene oxide: effects and control, *J. Environ. Health,* 43, 254, 1981.
214. **Salinas, E., Sasich, L., Hall, D. H., Kennedy, R. M., and Morriss, H.,** Acute ethylene oxide intoxication, *Drug Intell. Clin. Pharm.,* 15, 384, 1981.
215. **Ficarra, B. J.,** Toxicologic states treated in an emergency department, *Clin. Toxicol.,* 17, 1, 1980.
216. **Troisi, F. M.,** Late technopathic aphony due to ethylene oxide, *Med. Lav.,* 56, 373, 1965.
217. **Spasovski, M., Khristeva, V., Pernov, K., Kirkov, V., Dryanovska, T., Panova, Z., Bobev, G., Gincheva, N., and Ivanova, S.,** Health status of workers producing ethylene and ethylene oxide, *Khig. Zdraveopaz.,* 23, 41, 1980.
218. **Joyner, R. E.,** Chronic toxicity of ethylene oxide, *Arch. Environ. Health,* 8, 700, 1964.
219. **Jacobson, K. H., Hackely, E. B., and Feilsilver, L.,** The toxicity of inhaled ethylene oxide and propylene oxide vapors, *Arch. Ind. Health,* 13, 237, 1956.
220. **Trommer, A.,** Concerning the effects of inhaled ethylene oxide, thesis Wurzburg, 1931, cited in *Azch. Ind. Health,* 13, 218, 1956.
221. **Koelsch, F. and Lederer, E.,** Toxicity of ethylene oxide, *Zentralbl. Gewerehyg.,* 7, 264, 1930.
222. **Hollingsworth, R. L., Rowe, V. K., Oyen, F., McCollister, D. D., and Spencer, H. C.,** Toxicity of ethylene oxide determined on experimental animals, *Arch. Ind. Health,* 13, 217, 1956.

223. **Baer, F. and Griepentrog, F.,** Long-term feeding experiment with rats using ethylene oxide impregnated feed, *Bundesgesundheitsblatt,* 12, 106, 1969.
224. **Oser, B. L., Morgareidge, K., Cox, G. E., and Carson, S.,** Short term toxicity of ethylene chlorohydrin (ECH) in rats, dogs and monkeys, *Food Cosmet. Toxicol.,* 13, 313, 1975.
225. **Hunter, C. G., Stevenson, D. E., and Chambers, P. L.,** Acute and short-term oral toxicity in rats of RD 025, a propylene glycolethylene oxide copolymer, *Food Cosmet. Toxicol.,* 5, 195, 1967.
226. **Rowe, V. K., Hollingsworth, R. L., Oyen, R., McColliser, D. D., and Spencer, H. C.,** Toxicity of propylene oxide determined on experimental animals, *Arch. Ind. Health,* 13, 228, 1956.
227. **Johnson, B.,** Neurotoxicity of ethylene and propylene oxide, *Toxicol. Res. Proj. Dir.,* 5, 3, 1980.

Chapter 5

ALIPHATIC HALOGENATED HYDROCARBONS, ALCOHOLS, AND ACIDS AND THIOACIDS

John L. O'Donoghue

TABLE OF CONTENTS

I.	General Comments	100
II.	Methyl Chloride	101
	A. Introduction	101
	B. Human Neurotoxicity	101
	C. Experimental Neurotoxicity	102
	D. Effects on Other Organs	102
III.	Methyl Bromide	102
	A. Introduction	102
	B. Human Neurotoxicity	103
	1. Toxicity at High Vapor Concentrations	103
	2. Toxicity at Moderate Vapor Concentrations	103
	3. Toxicity at Low Vapor Concentrations	104
	C. Experimental Neurotoxicity	104
	D. Exposure to Fumigated Crops or Feeds	104
	E. Metabolism	106
	F. Mechanism of Action	106
IV.	Methyl Iodide	107
V.	Methylene Chloride	107
VI.	Carbon Tetrachloride	108
	A. Introduction	108
	B. Human Neurotoxicity	108
	C. Experimental Neurotoxicity	108
VII.	Ethyl Chloride	109
	A. Introduction	109
	B. Neurotoxicity	109
	C. Metabolism	110
	D. Other Effects	110
VIII.	Trichlorotrifluoroethane	110
IX.	Vinyl Chloride	110
	A. Introduction	110
	B. Human Neurotoxicity	110
	C. Experimental Neurotoxicity	111

X.	Trichloroethylene		111
	A.	Introduction	111
	B.	Human Neurotoxicity	112
	C.	Experimental Neurotoxicity	112
	D.	Trichloroethylene Decomposition Product — Dichloroacetylene	113
	E.	Neuropathology	113
	F.	Metabolism	113
XI.	Allyl Chloride		113
	A.	Introduction	113
	B.	Human Neurotoxicity	114
	C.	Experimental Neurotoxicity	114
XII.	Dichloroacetic Acid		114
	A.	Introduction	114
	B.	Neurotoxicity	114
	C.	Neuropathology	114
	D.	Mechanism of Action	115
XIII.	2-Chloropropionic Acid		115
	A.	Introduction	115
	B.	Experimental Neurotoxicity	115
	C.	Metabolism	116
XIV.	2-Bromopropionic Acid		116
XV.	2-Bromobutyric Acid		116
XVI.	1-Bromopentane		117
XVII.	2-Mercaptopropionic Acid		117
XVIII.	3-Mercaptopropionic Acid		117
XIX.	Other Aliphatic Chlorinated Substances		118
References			120

I. GENERAL COMMENTS

Many of the chemicals reviewed in this chapter appear to have a special affinity for the cerebellum. The reasons of this affinity are unknown but may relate to common metabolic products among the chemicals or intrinsic susceptibility of cerebellar neurons to halogenated materials. 2-Chloropropionic acid, 2-bromopropionic acid, 2-bromobutyric acid, 1-bromopropionic acid, 2-bromobutyric acid, and 1-bromopentane produce identical cerebellar granule cell necrosis and similar toxicity in other organs. Similarity of chemical structure and neuronal toxicity suggest a common metabolite or mechanism of neurotoxicity. 2-Mercaptopropionic acid and thiophene (Chapter 3) also produce identical granule cell damage implying that halogenation may not be necessary for neurotoxicity but metabolism leading to dehalogenation may result in reactive metabolites. Cerebellar signs, symptoms, or lesions are also observed in humans or experimental animals exposed to methyl chloride, methyl bromide, methyl iodide, and ethyl chloride. Toxic relationships among these materials have not been investigated.

II. METHYL CHLORIDE

A. Introduction

Methyl chloride or monochloromethane (CH_3Cl) is a nonirritating, sweet-smelling gas at room temperature. Primary use is as a chemical intermediate, particularly in the production of silicones, tetramethyl lead, synthetic rubber, and methyl cellulose and as a general methylating agent. It is also used as an extractant and as a blowing agent in polystyrene foams. Former uses include local and general anesthesia and refrigeration. Refrigeration use today is rare but reports occasionally appear on exposure due to leaking refrigerant.[1]

Industrial exposure is mainly by inhalation during the cutting of polystyrene foamed plastics.[2] Intentional ''recreational'' inhalation of methyl chloride[3] and absorption through the skin[4] may also occur.

B. Human Neurotoxicity

The toxicity of methyl chloride has recently been reviewed.[2,4-6] The primary toxic effects of this material involve the central nervous system. Involvement of the peripheral nervous system is unusual.[5,7,8] Severe acute methyl chloride toxicity results in central nervous system signs of depression and unconsciousness, and gastrointestinal signs and symptoms including vomiting, nausea, diarrhea, abdominal pain, and anorexia.[4-13] Deaths are infrequent and sometimes have been related to consumption of alcohol prior to exposure.[14]

More commonly, acute exposures result in staggering, ataxia, vertigo, weakness, emotional instability including depression, euphoria, and suicidal tendencies, abnormalities of vision including difficulty focusing, blurring and paralysis of accommodation, memory loss, speech impairment, headache, and sleep disturbances.[4-13] Less common or unusual signs include numbness, muscle atrophy, fasciculations, and myoclonia.[7,8] Scharnweber et al.[8] reported the cases of six men exposed to 200 to 400 ppm of methyl chloride for 2 to 3 weeks who showed many of these symptoms.

Mild overexposures to methyl chloride result in a ''drunken or inebriated state'' with incoordination and impaired judgment.[2,15]

Workplace studies or controlled human exposure studies are limited. Repko et al.[16] studied 122 methyl chloride workers exposed to a mean airborne level of 34 ppm using 73 behavioral measures of performance, electroencephalograms, neurological function studies, and demographic data. ''There seemed to be a decrease in speed and accuracy of simple tasks and a diminution of the ability to time-share in more complex performance situations.''[16] Other neural tests were unaffected by exposure to methyl chloride.

Hake et al.[17] and Stewart et al.[18] found no effects on behavioral, neurologic, electromyographic, or blood chemistry tests conducted on male and female volunteers exposed to 20, 100, or 150 ppm of methyl chloride for 1, 3, or 7.5 hr/day repeatedly for 5 days/week. Putz-Anderson et al.[19] found that volunteers inhaling 100 ppm of methyl chloride for 3 hr showed no effect on performance and that at levels of 200 ppm for 3 hr there was a marginally significant performance impairment of 4.5%. Methyl chloride inhalation (200 ppm/3 hr) did not potentiate the performance impairment caused by ingestion of 10 mg of diazepam, but the effects of combined exposure to both appeared additive.[19] A common finding has been large interindividual variation in blood and breath levels following methyl chloride inhalation.[17,19]

Recovery from acute intoxication usually occurs in a matter of several hours although in some cases several months are required. Symptoms following sublethal exposures have been reported as long as 13 years later to include mild but permanent neurologic and psychiatric sequelae.[7-10]

Autopsy reports following severe intoxication suggest diffuse damage to the central nervous system including both cerebrocortical and spinal damage.[5,11,13,20]

C. Experimental Neurotoxicity

Smith and von Oettingen[21] studied the lethality of methyl chloride in a variety of animal species. Doses ranged from 300 to 400 ppm, 6 hr/day, 6 days/week repeatedly for some species. In order of increasing sensitivity to 200 ppm of methyl chloride were frogs > chicken > cat > rabbit > rat > monkey > dog > guinea pigs = mice = goats. At 300 ppm for 64 weeks of exposure, no effects were seen in guinea pigs, mice, dogs, monkeys, rabbits, or rats. At 500 ppm, one dog developed "irreversible neuromuscular damage" after 29 weeks of exposure, two monkeys died after 16 or 17 weeks of exposure showing signs of "progressive debility and terminal persisting unconsciousness," and a rabbit born during the exposures showed slight neuromuscular signs as did older rabbits. Monkeys exposed to 2000 ppm for six to nine exposures had convulsions and prolonged periods of unconsciousness or prostration.

Kolkmann and Volk[22] reported the cerebellar effects of exposing guinea pigs to atmospheres of approximately 20,000 ppm for 10 min 6 times a week for total exposures of from 6 to 61. The earliest abnormalities were ataxia and hindlimb paresis after 17 exposures. Four animals developed ataxia, hindlimb paresis, ataxic head movements, and retardation of spontaneous reactions after 25 days. Morphological changes after 10 days of exposure included edema of the internal granular layer of the cerebellum which initially occurred in the vermis but later spread to other regions of the cerebellum. Accompanying the edema, there was pyknosis of cerebellar granule cells and by 21 days, focal necrosis of the internal granular layer. The authors suggested that methyl chloride has a special affinity for cerebellar granule cells.[22]

D. Effects on Other Organs

The primary concern in methyl chloride intoxication is due to disturbances in the central nervous system although gastrointestinal symptoms are common in acute intoxications. Massive methyl chloride exposures to people have been reported to lead to electrocardiogram changes,[4] heart rate changes,[1,24] hypotension, and liver[1,23] or kidney injury.[1,13]

In experimental studies, methyl chloride levels of 1500 ppm for 90 days produced moderate hepatic effects in rats and mice and exposures of 1000 ppm for 6 and 12 months resulted in testicular degeneration in rats.[2] In a bacterial mutagenesis assay, methyl chloride has been reported to cause base pair substitutions in *Salmonella typhimurim* strain TA 1535.[25]

III. METHYL BROMIDE

A. Introduction

Methyl bromide (monobromomethane, CH_3Br_3) is a colorless, nonflammable gas which is odorless at low concentrations and has a sweet odor at higher concentrations. The primary use of methyl bromide is as a fumigant for soil, warehouses, rail cars, fruits, and produce. It is also used to fumigate houses and as a methylating agent in the chemical industry. Its

Table 1
ACCIDENTAL HUMAN INHALATION EXPOSURES TO METHYL BROMIDE

Estimated exposure (ppm)	No. of individuals involved	Outcome	Ref.
60,000	1	Vomiting, giddy, delirium, convulsions, death, severe lung damage	26
10,000	1	Lethal	47
8,000	1	Found moribund	47
220, several hours	—	Without serious effect	47
50, 8 hr	—	Mild symptoms	47
≤35, 2 weeks	33 out of 90 exposed	Dermatitis due to contact; systemic illness included vomiting, anorexia, headache, vertigo, fainting, and abnormal vision	49

Note: Exposure concentrations in cases of methyl bromide poisoning have generally not been reported. The levels reported on this table are the best estimates of exposure available.

use as a refrigerant has been discontinued and as a fire extinguisher, particularly for ships, has largely been discontinued.

Methyl bromide has several characteristics which make it a particularly hazardous substance. Being a colorless and odorless gas, people may not be aware of its presence. Exposure does not produce immediate irritation or illness; therefore, people may voluntarily continue exposure. When signs of toxicity appear after a delay period, they frequently are followed by death or chronic central nervous system damage.

B. Human Neurotoxicity

The first reported case of methyl bromide poisoning occurred in 1893.[26] Between 1899 and 1952, 174 nonfatal and 47 fatal cases were reported.[27,28] Between 1957 and 1964, Hine[29] noted 166 cases of methyl bromide poisoning. At least 62 additional cases[29-46] have been reported. In most of these exposures, methyl bromide concentrations were not determined. A few cases in which estimates of the methyl bromide exposure concentration were made are listed in Table 1.

Symptomatology following methyl bromide exposure depends on the exposure concentration, the length of exposure, and individual susceptibility (Table 1). A few individuals have been reported to have sustained multiple acute poisonings with little or no sequela but most fit into the pattern described by Clarke et al.[47]

1. Toxicity at High Vapor Concentrations

At high atmospheric concentrations, methyl bromide is an irritant to the eyes and respiratory tract and has an unpleasant musty odor. If the exposure ends quickly, symptoms may include eye irritation, coughing, pneumonitis, headache, and sensory neuropathy including numbness and paresthesias. Residual sensory abnormalities may last for months. If escape does not occur, coma, severe repeated grand mal seizures or status epilepticus, pulmonary edema, or death may occur.

2. Toxicity at Moderate Vapor Concentrations

Exposures of the order of 100 to 500 ppm may be more dangerous because warning signs, such as eye irritation, nausea, and headache, are minor and often overlooked or misinter-

abnormalities included ataxia, nystagmus, tremors, scanning dysarthria, dysdiadochokinesis of the arms, sluggish lower limb reflexes, and hallucinations. Complete recovery occurred in 1 month. Seven other women and a man sniffing ethyl chloride for its hallucinogenic properties exhibited disorientation and paranoia but no cerebellar abnormalities.

C. Metabolism
Ethyl chloride appears to be eliminated rapidly by the lungs and is not metabolized to ethanol.[62]

D. Other Effects
Ethyl chloride at very high atmospheric concentrations for prolonged periods of time may result in damage to the kidneys and liver. Like other chlorinated solvents, ethyl chloride may sensitize the myocardium to beta-adrenergic-stimulating substances such as epinephrine resulting in ventricular fibrillation.[64,96,97]

VIII. TRICHLOROTRIFLUOROETHANE

Raffi and Violante[98] reported a single case of sensorimotor neuropathy in a woman who had worked as a laundress with trichlorotrifluoroethane (Freon 113) for several years. Weakness, pain, and paresthesias were most severe distally in the legs. Electrodiagnostic testing was consistent with axonal damage. Removal from exposure to trichlorotrifluroethane resulted in gradual recovery.

IX. VINYL CHLORIDE

A. Introduction
Vinyl chloride (chlorethene, chlorethylene, $CH_2=CHCl$) is a gas used in the production of polyvinylchloride and as a copolymer for other plastics. Other major uses have included solvent and aerosol propellant and a short-lived trial as an anesthetic.

B. Human Neurotoxicity
Industrial exposure to vinyl chloride and the production of polyvinyl chloride polymers have been associated with a variety of disorders such as acro-osteolysis, Raynaud's phenomenon, nonmalignant disorders of the liver and hematopoetic system, angiosarcoma and carcinoma of the liver, tumors of the brain, lungs, lymphoid, and hematopoietic tissues, and pulmonary function changes. There are several reviews of the toxicity of vinyl chloride and related substances.[99-107]

Studies of the toxicity of vinyl chloride have focused on the hepatic and carcinogenic effects of this material and there is relatively little information available on the much less frequently observed neurologic effects. Exposures resulting in toxicity to vinyl chloride have been related to inhalation of the monomer and working in polymer production areas.

High concentrations of vinyl chloride produce a pleasant taste in the mouth, euphoria, and giddiness.[108] Movements become slowed and apparently inebriated workers may easily fall asleep. Other abnormalities include headache, irritability, memory loss, paresthesia, tingling in extremities, and weight loss.[108,109]

Spirtas and colleagues[110] conducted a study of the symptoms reported by a group of vinyl chloride polymer workers and found a dose-related response pattern for symptoms of dizziness, lightheadedness, nausea, fatigue, and "pins and needles" feelings in the arms and legs. Waxweiler et al.[111] reported an epidemiologic survey on workers in this same plant and found only minimal symptoms related to job category. These included slightly higher prevalence of headaches, more common reports of having lost consciousness, and slightly diminished reflexes in one group.

Persistent signs of peripheral neuropathy have been associated with acro-osteolysis and Raynaud's phenomenon in workers who have been routinely assigned the job of reactor cleaner. The incidence of acro-osteolysis has been estimated to be between 1 and 6% of reactor cleaners.[112] Takeuchi and Mabuchi[113] described residual effects seen in a man who had worked as a reactor cleaner for 4.5 years. Nine years after giving up this job because of acro-osteolysis, paresthesia and abnormal electroencephalogram remained.

Malignant tumors of the brain have been associated with exposure to vinyl chloride.[114-120] Although the number of vinyl chloride-associated brain cancers is small, the occurrence is consistent and the preponderance of glioblastoma multiforme is unusual. Monson et al.[115] identified 5 brain tumors in vinyl chloride workers among 161 deaths when 1.2 brain tumors were expected. Three of the tumors were glioblastoma multiforme.

Nine out of ten brain tumors identified by Waxweiler et al.[111] in vinyl chloride workers were glioblastoma multiforme. The tenth tumor was not histologically typed. This is in contrast with the findings of 33% glioblastoma multiforme in a Yale autopsy series.[111] The relative risk of brain tumors in vinyl chloride workers has been estimated to be between 3 and 6 times higher than expected.[120]

C. Experimental Neurotoxicity

A low but significant incidence of brain neoplasms has been observed in rodents exposed to high levels of vinyl chloride. In a series of studies, Maltoni et al.[121] found 10 neuroblastomas in 180 rats exposed to vinyl chloride levels of 2,500 to 10,000 ppm and none in controls. Feron and Kroes[122] reported one esthesioneuroepithelioma and one malignant ependymoma in Wistar rats exposed to 5000 ppm of vinyl chloride 7 hr/day, 5 days/week for 52 weeks. Other tumors were found in the nasal passages, ceruminous glands, and lung.

X. TRICHLOROETHYLENE

A. Introduction

Trichloroethylene (CCl_2=CHCl) is a nonflammable, colorless liquid with a sweet odor similar to chloroform. It is also known as ethylene trichloride, Trilene, and tri. Trichloroethylene (TCE) is mainly used (90%) as a vapor degreasing agent and solvent in industry.[123] Small quantities are used as a dry cleaning agent, a chemical intermediate, an anesthetic, and an analgesic. Formerly, it was used as an extractant for foods such as decaffeinated coffee. Although not reactive, TCE decomposes in the presence of heat or light, producing dichloroacetylene, phosgene, carbon monoxide, and hydrogen chloride. Decomposition is catalyzed by moisture, metals, rubber acids, and resins.[124] Dichloroacetylene is a hazard when TCE is in contact with strong alkaline materials as in anesthesia equipment with soda lime carbon dioxide absorbers[124] or life support systems[125] for space travel and submarines. Industrial TCE is usually inhibited with triethylamine, triethanolamine, epichlorohydrin, or stearates.[123] Anesthetic grades are inhibited with thymol blue.

Production methods have changed over many years resulting in a purer industrial grade of TCE and different potential by-products. Over the last 15 years, production methods have been changing from a method which started with acetylene and produced 1,1,2,2-tetrachloroethane as an intermediate to predominantly one which uses ethylene as a starting material and produces ethylene dichloride as an intermediate.[123]

More importantly, much of the earlier literature on TCE includes observations which were probably due to TCE decomposition products, particularly dichloroacetylene, and not TCE itself.[124]

B. Human Neurotoxicity

The most prominent effects of TCE are those involving the nervous system although hepatic, renal, and hematopoietic changes, and cardiac arrythmias have been reported. Several reviews on TCE toxicity are available.[61,123,124,126-129]

TCE gained wide use as an industrial degreaser during World War I and soon afterward Plessner[130] reported four cases of trigeminal nerve damage due to TCE exposure. This led to unsuccessful trials of TCE for treating trigeminal neuralgia or tic douloureux. A contaminant was apparently responsible for the effects on the trigeminal nerve.[131,132]

During World War II, TCE was developed as a nonexplosive anesthetic to replace chloroform. Use of TCE in rebreathing anesthetic machines equipped with soda lime absorbers led to cranial nerve palsies, recurrence of facial herpes infections, and death, apparently due to TCE decomposition products.[133-135]

As an anesthetic, highly purified TCE has had considerable use in Europe. At levels of 500 ppm, psychomotor functions deteriorate and levels of 3,000 to 25,000 ppm are anesthetic.[126] Analgesia is present at levels of 5000 ppm.[124] Deep planes of anesthesia have been associated with cardiac and respiratory complications.[124]

Nervous system sequelae of TCE anesthesia are rare although cases of headache, prolonged drowsiness, mental aberrations, and psychosis have been reported.[124]

In the industrial setting, the situation has been very different because of differences in TCE purity and the presence of toxic decomposition products during use.

Acute exposure to high ambient levels or ingestion of TCE may lead to inebriation, gastrointestinal upset, coma, and death. Recovery from acute exposures may be incomplete with cranial nerve palsies, sensorimotor peripheral neuropathies, headaches, or psychoses.[136-145] Chronic exposures have been associated with similar effects and autonomic and cerebellar dysfunction.[124]

The most characteristic syndrome involves cranial nerve damage particularly to the trigeminal nerve. Damage to other cranial nerves may result in loss of vision due to optic nerve atrophy or retrobular neuritis,[144,151] loss of taste,[140] facial weakness due to VII nerve damage,[145] or hearing loss due to VIII nerve damage.[147] Lawrence and Partyka[147] described an unusual case of trigeminal anesthesia and permanent bulbar palsy in a man who had worked shoveling metal shavings cleaned in TCE. Saihan et al.[148] also described an unusual case with Raynaud's phenomenon and peripheral neuropathy but the association with TCE exposure was ambiguous. Spinal lesions associated with TCE exposure are very rare.[149,150]

Although subjective complaints of headache, fatigue, and neurologic abnormalities have been reported by people exposed to 100 ppm or less of TCE,[151-155] other studies including those using electrophysiological techniques have been negative at similar levels.[156,162,163]

C. Experimental Neurotoxicity

The neurotoxicity associated with industrial use of TCE has not been reproduced experimentally with pure grades of TCE.[123,124,164,165,166-168] Minor behavioral deficits of uncertain significance have been reported with experimental exposures.

Battig and Grandjean[169] found both enhancement and inhibition of rat behavior following exposure to 400 ppm, 8 hr/day, 5 days/week for 10 months depending on which task was performed. Activity in familiar surroundings was decreased and in unfamiliar surroundings was increased in rats exposed to 100 ppm repetitively.[170] Ikeda,[171] and Ikeda et al.[172] found only slight learning impairment in rats exposed repetitively to 2600 to 8000 ppm of TCE, 30 min/day, and no effect on other behavior.

Gerbils exposed continuously to 150 ppm TCE for 71 and 106 days with or without a 40-day recovery period had very slight changes in maze activity.[173] Effects following 320 ppm of continuous TCE exposure for months were slight only after exposure to 1,1,1-trichloroethane.[174,175] Slight neurochemical effects occur with similar exposures.[175-177]

D. Trichloroethylene Decomposition Product — Dichloroacetylene

Dichloroacetylene or DCA (ClC≡CCl) is much more toxic than most other chlorinated hydrocarbons. The LC_{50} for a DCA exposure of 6 hr was 19 ppm.[178] Lethal levels of DCA produced extensive renal necrosis and edema and nonspecific neuronal degeneration in brain stem nuclei, Purkinje cells, and cerebral ganglion cells.[178] Humphrey and McClelland[133] identified foci of necrosis with glial proliferation in the brain and meningeal inflammatory reactions in two rabbits exposed to DCA with trichloroethylene or ether.

Lethal and sublethal airborne levels of DCA from 17 to 307 ppm given to rabbits resulted in a temperature sensory loss in the distribution of the trigeminal nerve and neuronal degeneration in several brain stem nuclei and cerebellar Purkinje cells. The severest changes occurred in the sensory trigeminal nucleus and then the facial nerve with milder changes in the oculomotor nuclei, motor trigeminal nucleus, and eighth nerve nuclei.

When rats were continuously exposed to 2.8 ppm of a DCA mixture by inhalation, two of eight rats had weakness and difficulty walking and one rat appeared blind.[180]

E. Neuropathology

Little data exist on the neuropathology of TCE. The best account is that of Buxton and Hayward[150] who described autopsy findings in a man exposed to TCE and TCE decomposition products. Lesions were bilaterally symmetrical and were most severe in the trigeminal nerve, tracts, and nuclei, particularly its sensory divisions. Also affected were the nucleus solitarius and its tract and, to a lesser extent, other dorsal brain stem nuclei and the reticular formation. Secondary ischemic changes involved the cerebral and cerebellar cortices. The clinical course and outcome of this case suggest that a TCE decomposition product, most likely dichloroacetylene, was responsible.

Dogs exposed to TCE vapors chronically for up to 162 hr developed ataxia and showed damage to cerebellar Purkinje cells.[181] Rabbits given 2 mℓ TCE, twice a week for 41 to 247 days or 3 mℓ TCE, 3 times a week for 29 days by i.m. injection did not develop neurologic deficits.[182] They were reported to have moderately diffuse neuronal damage in the majority of cranial nerve nuclei and the cerebellum. The similarity of the neuronal changes to artifactually inducible ischemic changes and the presence of inflammatory changes which in some instances are caused by common rabbit protozoal infections make interpretation of this study difficult.

F. Metabolism

The metabolic conversion of trichloroethylene to trichloroacetic acid (TCA) and excretion of TCA in the urine following inhalation of TCE has been well-known and used as a basis for monitoring working exposure to TCE.[124] Other urinary metabolites in man include trichloroethanol conjugated to glucuronic acid and monochloroacetic acid. Metabolism from TCE to TCA may proceed through conversion of TCE to trichloroethylene epoxide and chlorol hydrate.[183,184] The formation of reactive intermediates may explain the binding of TCE metabolites to protein.[184]

The metabolism of TCE is inhibited by combined exposure to ethanol leading to potential serious effects on cardiac rhythm and death.[185] Storage of TCE and its metabolites in body lipids may allow the accumulation of a toxic dose.[186]

XI. ALLYL CHLORIDE

A. Introduction

Allyl chloride (3-chloro-1-propene, 3-chloropropene, 2-propenylchloride, chloropropylene, $CH_2=CH-CH_2-Cl$) is the most commercially important allyl compound because it is the starting material for a number of important derivatives including epichlorohydrin, glycerol, allyl alcohol, allyl amines, barbiturates, diuretics, and cyclopropane.[187] It is a flammable liquid of relatively high toxicity.[2,188]

B. Human Neurotoxicity

He et al.[189] reported polyneuropathy in 17 Chinese women exposed to allyl chloride form 7 months to 5.2 years in a plant manufacturing sodium allyl sulfonate. The predominant clinical presentation was a symmetrical, distal sensorimotor neuropathy with paresthesias, weakness in the legs and hands, and sensory deficits to light touch, vibration, and superficial pain. Recovery began 2 to 4 months after treatment but it took 9 to 11 months for completion. Relapses occurred on reexposure. Functional changes in the nervous system of people with prolonged exposure to allyl chloride has been reported in the Russian literature.[190]

C. Experimental Neurotoxicity

Rabbits given allyl chloride by inhalation or s.c. injection and mice given it by mouth develop a clinical peripheral neuropathy.[189,191,192] Axonal damage is of the "dying-back" type with proliferation of neurofilaments. Degeneration was most severe in distal regions of the peripheral nerves and was also present in the spinal cord following the pattern of a central-peripheral distal axonopathy.[189,192]

XII. DICHLOROACETIC ACID

A. Introduction

Dichloroacetic acid is used as an intermediate in organic chemistry, in the synthesis of pharmaceuticals, and at one time was proposed for use in antiseptics and as a preservative. Dichloroacetic acid is metabolic product of 1,1,2,2-tetrachloroethane in vitro.[193] In commercial products, contact is more likely to be to a salt of dichloroacetic acid and toxicity studies generally refer to the sodium salt. Sodium dichloroacetate has been used in clinical trials because of its ability to reduce blood glucose, lactate, alanine, triglycerides, free fatty acids, and cholesterol in human diabetics.[194]

B. Neurotoxicity

Woodard et al.[195] reported that the acute toxicity of sodium dichloroacetate (LD_{50} rat 4.48 g/kg, mouse 5.52 g/kg) was much less than sodium chloroacetate (LD_{50} rat 0.076 g/kg, mouse 0.225 g/kg). The only report of human neurotoxicity related to sodium dichloroacetate involved a 21-year-old man with a genetic hypercholesterolemia who was treated with 50 mg/kg for 16 weeks.[196] Symptoms included weakness in facial, hand, and leg muscles, decreased deep tendon reflexes, and decreased nerve conduction velocity. Improvement followed stopping treatment. 26.7% of a group of rats dosed orally with 2000 mg/kg of sodium dichloroacetate (DCA) developed hindlimb paralysis during a 3-month treatment period.[194] The first case was observed after 2 months of treatment. A few rats allowed a 4-week recovery period returned to normal. Rats dosed orally with 125 or 500 mg/kg for 3 months did not develop a neuropathy. Dogs appear much more sensitive to the acute toxicity and neurotoxicity of DCA.[194] Doses of 75 mg/kg or 100 mg/kg resulted in general weakness, hindlimb weakness, reduced activity, or ataxia in two of seven dogs prior to death. Other dogs receiving 100 mg/kg were paralyzed.

C. Neuropathology

Both rats (125, 500, and 2000 mg/kg) and dogs (50, 75, and 100 mg/kg) receiving DCA orally had lesions consisting of vacuoles or edema-like areas in myelinated areas of the cerebrum and to a lesser extent the cerebellum.[194] After recovery periods of 4 weeks for rats and 5 weeks for dogs, vacuoles remained even though rats appeared clinically recovered. Examination of optic and sciatic nerves did not reveal myelin vacuoles.

Spencer et al.[197] have reported that rats consuming water containing 2 g/ℓ of DCA for up to 20 weeks have widespread vacuolization of the brain and spinal cord and minimal vacuolization of peripheral nerves. Scattered degenerative changes were found in axons.

D. Mechanism of Action

DCA has effects on pyruvate metabolism in laboratory species and probably in man which result in reduction in blood values of glucose, lactate, alanine, triglycerides, free fatty acids, and cholesterol in diabetics but not nondiabetics.[194,198] Effects on enzymes responsible for gluconeogenesis, including decreased glucokinase and pyruvate kinase and increased glucose-6-phosphate dehydrogenase and malic enzymes, occur in both diabetic and normal rats.[198] Enzyme effects are tissue specific (present in the liver, absent in the jejunum) and may require metabolic activation of DCA since hepatic enzymes are not affected in vitro.[198]

The mechanism of neurotoxicity is unknown, but effects on mitochondrial enzymes involving pyruvate metabolism and the similarity of myelin vacuolization between DCA and hexachlorophene and cuprizone which inhibit mitochondrial oxidative phosphorylation suggest that mitochondrial abnormalities may be involved.

XIII. 2-CHLOROPROPIONIC ACID

A. Introduction

2-chloropropionic acid (α-chloropropionic acid, 2-chloropropanoic acid, $C_3H_5ClO_2$) is a highly irritating liquid used in the manufacture of herbicides. Mild to moderate skin burns have occurred in workers handling it.[199] It has a proposed threshold limit value of 2 ppm.[199]

B. Experimental Neurotoxicity

The oral LC_{50} for 2-chloropropionic acid was about 800 mg/kg in male rats.[200] Survivors of the LD_{50} test developed tremors which may have been the result of cerebellar damage.[200]

Sprague-Dawley rats fed diets containing 1.0, 0.5, 0.25, or 0.1% of 2-chloropropionic acid consumed 207 to 320, 220, 171, or 78 mg/kg/day of the acid, respectively.[200] After 3 to 4 days on a 1.0% 2-chloropropionic acid diet, rats (10) began to have difficulty walking and developed gross ataxia and tremors. Five rats given diets containing 0.5% 2-chloropropionic acid for 6 days were comparable to animals given 1.0% for 3 to 4 days. After 6 days consuming diets containing 0.25% of 2-chloropropionic acid, rats showed increased sensitivity to stimulation, particularly sound, and ataxia. The 0.25% diet was continued for an additional 36 days with little change in clinical signs. Rats given 0.1% diets showed no clinical abnormalities after 38 days of exposure.

Rats fed 1.0% 2-chloropropionic acid for 8 days showed little or no recovery from ataxia during a 36-day period on control diets. Neutralization of the acid with sodium hydroxide delayed the onset of ataxia in rats given 1.0% 2-chloropropionic acid from 3 to 4 days to 10 days but had no effect on the eventual outcome.

The typical lesion found in rats receiving 1.0 to 0.25% 2-chloropropionic acid was foci of necrotic cells in the inner granule cell layer of the cerebellum. In more severely affected animals, the foci coalesced into large areas of granule cell necrosis. The remainder of the brain was unaffected. Other lesions included atrophy of the testicular germinal epithelium in rats receiving 1.0 to 0.1% diets and thymic lymphoid necrosis in rats given 1.0%.[200] Rats fed diets containing 0.04 mol/kg of feed of 2-chloropropionate consumed approximately 4 to 2.5 mmol/kg/day of 2-chloropropionate for 12 weeks.[201] Within 2 to 4 weeks on this diet, they developed hindlimb weakness and abnormal gaits which were associated with a slight reduction in motor nerve conduction velocity and slightly smaller diameter of the tibial nerve.

Growth retardation and testicular atrophy also were observed. The cause of the neurological deficits were assumed to be due to PNS damage although the CNS was not examined.

C. Metabolism

The sodium salt of 2-chloropropionic acid, 2-chloropropionate, like dichloroacetate is an activator of the pyruvate dehydrogenase complex in both rats and dogs.[201-205] Activation of the pyruvate dehydrogenase complex results in lower blood glucose, lactate, and pyruvate in fasted rats.[203] In dogs given 125 mg/kg p.o., lactate and pyruvate were lowered but glucose, β-hydroxybutyrate, and acetoacetate were unchanged.[205] Both dichloroacetate and 2-chloropropionate stimulate leucine metabolism in the perfused rat heart.[204]

XIV. 2-BROMOPROPIONIC ACID

2-Bromopropionic acid (α-bromopropionic acid, 2-bromopropanoic acid) has limited use in synthetic organic chemistry. No references to human toxicity or animal neurotoxicity were found. It is a strong irritant when applied to guinea pig skin under a rubber dam for 24 hr and may be absorbed through the skin (dermal LD_{50} between 0.01 and 0.1 mℓ/kg).[200] The approximate oral LD_{50} in male rats and mice was 237 or 282 mg/kg, respectively.[200] LePoidevin[206] reported that the oral LD_{50} was 200 mg/kg for mice and that the LD_{50} of 3-bromobutyric acid was >1000 mg/kg indicating the importance of 2-substitution on acute toxicity. Following single oral doses of 2-bromopropionic acid, surviving rats had tremors and ataxia.[200] Mice developed weakness and convulsions.[200] Neurologic signs in rats suggest that single sublethal doses of 2-bromopropionic acid may result in irreversible cerebellar damage.

Cerebellar damage resulted from feeding five male rats diets containing 1.0% 2-bromopropionic acid.[200] The average dose was 315 mg/kg/day. After 4 to 6 days, all rats showed tremors of the entire body, ataxia, convulsions, increased activity, and increased sensitivity to external stimuli such as touch and noise. Widespread degeneration and necrosis of cerebellar granule cells were found microscopically in four rats. No lesions were identified in the cerebrum.

Other effects included thymic atrophy in one rat and abnormal epididymal sperm in two rats. Another group of rats was unaffected by the presence of 0.1% of 2-bromopropionic acid (90 mg/kg/day) in their diets for 12 days.

XV. 2-BROMOBUTYRIC ACID

2-Bromobutyric (α-bromobutyric acid, 2-bromobutanoic acid) acid is an oily liquid with limited use as an intermediate for synthetic organic chemistry. No human toxicity or animal neurotoxicity data were found in the literature. The approximate oral LD_{50} of this compound as a 10% solution in corn oil was 181 or 70 mg/kg for male rats and mice, respectively.[200] LePoidevin reported that the oral LD_{50} was 325 mg/kg for mice and that 2-substitution had significant effects on acute toxicity since substitution on the 3 or 4 carbons produced $LD_{50}s$ of >1000 mg/kg.[206] 2-Bromobutyric acid is corrosive to guinea pig skin and is absorbed through the skin (dermal LD_{50} < 0.1 mℓ/kg).[200] Repeated doses as low as 50 mg/kg/day for 5 to 7 days were corrosive to the gastrointestinal tract.[200] To test the toxicity of the 2-substituted butyrate portion of the molecule, the acid was neutralized with sodium hydroxide.

One dose of 1000 mg/kg of the sodium salt by gavage killed all five dosed rats.[200] One dose of 500 mg/kg killed one of five rats and after 2 to 3 doses, the survivors were weak, ataxic, had labored breathing, and were in poor physical condition. Microscopically, all rats showed a moderate degree of pyknosis of cerebellar granule cells. One animal also developed atrophy of the thymus. Fifteen doses of 100 mg/kg of the sodium salt given to five male rats did not have any apparent clinical effect. Microscopically four rats had pyknosis of cerebellar granule cells and three had abnormal giant cells among epididymal spermatozoa.

XVI. 1-BROMOPENTANE

Synonyms   n-Amylbromide
n-Pentylbromide
1-Pentylbromide

1-Bromopentane is a liquid used as an intermediate in synthetic organic chemistry. No human toxicity or animal neurotoxicity data have been reported.

The approximate oral LD_{50} in male rats and mice was 3200 and >3200 mg/kg, respectively.[200] When held under a rubber dam against the skin of guinea pigs, it was a strong skin irritant.[200] After three doses of 1000 mg/kg by gavage, five rats had generalized muscle weakness, ataxia, and tremors involving the entire body.[200] Body weight gain and feed consumption were severely depressed. Rats receiving 15 doses of 100, 10, or 1 mg/kg of 1-bromopentane appeared clinically normal. Cerebellar granule cell necrosis was evident histologically in rats from the 1000 and 100 mg/kg groups. Testicular germinal cell atrophy and thymic cortical atrophy were evident in tissues from rats from the 1000 mg/kg group.

XVII. 2-MERCAPTOPROPIONIC ACID

Synonyms   2-Mercaptopropanoic Acid
2-Thiolactic Acid
α-Mercaptopropionic Acid

2-Mercaptopropionic acid is a foul-smelling, combustible, corrosive liquid which has been used in depilatories, hair waving solutions, and as an intermediate in synthetic organic chemistry. No references to human or animal neurotoxicity were found. Human sensitization has been reported.[207]

The oral LD_{50} was approximately 1131 and 672 mg/kg for male rats and mice, respectively.[200] Both rats and mice had tremors and convulsions before death. Among a group of five rats given 1000 mg/kg of 2-mercaptopropionic acid by gavage, two had grand mal seizures after two doses.[200] Three rats which survived two doses of 1000 mg/kg were ataxic, staggered, had tremors, and had a base-wide stance. Groups of ten rats given 500 or 100 mg/kg did not gain weight normally and showed slight muscle weakness but did not have seizures or ataxia. On routine histopathology examination, cerebellar granule cells were pyknotic in animals receiving 1000 mg/kg but not in those receiving 500 or 100 mg/kg. Other histologic effects included acute inflammation of the gastrointestinal tract and thymic cortical degeneration.

XVIII. 3-MERCAPTOPROPIONIC ACID

3-Mercaptopropionic acid (β-mercaptopropionic acid) has limited use in synthetic organic chemistry and may have been used in hair waving preparations. Human toxicity has not been reported, but this chemical has been used as a model for epilepsy and as an inhibitor of glutamate decarboxylase in vivo and in vitro.[208-212]

The oral LD_{50} in rats has been reported to be between 50 and 400 mg/kg.[207] Weakness and convulsions similar to those produced by thioglycolic acid were observed. Intraperitoneal injection (35 to 180 mg/kg) of 3-mercaptopropionic acid produces sudden severe convulsions in rats about 7 min following injection.[210] Doses of 35 and 90 mg/kg result in recovery in

Table 4
CHLORINATED ALCOHOLS AND SUGARS REPORTED NEUROTOXIC

Chemical	Species	Route	Lowest effective dose reported	Effect	Ref.
α-chlorohydrin (3-chloro-1,2-propanediol)	Mouse	Gavage	50 mg/kg/d, 4 days	Paralysis	216
1-amino-3-chloro-2-propanol	Rhesus monkey	Gavage	≥50 mg/kg/d, 7 d/wk, 13 wk	Focal brain edema; glial cell damage	218
6-chloro-6-deoxyglucose	Mouse	Gavage	240 mg/kg/d, 8 days	Paralysis	216
6-chloro-6-deoxyglucitol	Mouse	Gavage	480 mg/kg/d, 11 days	Paralysis	216
6-chloro-6-deoxyfructose	Mouse	Gavage	240 mg/kg/d, 6 days	Paralysis	216
6-chloro-6-deoxymannose	Mouse	Gavage	1200 mg/kg/d, 16 days	Paralysis	216
6-chloro-6-deoxygalactose	Mouse	Gavage	1200 mg/kg/d, 6 days	Paralysis	216

15 to 30 min. Doses of 150 and 180 mg/kg result in death within 30 min. Karlsson et al.[210] have shown that 3-mercaptopropionic acid produces a large reversible inhibition of glutamate decarboxylase in the brain prior to the onset of convulsions. Glutamate decarboxylase converts the putative excitatory neurotransmitter glutamate to the putative inhibitory neurotransmitter γ-aminobutyric acid (GABA). GABA levels decrease 20 to 35% 4 min after the injection of 3-mercaptopropionic acid, and seizures occur within 3 min. Aminooxyacetic acid, a drug which blocks the catabolism of GABA, producing increased GABA levels, delays or prevents the effects of 3-mercaptopropionic acid.[210,212] Convulsions have also been produced in baboons with photosensitive epilepsy and in mice.[209] Alterations in amino acid levels other than GABA and in mitochondrial oxidation appear to be secondary responses to reduce GABA levels and convulsions.[208,210,213]

Degenerative changes have been described in cerebellar Purkinje cells and in nerve terminals of treated rats.[214] Morphological abnormalities were not distributed in GABA neurons of the cerebral cortex, and therefore, the abnormalities may be specific for Purkinje neurons and not all GABA neurons.

XIX. OTHER ALIPHATIC CHLORINATED SUBSTANCES

α-Chlorohydrin or 3-chloro-1,2-propanediol has been used in the preparation of dyes and in dynamite. While testing for antifertility drugs, Samojlik and Chang,[215] and Ford and Waites[216] found that it caused paralysis in rats and mice. The effect in mice is believed to be similar to the effects of chlorinated sugars[217] (Table 4) and 1-amino-3-chloro-2-propanol.[218] Other nonhalogenated substances including 6-amino-nicotinamide,[219] metronidazole,[220] and misonidazole[221] produced similar brain lesions.

REFERENCES

1. **Spevak, L., Nadj, V., and Fellé,D.**, Methyl chloride poisoning in four members of a family, *Br. J. Ind. Med.*, 33, 272, 1976.
2. **Torkelson, T. R. and Rowe, V. K.**, Halogenated aliphatic hydrocarbons containing chlorine, bromine and iodine, in *Patty's Industrial Hygiene and Toxicology*, Vol. 2B, 3rd rev. ed., Clayton, G. and Clayton, F., Eds., John Wiley-Interscience, New York, 1981, 3433.
3. **Barnes, G. E.**, Solvent abuse: a review, *Int. J. Addict.*, 14, 1, 1979.
4. Methyl chloride, in *Documentation of Threshold Limit Values*, 4th ed., American Conference of Governmental Industrial Hygienists, Cincinnati, 1980, 268.
5. **Repko, J. D. and Lasley, S. M.**, Behavioral, neurological, and toxic effects of methyl chloride: a review of the literature, *Crit. Rev. Toxicol.*, 6, 283, 1979.
6. **Repko, J. D.**, Neurotoxicity of methyl chloride, *Neurobehav. Toxicol. Teratol.*, 3, 425, 1981.
7. **Gudmundsson, G.**, Methyl chloride poisoning 13 years later, *Arch. Environ. Health*, 32, 236, 1977.
8. **Scharnweber, H. C., Spears, G. N., and Cowles, S. R.**, Chronic methyl chloride intoxication in six industrial workers, *J. Occup. Med.*, 16, 112, 1974.
9. **MacDonald, J. D. C.**, Methyl chloride intoxication: report of 8 cases, *J. Occup. Med.*, 6, 81, 1964.
10. **Noro, L. and Petterson, T.**, Metyl klorid förgifning, *Nord. Med.*, 64, 881, 1960.
11. **Baker, H. M.**, Intoxication with commercial methyl chloride, *JAMA*, 88, 1137, 1927.
12. **Hansen, H., Weaver, N. K., and Venable, F. S.**, Methyl chloride intoxication, *Arch. Ind. Hyg. Occp. Med.*, 8, 328, 1953.
13. **Kegel, A. H., McNally, W. D., and Pope, A. S.**, Methyl chloride poisoning from domestic refrigerators, *JAMA* 93, 353, 1929.
14. **Borovska, D., Jindrichova, J., and Klima, M.**, Methyl chloride poisoning in the country of East Bohemia, *Z. Gesamte Hyg.*, 22, 241, 1976.
15. **Eckardt, R. E.**, Industrial intoxications which may simulate ethyl alcohol intake, *Ind. Med. Surg.*, 40, 30, 1971.
16. **Repko, J. D., Jones, P. D., Garcia, L. S., Jr., Schneider, E. J., Roseman, E., and Corum, C. R.**, Behavioral and Neurological Effects of Methyl Chloride, Department of Health, Education, and Welfare (NIOSH) Publ. No. 77-125, Washington, D.C., 1976.
17. **Hake, C. L., Stewart, R. D., Wu, A., Forster, H. V., and Newton, P. E.**, Experimental human exposures to methyl chloride at industrial environment levels, *Toxicol. Appl. Pharmacol.*, 41, 198, 1977.
18. **Stewart, R. D., Hake, C. L., Wu, A., Graff, S. A., Forster, H. V., Keeler. W. H., Lebrun, A. J., Newton, P. E., and Soto, R. J.**, Methyl Chloride: Development of a Biologic Standard for the Industrial Worker by Breath Analysis, Department of Health, Education, and Welfare (NIOSH) Publ. No. 77-1, Washington, D.C., 1977.
19. **Putz-Anderson, V., Setzer, J. V., Croxton, J. S., and Phipps, F. C.**, Methyl chloride and diazepam effects on performance, *Scand. J. Work Environ. Health*, 7, 8, 1981.
20. **Van der Klout, A.**, Methyl chloride poisoning, *Ill. Med. J.*, 65, 508, 1934.
21. **Smith, W. W. and von Oettingen, W. F.**, The acute and chronic toxicity of methyl chloride. I. Mortality resulting from exposures to methyl chloride in concentrations of 4,000 to 300 parts per million, *J. Ind. Hyg. Toxicol.*, 29, 47, 1947.
22. **Kolkmann, F. W. and Volk, B.**, Necroses in the granular cell layer of the cerebellum due to methyl chloride intoxication in guinea pigs, *Exp. Pathol.*, 10, 298, 1975.
23. **Wood, M. W. W.**, Cirrhosis of liver in a refrigerator engineer attributed to methyl chloride, *Lancet*, 260, 508, 1951.
24. **McNally, W. D.**, Eight cases of methyl chloride poisoning with three deaths, *J. Ind. Hyg. Toxicol.*, 20, 94, 1946.
25. **Andrews, A. W., Zawistowski, E. S., and Valentine, C. R.**, A comparison of the mutagenic properties of vinyl chloride and methyl chloride, *Mutat. Res.*, 40, 273, 1976.
26. **Wyers, H.**, Methyl bromide intoxication, *Br. J. Ind. Med.*, 2, 24, 1945.
27. **Von Oettingin, W. F.**, *Halogenated Hydrocarbons, Toxicity and Potential Dangers*, U.S. Public Health Serv. Publ. No. 414, Washington, D.C., 1955, 15.
28. Methyl bromide, in *Documentation of Threshold Limit Values*, 4th ed., American Conference of Governmental Industrial Hygienists, Cincinnati, 1980, 265.
29. **Hine, C. H.**, Methyl bromide poisoning: a review of ten cases, *J. Occup. Med.*, 11, 1, 1969.
30. **Rathus, E. M. and Landry, P. J.**, Methyl bromide poisoning, *Br. J. Ind. Med.*, 18, 53, 1961.
31. **Kantarjian, A. D. and Shaheen, A. S.**, Methyl bromide poisoning with nervous system manifestations resembling polyneuropathy, *Neurology*, 13, 1054, 1963.
32. **Longly, E. O. and Simpson, G. R.**, Acute Methyl Bromide Poisoning, *Proc. 4th Int. Congr. Rural Med.-Whithe*, 1970, 32.

33. **Araki, S., Ushio, K., Suwa, K., Abe, A., and Uehara, K.,** Methyl bromide poisoning: a report based on fourteen cases, *Sangyo Igaku,* 13, 507, 1971.
34. **Takahashi, S., Moroji, T., and Yamauchi, T.,** A case of methyl bromide intoxication which showed paroxysmal abnormal EEG, *Rinsho Noha,* 15, 725, 1973.
35. **Yamauchi, T., Moroji, T., and Takahashi, S.,** Bromomethane poisoning: a case of neurologic disturbances practically restricted to the sensory system, *Brain Nerve (Tokyo),* 26, 1005, 1974.
36. **Mellerio, F., Gaultier, M., and Bismut, C.,** Electroencephalogram during acute methyl bromide poisoning, *Eur. J. Toxicol.,* 7, 119, 1974.
37. **Goulon, M., Nouailhat, F., Escourolle, R., Zarranz-Imirizaldu, J. J., Grosbus, S., and Lévy-Alcover, M. A.,** Methyl bromide poisoning. Report of three cases with one death. An anatomical study in one case of stupor and myoclonus with a five years survey, *Rev. Neurol.,* 131, 445, 1975.
38. **Ishitsu, S., Momotani, H., Sato, M., Nakayama, E., and Minami, M.,** Methyl Bromide Intoxication, *Proc. 18th Annu. Meet. Jpn. Soc. Ind. Med.,* 62, 1975.
39. **Shield, L. K., Coleman, T. L., and Markesbery, W. R.,** Methyl bromide intoxication: neurologic features, including simulation of Reye syndrome, *Neurology,* 27, 959, 1977.
40. **Ushio, K. and Osozuka, R.,** A case of severe intoxication due to methyl bromide, *Sangyo Igaku,* 19, 355, 1977.
41. **Van Den Oever, R., Van De Mitrap, L., and Lahaye, D.,** Occupational poisoning with bromomethane, *Arch. Belg. Med. Soc. Hyg. Med. Trav. Med. Leg.,* 36, 353, 1978.
42. **Gil-Peralta, A., Bautista-Lorite, J., and Alberca, R.,** Peripheral syndromes in methyl bromide poisoning, *Rev. Neurol.,* 7, 13, 1979.
43. **Verberk, M. M., Rooyakkers-Beemster, T., De Vlieger, M., and Van Vliet, A. G. M.,** Bromine in blood, EEG and transaminases in methyl bromide workers *Br. J. Ind. Med.,* 36, 59, 1979.
44. **Zatuchni, J. and Hong, K.,** Methyl bromide poisoning seen initially as psychosis, *Arch. Neurol.,* 38, 529, 1981.
45. **Takahaski, S., Moroji, T., and Yamauchi, T.,** A case of methyl bromide intoxication which showed paroxysmal abnormal EEG, *Rinsho Noha,* 15, 725, 1973.
46. **Simons, A. J. R.,** Myoclonus following methyl bromide intoxication, *Electroenceph. Clin. Neurophysiol.,* 31, 108, 1971.
47. **Clarke, C. A., Roworth, C. G., and Holling, H. E.,** Methyl bromide poisoning. An account of four recent cases met with in one of H. M. Ships, *Br. J. Ind. Med.,* 2, 17, 1945.
48. **Johnstone, R. T.,** Methyl bromide intoxication of a large group of workers, *Ind. Med.,* 14, 495, 1945.
49. **Watrous, R. M.,** Methyl bromide — local and mild systemic toxic effects, *Ind. Med.,* 11, 575, 1942.
50. **Irish, D. D., Adams, E. M., Spencer, H. C., and Rowe, V. K.,** The response attending exposure of laboratory animals to vapors of methyl bromide, *J. Ind. Hyg.,* 22, 218, 1940.
51. **Sato, M., Uchiyama, H., and Marinobu, S.,** Experimental studies of intoxication due to methyl bromide centered on the persistence of bromine in organs, *Proc. Annu. Meet. Jpn. Soc. Ind. Hyg.,* 53, 305, 1980.
52. **Ikeda, T., Kishi, R., Yammaura, K., Miyake, H., Sato, M., and Ishizu, S.,** Behavioral effects in rats following repeated exposure to methyl bromide, *Toxicol. Lett.,* 6, 293, 1980.
53. **Miyakawa, M., Hasegawa, H., Sato, M., Sudo, A., Homma, T., and Okonogi, K.,** Studies on intoxication due to methyl bromide, *Proc. Annu. Meet. Jpn. Soc. Ind. Hyg.,* 53, 309, 1980.
54. **Ishizu, S., Momotani, H., Sato, M., Uchiyama, H., and Mori, N.,** Experimental studies of intoxication due to methyl bromide on the biochemical change of the blood, *Proc. Annu. Meet. Jpn. Soc. Ind. Hyg.,* 53, 303, 1980.
55. **Anger, W. K., Setzer, J. V., Russo, J. M., Brightwell, W. S., Wait, R. G., and Johnson, B. L.,** Neurobehavioral effects of methyl bromide inhalation expsoures, *Scand. J. Work Environ. Health,* 7(Suppl. 4), 40, 1981.
56. **Knight, H. D. and Costner, G. C.,** Bromide intoxication of horses, goats, and cattle, *J. Am. Vet. Med. Assoc.,* 171, 446, 1977.
57. **Knight, H. D. and Reina-Guerra, M.,** Intoxication of cattle with sodium bromide-contaminated feed, *Am. J. Vet. Res.,* 38, 407, 1977.
58. **Shibata, N., Yomura, Y., and Hashimoto, K.,** Studies on intoxication due to intake of food fumigated with methyl bromide. II. The first experiment of maintaining rats on diets fumigated with methyl bromide, *St. Marianna Ikadaigaku Zasshi,* 7, 40, 1979.
59. **Yomura, Y. and Shibata, N.,** Studies on intoxication due to intake of food fumigated with methyl bromide. III. The second experiment of maintaining infantile rats on synthetic diet fumigated with methyl bromide, *St. Marianna Ikadaigaku Zasshi,* 17, 342, 1979.
60. **Yomura, Y. and Shibata, N.,** Studies on the intoxication due to the intake of foods fumigated with methyl bromide. I. A case of intoxication due to methyl bromide, which started the studies, *St. Marianna Ikadaigaku Zasshi,* 6, 389, 1978.

61. **Torkelson, T. R. and Rowe, V. K.,** Halogenated aliphatic hydrocarbons containing chlorine, bromine, and iodine, in *Patty's Industrial Hygiene and Toxicology,* Vol. 2B, 3rd rev. ed., Clayton, G. D. and Clayton, F. E., Eds., John Wiley-Interscience, New York, 1981, 3442.
62. **Williams, R. T.,** *Detoxification Mechanisms,* John Wiley-Interscience, New York, 1959, 26.
63. **Sharpless, S. K.,** Hypnotics and sedatives, in *The Pharmacological Basis of Therapeutics,* 3rd ed., Goodman, L. S. and Gilman, A., Eds., Macmillan, New York, 1965, 130.
64. **Irish, D. D.,** Halogenated hydrocarbons. I. Aliphatic, in *Patty's Industrial Hygiene and Toxicology,* 2nd ed., John Wiley-Interscience, New York, 1963, 1241.
65. **Garland, A. and Camps, F. E.,** Methyl iodide poisoning, *Br. J. Ind. Med.,* 2, 209, 1945.
66. **Appel, G. B., Galen, R., O'Brien, J., and Schoenfeldt, R.,** Methyl iodide intoxication. A case report, *Ann. Intern. Med.,* 82, 534, 1975.
67. **Cherry, N., Venables, H., Waldron, H. A., and Wells, G. G.,** Some observations on workers exposed to methylene chloride, *Br. J. Ind. Med.,* 38, 351, 1981.
68. **Pankow, D., Gutewort, R., Glatzel, W., and Tietze, K.,** Effect of dichloromethane on carboxyhemoglobin and motor nerve conduction velocity in rats, *Toxikol. Anal. Phobl. Loesungsmittelexpo.* (Vortr. Minisymp.), 60, 1980.
69. **Winneke, G.,** The neurotoxicity of dichloromethane, *Neurobehav. Toxicol. Teratol.,* 3, 391, 1981.
70. **Savolainen, H., Kurppa, K., Pfäffli, P., and Kivisto, H.,** Dose-related effects of dichloromethane on rat brain in short-term inhalation exposure, *Chem.-Biol. Interact.,* 34, 315, 1981.
71. **Kurppa, K., Kivisto, H., and Vainio, H.,** Dichloromethane and carbon monoxide inhalation: carboxyhemoglobin addition, and drug metabolizing enzymes in rat, *Int. Arch. Occup. Environ. Health,* 49, 83, 1981.
72. **Stevens, H. and Foster, F. M.,** Effect of carbon tetrachloride on the nervous system, *Arch. Neurol. Psychiatry,* 70, 635, 1953.
73. National Institute for Occupational Safety and Health, Criteria for a Recommended Standard: Occupational Exposure to Carbon Tetrachloride, Department of Health, Education, and Welfare, Publ. No. 76-133, Washington, D. C., 1975.
74. **Cohen, M. M.,** Central nervous system in carbon tetrachloride intoxication, *Neurology,* 7, 238, 1957.
75. **Luse, S. A. and Wood, W. G.,** The brain in fatal carbon tetrachloride poisoning, *Arch. Neurol.,* 17, 304, 1967.
76. **Stevens, H.,** Neurotoxicity of some common halogenated hydrocarbons, in *Laboratory Diagnosis of Diseases caused by Toxic Agents,* Sunderman, F., Sr. and Sunderman, F., Jr., Eds., Warren H. Green, St. Louis, Mo., 1970, 193.
77. **Lahl, R.,** Carbon tetrachloride poisoning and the CNS: review of neurological and mental symptomatology in man, *Psychiatr. Neurol. Med. Psychol.,* 25, 1, 1973.
78. **Farrell, C. and Senseman, L.,** Carbon tetrachloride polyneuritis: a case report, *R. I. Med. J.,* 27, 334, 1944.
79. **Wirtschafter, Z. T.,** Toxic amblyopia and accompanying physiological disturbances in carbon tetrachloride intoxication, *Am. J. Public Health,* 23, 1035, 1933.
80. **Smyth, H. F., Smyth, H. F., Jr., and Carpenter, C. P.,** The chronic toxicity of carbon tetrachloride; animal exposures and field studies, *J. Ind. Hyg. Toxicol.,* 18, 277, 1936.
81. **Gray, I.,** Carbon tetrachloride poisoning: report of seven cases with two deaths, *N.Y. State J. Med.,* 47, 2311, 1947.
82. **Smith, A. R.,** Optic atrophy following inhalation of carbon tetrachloride, *Arch. Ind. Hyg. Occup. Med.,* 1, 348, 1950.
83. **Korenke, H. D. and Pribilla, O.,** Suicide by a single inhalation of carbon tetrachloride accompanied by leukoencephalopathy, *Arch. Toxicol.,* 25, 109, 1969.
84. **Tanohata, K. and Tagawa, D.,** Histopathologic study of central nervous system in experimental carbon tetrachloride poisoning, *Nagasaki Igakkai Zasshi,* 10, 1505, 1932.
85. **Biancalani, A.,** Richerche sperimentali sulle alterazioni del sistema nervosa centrale nella into sicazione da tetrachoruro di carbonio, *Riv. Pat. erv. Ment.,* 44, 352, 1934.
86. **Lahl, R.,** The pathomorphology of the CNS in carbon tetrachloride poisoning. IV. The histopathology of the rhombencephalon in experimental studies on random bred rabbits, *Zentralbl. Allg. Pathol. Pathol. Anat.,* 118, 305, 1974.
87. **Lahl, R.,** The pathomorphology of the CNS in carbon tetrachloride poisoning. V. The histopathology of the spinal cord in experimental studies on random bred rabbits, *Acta Morphol. Acad. Sci. Hung.,* 22, 47, 1974.
88. **Lahl, R.,** The pathomorphology of the CNS in carbon tetrachloride poisoning. VII. The histopathology of lumbar spinal ganglia in experimental studies, *Zentralbl. Allg. Pathol. Pathol. Anat.,* 119, 276, 1975.
89. **Diemer, N. H.,** Size and density of oligodendroglial nuclei in rats with CCl_4-induced liver disease, *Neurobiology,* 5, 197, 1975.

90. **Diemer, N. H.,** Number of Purkinje cells and Bergmann astrocytes in rats with CCl_4-induced liver disease, *Acta Neurol. Scand.,* 55, 1, 1976.
91. **Diemer, N. H.,** Glial and neuronal alterations in the corpus striatum of rats with CCl_4-induced liver disease, *Acta Neurol. Scand.,* 55, 16, 1976.
92. **Almeida, V. M.,** Use and abuse of alcohol and drugs — a clinical study of certain aspects of their interrelationships, *Bol. Sanit. Panam.,* 80, 45, 1980.
93. **Hess, J. P., Cohn, D. F., and Streifler, M.,** Ethyl chloride sniffing and cerebellar dysfunction, *Isr. Ann. Psychiatry,* 17, 122, 1979.
94. **Anon.,** A huffin' and a puffin', a sniffin' and a suckin', *Lancet,* 2, 876, 1974.
95. **Morris, T. E.,** Ethyl chloride, in *Kirk-Othmer Encyclopedia of Chemical Technology,* Vol. 5, 3rd ed., Kirk, R. E., Othmer, D. F., Grayson, M., and Eckroth, D., Eds., John Wiley-Interscience, 1979, 714.
96. Ethyl chloride, in *Hygienic Guide Series,* Vol. 1, American Industrial Hygiene Assoc., Akron, Ohio, 1978.
97. Ethyl Chloride, in *Documentation of the Threshold Limit Values,* 4th ed., American Conference of Governmental Industrial Hygienists, Cincinnati, 1980.
98. **Raffi, G. B. and Violante, F. S.,** Is Freon 113 neurotoxic? A case, *Int. Arch. Occup. Environ. Health,* 49, 125, 1981.
99. **Selikoff, I. and Hammond, E. C., Eds.,** Toxicity of vinyl chloride-polyvinyl chloride, *Ann. N.Y. Acad. Sci.,* 246, 1, 1975.
100. **Walker, J. R.,** Vinyl Chloride and Health; A Select Bibliography, Central Libraries, Birmingham, England, 1976.
101. **Milby, T. M.,** Vinyl Chloride: An Information Resource, Department of Health, Education, and Welfare (NCI) Publ. No. 78-1599, Bethesda, Md., 1978.
102. **Warren, H. S., Hugg, J. E., and Gerstner, H. B.,** Vinyl Chloride, a Review 1835—1975; A Literature Compilation, 1976—1977, ORNL/TIRC-78-3, Oak Ridge National Laboratories, Oak Ridge, Tenn., 1978.
103. Vinyl chloride, polyvinyl chloride and vinyl chloride-vinyl acetate copolymers, *IARC Monogr. on the Evaluation of Carcinogenic Risks of Chemicals to Man,* 377, 1979.
104. **Jackson, S.,** Vinyl Chloride Toxicology: Literature Search, January 1977 to October 1980, Publ. No. 80-8, National Library of Medicine, Bethesda, Md., 1980.
105. **VanDuuren, B. L.,** Selected Abstracts on Carcinogenicity of Vinyl Chloride and Related Compounds, Department of Health, Education, and Welfare (NCI), Bethesda, Md., 1980.
106. Conference to reevaluate the toxicity of vinyl chloride, poly (vinyl chloride) and structural analogs, *Environ. Health Perspect.,* 41, 1, 1981.
107. **Binns, C. H.,** Vinyl chloride: a review, *J. Soc. Occup. Med.,* 29, 134, 1979.
108. **Suciu, I., Prodan, L., Ilea, E., Pauduraru, A., and Pascu, L.,** Clinical manifestations in vinyl chloride poisoning, *Ann. N.Y. Acad. Sci.,* 246, 53, 1975.
109. **Veltman, G., Lange, C. E., Jühe, S., Stein, G., and Bachner, U.,** Clinical manifestations and course of vinyl chloride disease, *Ann. N.Y. Acad. Sci.,* 246, 6, 1975.
110. **Spirtas, R., McMichal, A. S., Gamble, J., and VanErt, M.,** The association of vinyl chloride exposures with morbidity symptoms, *Am. Ind. Hyg. Assoc. J.,* 36, 779, 1975.
111. **Waxweiler, R. J., Falk, H., McMichael, A., Mallov, J. S., Grivas, A. S., and Stringer, W. T.,** A Cross-sectional Epidemiological Survey of Vinyl Chloride Workers, Department of Health, Education, and Welfare (NIOSH) Publ. No. 77-177, Cincinnati, 1977.
112. **Walker, A.,** Occupational acro-osteolysis, *Proc. R. Soc. Med.,* 68, 343, 1975.
113. **Takeuchi, Y. and Mabuchi, C.,** A case of occupational acro-osteolysis, presumably caused by vinyl chloride, *Jpn. J. Ind. Health,* 15, 385, 1973.
114. **Tabershaw, I. R. and Gaffey, W. R.,** Mortality studies of workers in the manufacture of vinyl chloride and its polymers, *J. Occup. Med.,* 16, 509, 1974.
115. **Monson, R. R., Peters, J. M., and Johnson, M. N.,** Proportional mortality among vinyl-chloride workers, *Lancet,* 2, 397, 1974.
116. **Waxweiler, R. J., Stringer, W., Wagoner, J. K., Jones, J., Falk, H., and Carter, C.,** Neoplastic risk among workers exposed to vinyl chloride, *Ann. N.Y. Acad. Sci.,* 271, 40, 1976.
117. **Fox, A. J. and Collier, P. F.,** Mortality experience of workers exposed to vinyl chloride monomer in the manufacture of polyvinyl chloride in Great Britain, *Br. J. Ind. Med.,* 34, 1, 1977.
118. **Wagoner, J. K., Infante, P. F., and Apfeldorf, R. B.,** Toxicity of vinyl chloride and polyvinyl chloride as seen through epidemiologic observations, *J. Toxicol. Environ. Health,* 6, 1101, 1980.
119. **Emmerich, K. H. and Norpoth, K.,** Malignant tumors after chronic exposure to vinyl chloride, *J. Cancer Res. Clin. Oncol.,* 102, 1, 1981.
120. **Infante, P. F.,** Observations of the site-specific carcinogenicity of vinyl chloride to humans, *Environ. Health Perspect.,* 41, 89, 1981.
121. **Maltoni, C., Lefemine, G., Chieco, P., and Carretti, D.,** Vinyl chloride carcinogenesis. Current results and perspectives, *Med. Lav.,* 65, 421, 1974.

122. **Feron, V. J. and Kroes, R.,** One-year time sequence inhalation toxicity study of vinyl chloride in rats. II. Morphological changes in the respiratory tract, ceruminous glands, brain, kidneys, heart and spleen, *Toxicology,* 13, 131, 1979.
123. National Institute for Occupational Safety and Health, Criteria for a Recommended Standard to Occupational Exposure to Trichloroethylene, Department of Health, Education, and Welfare, Washington, D.C., 1973.
124. **Defalque, R. J.,** Pharmacology and toxicology of trichloroethylene. A critical review of the world literature, *Clin. Pharmacol. Ther.,* 2, 665, 1961.
125. **Saunders, R. A.,** A new hazard in closed environmental atmospheres, *Arch. Environ. Helth,* 14, 380, 1967.
126. **Huff, J. E.,** New evidence on the old problems of trichloroethylene, *Ind. Med. Surg.,* 40, 25, 1971.
127. **Waters, E. M., Gerstner, H. B., and Huff, J. E.,** Trichloroethylene. I. An overview, *J. Toxicol. Environ. Health,* 2, 671, 1977.
128. **Feldman, R. G.,** Trichloroethylene, in *Handbook of Clinical Neurology,* Vol. 36, Vinken, P. J. and Bruyn, G. W., Eds., Elsevier/North-Holland, Amsterdam, 1979, 457.
129. **Annau, Z.,** The neurobehavioral toxicity of trichloroethylene, *Neurobehav. Toxicol. Teratol.,* 3, 417, 1981.
130. **Plessner, W.,** Ueber Trigeminuser Krankung Infolge von Trichloraethylenvergiftung, *Neurol. Zentralbl.,* 34, 916, 1915.
131. **Oljenick, I.,** Trichloroethylene treatment of trigeminal neuralgia, *JAMA,* 91, 1085, 1928.
132. **Glaser, M. A.,** Treatment of trigeminal neuralgia with trichloroethylene, *JAMA,* 96, 916, 1931.
133. **Humphrey, J. H. and McClelland, M.,** Cranial nerve palsies with herpes following general anesthesia, *Br. Med. J.,* 1, 315, 1944.
134. **McClelland, M.,** Some toxic effects following trilene decomposition products, *Proc. R. Soc. Med.,* 37, 526, 1944.
135. **Firth, J. B. and Stuckey, R. E.,** Decomposition of trilene in closed circuit anesthesia, *Lancet,* 1, 814, 1945.
136. **Harenko, A.,** Two peculiar instances of psychotic disturbance in trichloroethylene poisoning, *Acta Neurol. Scand. Suppl.,* 31, 139, 1967.
137. **Meyer, H. J.,** Per oral poisoning with trichloroethylene, *Arch. Toxikol.,* 21, 225, 1966.
138. **Assmus, H.,** Irreversible sensory trigeminal lesion following handling of trichloroethylene, *Aktuel. Neurol.,* 2, 131, 1975.
139. **Hepple, N. V.,** Sniffing of a shoe cleaner, *Br. Med. J.,* 4, 387, 1968.
140. **Mitchell, A. B. S. and Parsons-Smith, B. G.,** Trichloroethylene neuropathy, *Br. Med. J.,* 1, 422, 1969.
141. **Gwynne, E. I.,** Trichloroethylene neuropathy, *Br. Med. J.,* 2, 315, 1969.
142. **Fra, L., Gandiglio, G., Riccio, A., and Sandigliano, G.,** Clinical observation and neurophysiological study of a case of chronic trichloroethylene poisoning, *Med. Lav.,* 57, 606, 1966.
143. **Todd, J.,** Trichloroethylene poisoning with paranoid psychosis and lilliputian hallucination, *Br. Med. J.,* 7, 439, 1954.
144. **Tabacchi, G., Corsico, R., and Gallinelli, R.,** Retrobulbar neuritis caused by suspected chronic trichloroethylene poisoning, *Ann. Otlalmol. Clin. Ocul.,* 92, 787, 1966.
145. **Feldman, R. G., Mayer, R. M., and Taub, A.,** Evidence for peripheral neurotoxic effect of trichloroethylene, *Neurology,* 20, 599, 1970.
146. **Tomasini, M. and Sartorelli, E.,** Chronic poisoning from inhalation of commercial trichloroethylene with impairment of the 8th pair of cranial nerves, *Med. Lav.,* 62, 277, 171.
147. **Lawrence, W. H. and Partyka, E. K.,** Chronic dysphagia and trigeminal anesthesia after trichloroethylene exposure, *Ann. Intern. Med.,* 95, 710, 1981.
148. **Saihan, E. M., Burton, J. L., and Heaton, K. W.,** A new syndrome with pigmentation, scleroderma, Gynaecomastia, Raynaud's phenomenon and peripheral neuropathy, *Br. J. Dermatol.,* 99, 437, 1978.
149. **Sagawa, K., Nishitani, H., Kawai, H., Kuge, Y., and Ikeda, M.,** Transverse lesion of spinal cord after accidental exposure to trichloroethylene, *Int. Arch. Arbeitsmed.,* 31, 257, 1973.
150. **Buxton, P. H. and Hayward, M.,** Polyneuritis cranialis associated with industrial trichloroethylene poisoning, *J. Neurol. Neurosurg. Psychiatry,* 30, 511, 1967.
151. **Granjean, E., Münchinger, R., Turrian, V., Haas, P. A., Knoepfel, H. K., and Rosenmund, H.,** Investigations into the effects of exposure to trichloroethylene in mechanical engineering, *Br. J. Ind. Med.,* 12, 131, 1955.
152. **Bardodej, Z. and Vyskocil, J.,** The problem of trichloroethylene in occupational medicine, *Arch. Ind. Health,* 13, 581, 1956.
153. **Lilis, R., Stanescu, D., Muica, N., and Roventa, A.,** Chronic effects of trichloroethylene exposure, *Med. Lav.,* 60, 595, 1969.
154. **Haas, P. A.,** Industriehygienische untersuchungen uber das Trichloroaethylen, Thesis 2979, Technische Hoch Schule, Zurich, 1960.

210. **Karlsson, A., Fonnum, F., Malthe-Sørenssen, D., and Storm-Mathisen, J.,** Effect of the convulsive agent 3-mercaptopropionic acid on the levels of GABA, other amino acids and glutamate decarboxylase in different regions of the rat brain, *Biochem. Pharmacol.,* 23, 3053, 1974.
211. **Lamar, C., Jr.,** Mercaptopropionic acid: a convulsant that inhibits glutamate decarboxylase, *J. Neurochem.,* 17, 165, 1970.
212. **Löscher, W.,** 3-Mercaptopropionic acid: convulsant properties, effects on enzymes of the γ-aminobutyrate system in mouse brain and antagonism by certain anticonvulsant drugs, aminooxyacetic acid and gabaculine, *Biochem. Pharmacol.,* 28, 1397, 1979.
213. **Rodriguez de Lores Arnaiz, G. and Robiolo de Esteves, B.,** Inhibition *in vitro* of protein synthesis in brain subcellular fractions by the convulsant 3-mercaptopropionic acid, *Biochem. Pharmacol.,* 24, 2307, 1975.
214. **Rodriguez de Lores Arnaiz, G., Alberici de Canal, M., and DeRobertis, E.,** Alteration of GABA system and Purkinje cells in rat cerebellum by the convulsant 3-mercaptopropionic acid, *J. Neurochem.,* 19, 1379, 1972.
215. **Samojlik, E. and Chang, M. C.,** Antifertility activity of 3-chloro-1,2-propanediol (U-5897) on male rats, *Biol. Reprod.,* 2, 299, 1970.
216. **Ford, W. C. L. and Waites, G. M. H.,** Activities of various 6-chloro-6-deoxysugars and (S) alpha-chlorohydrin in producing spermatocoeles in rats and paralysis in mice and inhibiting glucose metabolism in bull spermatozoa *in vitro, J. Reprod. Fertil.,* 65, 177, 1982.
217. **Jacobs, J. M. and Ford, W. C. L.,** The neurotoxicity and antifertility properties of 6-chloro-6-deoxyglucose in the mouse, *Neurotoxicology,* 2, 405, 1981.
218. **Heywood, R., Sortwell, R. J., and Prentice, D. E.,** The toxicity of 1-amino-3-chloro-2-propanol hydrochloride (CL 88,236) in the Rhesus monkeys, *Toxicology,* 9, 219, 1978.
219. **Sasaki, S.,** Brain edema and gliopathy induced by 6-aminonicotinamide intoxication in the central nervous system of rats, *Am. J. Vet. Res.,* 43, 1691, 1982.
220. **Rogulja, P. V., Kovac, W., and Schmid, H.,** Metronidazole-encephalophathic der Ratte, *Acta Neuropathol.,* 25, 36, 1973.
221. **Griffin, J. W., Price, D. L., Kuethe, D. O., and Goldberg, A. M.,** Neurotoxicity of misonidazole in rats. I. Neuropathology, *Neurotoxicology,* 1, 299, 1979.

Chapter 6

AROMATIC HYDROCARBONS

John L. O'Donoghue

TABLE OF CONTENTS

I. General Comments 130
II. Toluene 130
    A. Introduction 130
    B. Human Neurotoxicity 130
    C. Experimental Neurotoxicity 131
    D. Metabolism 132
    E. Effects on Other Organs 132

III. Styrene Monomer 132
    A. Synonyms 132
    B. Introduction 132
    C. Human Neurotoxicity 133
    D. Experimental Neurotoxicity 134

IV. Vinyl Toluene 134

References 136

I. GENERAL COMMENTS

Aromatic hydrocarbons make up a large class of substances with wide uses in industry and in commercial products. They are used as fuels, paint and lacquer solvents, in adhesives, polymer solvents and additives, and cleaners, as raw materials for the synthesis of other chemicals and medicines, and in many other uses requiring volatile solvents or carriers. They also result from incomplete combustion of carbonaceous materials.

As a group, their high lipid solubility results in accumulation in lipid rich tissues, particularly the brain, spinal cord, and peripheral nerves.[1] Accumulation of these materials in the nervous system may lead to functional impairment either acutely or chronically. Impairment may be seen as nonspecific complaints of nausea, vomiting, weakness, tiredness, vertigo, or in more severe cases inebriation or unconsciousness. With some materials, such as benzene, seizures may occur.[1]

Recent studies[2-6] on workers chronically exposed to mixtures of aliphatic and aromatic substances have found psychological and neurophysiological differences between solvent exposed and nonexposed workers, particularly painters, suggesting that aromatic hydrocarbons may have subtle long-lasting neurologic effects. For the most part, with the possible exception of the materials covered in this chapter, specific aromatic hydrocarbons causing long-lasting neural effects have not been identified.

II. TOLUENE

A. Introduction

Toluene (toluol, methacide, methyl benzene) is a widely used aromatic solvent. Commercially, it is converted into benzene and its derivatives: toluene diisocyanate, vinyl toluene, nitrotoluenes, and other derivatives.[7] Toluene is a component of gasolines and is extensively used in glues, adhesives, lacquers, paints, inks, and a variety of thinner products. In the past, toluene frequently contained benzene and this has caused more confusion in the literature between the effects of toluene and benzene.[7] Although pure (99%) toluene is available and widely used, commercial products may contain combinations of toluene with benzene, hexane, or other aromatic and aliphatic solvents.

B. Human Neurotoxicity

Routine use of toluene in industry and legitimate uses of commercial products have generally not been associated with lasting symptoms or signs of neurotoxicity although long term exposures to mixtures of solvents which contain toluene are reported to result in subtle neural effects.[8-12]

Based on a few reports of intentional exposures,[13-16] accidental exposure,[17] and industrial exposure,[18] the acute effects of toluene exposures may be summarized as follows:
At 50 to 200 ppm — subjective symptoms of fatigue and headache; between 200 to 400 ppm — headache, lassitude, slight incoordination, and slightly reduced reaction time; between 400 to 800 ppm — headache, dizziness, hilarity, incoordination, reduced time, weakness, and confusion; at 800 ppm — symptoms are more severe and as, the level increases, loss of consciousness occurs.

Capellini and Alessio[19] reported that a man working in an environment with a mean toluene level of 250 ppm had eye irritation, occasional stupor, insomnia, and nervousness. Other workers exposed to mean toluene levels of 125 ppm did not have nervous system effects. Matsushita et al.[20] reported subjective symptoms of weakness, dysmenorrhea and dermatitis, and abnormal tendon reflexes in women chronically exposed to 60 to 100 ppm of toluene. Boor and Hurtig[21] studied an optician exposed to pure toluene while cleaning lenses. Slurred speech, staggering, fatigue, and clumsiness were recoverable within a month after exposure. Chronic illnesses, were not found in a group of 46 workers exposed to varying (20 to 200 ppm) levels of toluene in a plant manufacturing truck tarpaulins.[23]

The predominant exposure associated with clinical reports of toluene neurotoxicity is intentional inhalation of glues, paint and lacquer thinners, and aerosolized paint products for euphoric effects.[21,23-37]

Since toluene is a common ingredient of these types of products, many reports refer to toluene without actual analyses of the products. From the number of reports on toluene abuse in the literature, toluene would appear to be the most commonly abused commercial solvent.[38]

The exposures involved in "sniffing or huffing" toluene-containing products is very high, on the order of several thousand ppm, for prolonged periods of time. Consumption of 0.5 to 1 ℓ of pure toluene per week is reported.[21,24,29]

Under these conditions, acute encephalopathy may occur, usually after years of near daily use. The most common clinical abnormality described is cerebellar ataxia but cerebral effects are also likely to occur. Clinical abnormalities may also include euphoria, personality changes, hallucinations, drowsiness, suicidal tendencies, dysarthria, nystagmus, positive Babinski sign, abnormal vision, and convulsions.[21,23-37]

Recovery from encephalopathy is common although lasting or possibly irreversible effects may occur.[21,28,29,32] Of 19 adolescent glue sniffers, King et al.[28] found that 13 recovered completely, one had cerebellar ataxia lasting one year, and five with psychological impairment were lost to follow-up. Kelly[25] reported improvement but persistence of slight ataxia in a girl who had sniffed paint containing toluene for a year and a half. Permanent cerebral cortical damage was found in a man who had sniffed toluene daily for about 14 years.[24] Even in this case, abstention for 6 days lessened ataxia considerably.[24] Cerebral cortical atrophy was found in 6 of 11 toluene sniffers with neurologic abnormalities.[31] Two of the six had cerebellar atrophy.[31]

No convincing reports of peripheral neuropathy due to pure toluene are available. In most instances, such as the report by Shirabe et al.,[39] other solvents including n-hexane have been combined with toluene and these other solvents appear responsible for peripheral nerve damage. Exerimentally, toluene decreases the neurotoxic effects of n-hexane.[40]

Hypokalemic paralysis has been associated with prolonged toluene sniffing in a few cases.[41-43] Following recovery from paralysis, resumption of toluene sniffing results in a recurrence of hypokalemia.[43] Affected individuals develop severe metabolic acidosis apparently due to impairment of renal tubular acidification resulting in hypokalemia and paralysis.[41] Acidosis and electrolyte imbalances may be very common in toluene sniffers. In a study on adult paint sniffers, 19 of 22 patients had acidosis and electrolyte imbalance.[32]

C. Experimental Neurotoxicity

Toluene, like many organic solvents, produces central nervous system depression and unconsciousness in laboratory animals. Toluene (0.07 mℓ) given i.v. to a dog every 3 to 5 days for a total of 10 doses resulted in rigidity and twitching with recovery in 5 to 10 min.[44] Histologic changes included shrinkage of cortical neurons and myelin pallor.[44] Rats and dogs inhaled 980, 480, or 240 ppm of a toluene concentrate containing about 50% toluene for 13 weeks, 6 hr/day without evidence of neurotoxicity.[45]

Rats exposed to 4000, 2000, or 1000 ppm of toluene for 4 hr showed disturbed sleep at 4000 and 2000 ppm and at 1000 ppm, difficulty entering the slow-wave phase of sleep and facilitation of entry into paradoxical sleep.[46] EEG changes were peculiar to each toluene concentration. Abnormal EEG patterns were induced in cats by intratracheal injection of toluene[47] and inhalation of toluene-containing thinner.[48] The convulsion threshold for Bemegride was lowered in rats exposed to 2000 ppm toluene 8 hr/day for 8 weeks.[49]

Horiguchi and Inoue[50] reported that toluene 1, 10, 100, or 1000 ppm for 6 hr/day for 20 days showed less wheel-turning activity after 6 to 10 days of exposure. Shigeta et al.[51]

studied responses of rats on lever pressing shock avoidance behavior. One-hour exposures to 3000 ppm toluene increased lever pressing but exposures to 1000 ppm did not. For 2-hr exposures, 1500 ppm toluene was a minimal effective concentration.[52] Repeated exposures of 800 ppm, 7 hr/day, 5 day/wk for 12 weeks did not effect avoidance behavior but 2000 ppm toluene for 4 weeks increased responses in some rats.[53] Learning was impaired in rats repeatedly exposed to 4000 ppm toluene, 2 hr/day for 60 days.[54] Although there have been several other studies of toluene, there have been no models which reproduce the conditions observed in men.[7,11,55,56]

D. Metabolism

Toluene may be absorbed through the skin and the gastrointestinal tract, although inhalation is the most frequent and most important route of exposure.[1] In rats, Carlsson and Lindquist[57] found that the largest amount of toluene and its metabolites accumulated in adipose tissue followed by kidneys, adrenal glands, liver, cerebrum, and cerebellum. One hour after exposure, brain levels of toluene had dropped sharply and by 6 hr were only one third of original levels.[57] In men, toluene uptake was dependent on body fat with the least amount of body fat associated with the lowest toluene uptake.[57]

Blood levels of 1.2 mg/ℓ have been reported in workers exposed to 100 ppm toluene.[58] Toluene levels in men who have abused toluene have been reported to range from 9.8 to 31.2 mg/ℓ.[59]

In the liver, toluene is metabolized to benzoic acid with benzyl alcohol and benzaldehyde as intermediates.[1,57-59] Benzoic acid is conjugated with glycine and excreted in the urine as hippuric acid or is conjugated with glucuronic acid and excreted as benzoylglucuronates.[1] About 75% of absorbed toluene follows this pathway while the remainder is eliminated through the lungs.[60] Urinary hippuric acid levels have been used to monitor toluene exposure.[61]

E. Effects on Other Organs

Reversible hepatic and renal changes and myocardial sensitization to catacholamines are more fully described in recent reviews of toluene toxicity.[1,7,11,55,56]

III. STYRENE MONOMER

A. Synonyms

Ethenyl benzene	Styrolene
Vinyl benzene	Styropol
Cinnamene	Phenylethylene
Styrol	Phenylethene

B. Introduction

Styrene ($C_6H_5CH=CH_2$) is a clear, colorless, volatile liquid with a strong pungent but tolerable odor at ambient air levels of 100 ppm.[62] Several million tons of styrene are used world-wide in the production of polystyrene, styrene-butadiene copolymer for synthetic rubber, styrene-acrylonitrile copolymer, acrylonitrile-butadiene-styrene copolymer, polyester resins for reinforced fiberglass products, coatings, and paints.

Occupational exposure to styrene occurs during monomer manufacture, transportation, and polymerization. Exposures may occur due to leaks in reactors, tubing, and other equipment, and during maintenance of production equipment, production of glass reinforced polyester plastics, and during use of styrene-containing resins and coatings.[63] Typical occupational styrene exposures are about 20 ppm except that the intensity and extent of exposure is much greater in the production of glass reinforced plastics.[63] Although skin exposure occurs, the primary concern is inhalation of styrene vapors.[63]

Table 1
EFFECTS OF STYRENE INHALATION ON VOLUNTEERS

Styrene air concentration (ppm)	Length of exposure	Effect(s)	Ref.
<10	—	Styrene odor is not detected	7
50—100	1—6 hr	Styrene odor is strong but not objectionable. Eye irritation is transient at 100 ppm; tests of coordination and dexterity and a modified Romberg test were unaffected	7, 10
200	—	Strong objectionable odor and nasal irritation occur	7, 10
350		Continuous exposure to 50, 150, 250, and 350 ppm styrene for 30 min at each exposure level impaired reaction time only at the 350 ppm level; perceptual speed and manual dexterity were unaffected	11
376	25 min	Unable to perform a modified Romberg test	10
	50 min	Nausea present; manual dexterity and coordination decreased	10
	60 min	Headache and an inebriated feeling were present	10
600	—	Very strong odor and strong eye and nasal irritation occurred	7
800	4 hr	Exposures at this level produce immediate eye, nose, and throat irritation, a pronounced, persistent metallic taste, listlessness, drowsiness, and impaired balance; after-effects included slight muscle weakness, unsteadiness, inertia, and depression	12

C. Human Neurotoxicity

Ocular and nasal irritation are common complaints due to styrene exposure. Nausea, loss of appetite, vomiting, and general weakness lasting a few hours have been reported as symptoms of "styrene sickness" in fiberglass production workers heavily exposed to styrene and other solvents.[64]

Effects of short term styrene exposure on volunteers are listed in Table 1. Inhalation of styrene vapors at around 350 to 375 ppm for short periods (0.5 to 1 hr) reduces manual dexterity and coordination[65] and impairs reaction time.[66] Increasing styrene concentration to around 800 ppm further depresses the nervous system. Thus styrene, under these conditions, acts like many other volatile organics in depressing central nervous system function.

Considering the large amount of styrene used and the large number of people exposed[63,68] to styrene, reports of clinical neurologic impairment are surprisingly few. These have included a single case of central retinal vein occlusion,[69] diminished night vision,[70] a single case of reversible retrobulbar neuritis,[71] a single case of skin atrophy, neurogenic muscular atrophy, anxiety reactions and abnormal electromyogram,[72] and two cases of abnormal tiredness, depression, and abnormal Rorschach tests.[73]

Field surveys of styrene workers have generally not described clinically apparent neurologic deficits even in the glass reinforced plastic industry where exposure is probably the greatest.[63] Subclinical effects associated with styrene exposure include electroencephalographic pattern changes, reduction in some conduction properties of peripheral nerves, fatigue, and reduced performance. Several field studies describing these effects are summarized in Table 2. The reversibility of these effects has not been studied but since reports of clinically evident impairment are rare, reversibility is likely.

Holmberg[84] reviewed files in the Finnish Register of Congenital Malformations and found single cases of anencephaly and hydrocephalus in children born to mothers who had worked in reinforced plastics production. He also mentioned a third case of anencephaly in a child born to a women who had exposure to styrene in the home. No causal relationship between styrene and nervous system malformations has been established.

D. Experimental Neurotoxicity

Although there are many studies on the toxicity of styrene in laboratory species,[1,68,87,88] there are few directed specifically at neurotoxicity. Persistent or irreversible clinical neurologic deficits have not been described in animals exposed to styrene.

Spencer et al.[89] recorded the sequence of clinical abnormalities due to styrene including incoordination, tremors, loss of equilibrium, and progression to loss of consciousness. Thus, styrene acts as a central nervous system depressant.

Styrene infused i.v. (3.1 to 12.6 mg/min) into rabbits produced a dose-correlated positional nystagmus as does alcohol, xylene, and methyl chloroform, indicating a direct effect of styrene on the vestibular system.[90] An unusual effect of styrene was reversal of the direction of rotation-induced nystagmus.

Interference in dopaminergic neural transmission has been suggested on the basis of increased brain serotonin and noradrenalin levels,[91] decreased monoamine oxidase activity,[91] and increased dopamine receptor binding of ^{3}H-spiroperidol[92] in rats.

Inhalation of 300 ppm styrene 6 hr/day for several weeks led to the accumulation of styrene in the brain and perirenal fat of rats.[93] Tissue styrene levels increased up to the fourth exposure week at which point they gradually decreased to half of the 4-week level. Neurochemical changes under these conditions included increased brain lysosomal acid proteinase levels and minor changes in spinal axonal proteins.[93] Glial cells had minor transitory changes in acid proteinase and RNA concentrations.[94] No behavioral effects on ambulation, rearing, preening frequency, or preening time was observed.[94]

IV. VINYL TOLUENE

Vinyltoluene (methyl styrene, ethenyl methyl benzene, tolylethene) is a combustible liquid with a disagreeable odor. It is used as a solvent and chemical intermediate. Industrial samples contain meta and para isomers of vinyltoluene and rarely the ortho isomer.

No reports of neurotoxicity in man due to vinyltoluene were found. Decreased motor nerve conduction velocity and evoked motor action potential amplitude indicative of axonal damage occurred in rats exposed to 100 or 300 ppm vinyltoluene, 6 hr/day, 5 days/week for 12 weeks.[95] No effects occurred at 50 ppm vinyltoluene. Changes in a group of neurochemical assays (acid proteinase, glutathione peroxidase, azoreductase, NADPH-diaphorase, RNA, and 2′,3′-cyclic nucleotide 3′-phosphohydrolase) were measured in rats exposed to vinyltoluene vapors for 1 or 2 weeks.[96] These changes were not associated with clinical symptoms and are difficult to evaluate since the statistics used in the study (Student's test) would lead to an increased notation of false positive results. No reports describing clinical signs or neuropathologic lesions due to vinyltoluene are available.

Table 2
FIELD STUDIES OF NEUROLOGIC FUNCTION IN STYRENE WORKERS

Populations surveyed	Job type	Findings	Ref.
494 styrene workers	Monomer manufacturing and polystyrene polymerization plant	Prenarcotic signs associated with higher styrene exposure; hypoesthesia and hypoactive reflexes associated with duration of styrene exposure; peroneal nerve conduction <40 m/sec associated with exposure duration; no spontaneous symptoms or disabilities	19, 20
345 styrene workers	Monomer manufacturing and polystyrene polymerization plant	22% of the population complained of eye irritation but conjunctival injection was rare; no retrobulbar neuritis or central retinal vein occlusion occurred	21
27 styrene workers and 27 control workers	Fiberglass boat factory	Tiredness more common in styrene workers; reaction time slower in men with blood styrene levels of ≥ 5.5 mol/ℓ	22
17 styrene workers and age-matched controls from a motor workshop	Fiberglass product plant	Reaction time in men exposed to >150 ppm styrene was prolonged	23
98 styrene workers and 43 concrete reinforcement workers	Reinforced polyester plastics production	Increased complaints of tiredness, concentration difficulties, and irritation without correlation to urinary mandelic acid levels; increased prevalance of abnormal EEG patterns associated with urinary mandelic acid levels >700 mg/ℓ; two of 20 psychological functions (visumotor accuracy and lowered psychomotor performance) were associated with high urinary mandelic acid; nerve conduction velocity was unrelated to urinary mandelic acid levels	24—27
33 styrene workers and 6 hospital transport service employees	Polystyrene production and polyester resin products	The duration of sensory nerve action potential (SAP) was increased and SAP amplitude tended to be decreased in styrene groups, but effects were not exposure related; other sensory and all motor nerve parameters showed no relationship to styrene exposure; the dominant alpha EEG frequency was normal but there was an increase in fast activity	28
122 styrene workers	—	Headache, sleep disorders, increased fatigue, weight changes, autonomic imbalance, decreased reflexes, and abnormal EEG patterns occurred	30
61 styrene workers	Plastic lifeboat production	Abnormal EEG patterns were found	31

REFERENCES

1. **Sandmeyer, E. E.**, Aromatic hydrocarbons, in *Patty's Industrial Hygiene and Toxicology*, Vol. 2B, 3rd. rev. ed., Clayton, G. D. and Clayton, F. E., Eds., John Wiley-Interscience, New York, 1981, 3253.
2. **Seppäläinen. A. M., Lindström, K., and Martelin, T.**, Neurophysiological picture of solvent poisoning, *Am. J. Ind. Med.*, 1, 31, 1980.
3. **Husman, K.**, Symptoms of car painters with long-term exposure to a mixture of organic solvents, *Scand. J. Work Environ. Health*, 6, 19, 1980.
4. **Juntunen, J., Hupli, V., Hernberg, S., and Luisto, M.**, Neurological picture of organic solvent poisoning in industry: a retrospective clinical study of 37 patients, *Int. Arch. Occup. Environ. Health*, 46, 219, 1980.
5. **Husman, K. and Karli, P.**, Clinical neurological findings among car painters exposed to a mixture of organic solvents, *Scand. J. Work Environ. Health*, 6, 33, 1980.
6. **Lindström, K.**, Changes in psychological performances of solvent-poisoned and solvent-exposed workers, *Am. J. Ind. Med.*, 1, 69, 1980.
7. National Institute for Occupational Safety and Health, Criteria for a Recommended Standard: Occupational Exposure to Toluene, Department of Health, Education, and Welfare, Washington, D.C., 1973.
8. **Elofsson, S. A., Gamberale, F., Hindmarsh, T., Iregren, A., Isaksson, A., Johnsson, I., Knave, B., Lydahl, E., Mindus, P., Persson, H. E., Philipson, B., Steby, M., Struwe, G., Söderman, E., Wennberg, A., and Widen, L.**, Exposure to organic solvents. A cross-sectional epidemiologic investigation on occupationally exposed car and industrial spray painters with special reference to the nervous system, *Scand. J. Work Environ. Health*, 6, 239, 1980.
9. **Lindström, K.**, Psychological performances of workers exposed to various solvents, *Work, Environ. Health*, 10, 151, 1973.
10. **Hänninen, H., Eskeline, L., Husman, K., and Nurminen, M.**, Behavioral effects of long-term exposure to a mixture of organic solvents, *Scand. J. Work Environ. Health*, 4, 240, 1976.
11. **Cohr, K.-H. and Stokholm, J.**, Toluene: a toxicologic review, *Scand. J. Work Environ. Health*, 5, 71, 1979.
12. **Seppäläinen, A. M., Husman, K., and Mattenson, G.**, Neurophysiological effects of long-term exposure to a mixture of organic solvents, *Scand. J. Work Environ. Health*, 4, 304, 1978.
13. **von Oettingen, W. F., Neal, P. A., and Donahue, D. D.**, The toxicity and potential dangers of toluene: preliminary report, *JAMA*, 118, 579, 1942.
14. **von Oettingen, W. F., Neal, P. A., Donahue, D. D., Svirbely, J. L., Baernstein, H. D., Monaco, A. R., Valaer, P. J., and Mitchell, J. L.**, The Toxicity and Potential Dangers of Toluene with Special Reference to its Maximal Permissible Concentration, Bull. 279, Public Health Serv. Publ., Washington, D.C., 1942.
15. **Carpenter, C. P., Shaffer, C. B., Weil, C. S., and Smyth, H. F., Jr.**, Studies on the inhalation of 1,3-butadiene; with a comparison of its narcotic effect with benzol, toluol, and styrene, and a note on the elimination of styrene by the human, *J. Ind. Hyg. Toxicol.*, 26, 69, 1944.
16. **Ogata, N., Tomokuni, K., and Takatsuka, Y.**, Urinary excretion of hippuric acid and m- or p-methylhippuric acid in the urine of persons exposed to vapours of toluene and m- or p-xylene as a test of exposure, *Br. J. Ind. Med.*, 27, 43, 1970.
17. **Longley, E. O., Jones, A. T., Welch, R., and Lomaev, O.**, Two acute toluene episodes in merchant ships, *Arch. Environ. Health*, 14, 481, 1967.
18. **Wilson, R. H.**, Toluene poisoning, *JAMA*, 123, 1106, 1943.
19. **Capellini, A. and Alessio, L.**, The urinary excretion of hippuric acid in workers exposed to toluene, *Med. Lav.*, 62, 196, 1971.
20. **Matsushita, T., Arimatsu, Y., Ueda, A., Satoh, K., and Nomura, S.**, Hematological and neuromuscular response of workers exposed to low concentrations of toluene vapor, *Ind. Health*, 13, 115, 1975.
21. **Boor, J. W. and Hurtig, H. I.**, Persistent cerebellar ataxia after exposure to toluene, *Ann. Neurol.*, 2, 440, 1977.
22. **Tähti, H., Kärkkäinen, S., Pyykko, K., Rintala, E., Kataja, M., and Vapaatalo, H.**, Chronic occupational exposure to toluene, *Int. Arch. Occup. Environ. Health*, 48, 61, 1981.
23. **Grabski, D. A.**, Toluene sniffing producing cerebellar degeneration, *Am. J. Psychiatry*, 118, 461, 1961.
24. **Knox, J. W. and Nelson, J. R.**, Permanent encephalopathy from toluene inhalation, *N. Engl. J. Med.*, 275, 1494, 1966.
25. **Kelly, T. W.**, Prolonged cerebellar dysfunction associated with paint sniffing, *Pediatrics*, 56, 605, 1975.
26. **Helliwell, M. and Murphy, M.**, Drug induced neurological disease, *Br. Med. J.*, 1, 1283, 1979.
27. **Lachapelle, J., Duplantis, F., Rousseau, S., Boileau, J., and Roy, L. E.**, Cerebellar degeneration due to toluene toxicity, *Union Med. Can.*, 3, 132, 1982.
28. **King, M. D., Day, R. E., Oliver, J. S., Lush, M., and Watson, J. M.**, Solvent encephalopathy, *Br. Med. J.*, 283, 663, 1981.

29. **Malm, G. and Lying-Tunell, Y.**, Cerebellar dysfunction related to toluene sniffing, *Acta Neurol. Scand.*, 62, 188, 1980.
30. **Takeuchi, Y., Hisanaga, N., Ono, Y., Ogawa, T., Hamaguchi, Y., and Okamoto, S.**, Cerebellar dysfunction caused by sniffing of toluene-containing thinner, *Ind. Health*, 19, 163, 1981.
31. **Schikler, K. N., Seitz, K., Rice, J. F., and Strader, T.**, Solvent abuse associated cortical atrophy, *J. Adolesc. Health Care*, 3, 37, 1982.
32. **Streicher, H. Z., Gabow, P. A., Moss, A. H., Kono, D., and Kaehny, W. D.**, Syndromes of toluene sniffing in adults, *Ann. Intern. Med.*, 94, 758, 1981.
33. **Akiguchi, I., Fujiwara, T., Iwai, N., and Kawai, C.**, A case of chronic thinner (toluene) intoxication with myeloneuropathy and EEG abnormality, *Rinsho Shinkeigaku*, 17, 586, 1977.
34. **Allister, C., Lush, M., Oliver, J. S., and Watson, J. M.**, Status epilepticus caused by solvent abuse, *Br. Med. J.*, 283, 1156, 1981.
35. **Keane, J. R.**, Toluene optic neuropathy, *Ann. Neurol.*, 4, 390, 1978.
36. **Weisenberger, B. L.**, Toluene habituation, *J. Occup. Med.*, 19, 569, 1977.
37. **Tarsh, M. J.**, Schizophreniform psychosis caused by sniffing toluene, *J. Soc. Occup. Med.*, 29, 131, 1979.
38. **Bruckner, J. V. and Peterson, R. G.**, Evaluation of toluene and acetone inhalant abuse. I. Pharmacology and pharmacodynamics, *Toxicol. Appl. Pharmacol.*, 61, 27, 1981.
39. **Shirabe, T., Tsuda, T., Terao, A., and Araki, S.**, Toxic polyneuropathy due to glue-sniffing. Report of two cases with a light and electron-microscopic study of the peripheral nerves and muscles, *J. Neurol. Sci.*, 21, 101, 1974.
40. **Takeuchi, Y., Ono, Y., and Hisanaga, N.**, An experimental study in the combined effects of n-hexane and toluene on the peripheral nerve of the rat, *Br. J. Ind. Med.*, 38, 14, 1981.
41. **Taher, G., Anderson, R., McCartney, R., Popovtzer, M. M., and Schrier, R. W.**, Renal tubular acidosis associated with toluene "sniffing," *N. Engl. J. Med.*, 290, 765, 1974.
42. **Bennett, R. H. and Forman, H. R.**, Hypokalemic periodic paralysis in chronic toluene exposure, *Arch. Neurol.*, 37, 673, 1980.
43. **Kroeger, R. M., Moore, R. J., Lehman, T. H., Giesy, J. D., and Skeeters, C. E.**, Recurrent urinary calculi associated with toluene sniffing, *J. Urol.*, 123, 89, 1980.
44. **Baker, A. B. and Tichy, F. Y.**, Effects of organic solvents and industrial poisonings on central nervous system, *Res. Publ. Assoc. Nerv. Ment. Dis.*, 32, 475, 1953.
45. **Carpenter, C. P., Geary, D. L., Jr., Myers, R. C., Nachreiner, D. J., Sullivan, L. J., and King, J. M.**, Petroleum hydrocarbon toxicity studies. XIII. Animal and human response to vapors of toluene concentrate, *Toxicol. Appl. Pharmacol.*, 36, 473, 1976.
46. **Takeuchi, Y. and Hisanaga, N.**, The neurotoxicity of toluene: EEG changes in rats exposed to various concentrations, *Br. J. Ind. Med.*, 34, 314, 1977.
47. **Contreras, C. M., Gonzalez-Estrada, T., Zaraboza, D., and Fernandez-Guardiola, A.**, Petit mal and grand mal seizures produced by toluene or benzene intoxication in the cat, *Electroencephalogr. Clin. Neurophysiol.*, 46, 290, 1979.
48. **Izumi, T. and Sato, S.**, A neurophysiological study on thinner intoxication, *Psychiatr. Neurol. Jpn.*, 73, 99, 1971.
49. **Takeuchi, Y. and Susuki, H.**, Change of convulsion threshold in the rat exposed to toluene, *Ind. Health*, 13, 109, 1975.
50. **Horiguchi, S. and Inoue, K.**, Effects of toluene on the wheel-turning activity and peripheral blood findings in mice. An approach to the maximum allowable concentration of toluene, *J. Toxicol. Sci.*, 2, 363, 1977.
51. **Shigeta, S., Aikawa, H., Misawa, T., and Kondo, A.**, Effect of single exposure to toluene on Sidman avoidance response, *J. Toxicol. Sci.*, 3, 305, 1978.
52. **Shigeta, S., Misawa, T., and Aikawa, H.**, Effects of concentration and duration of toluene exposure on Sidman avoidance in rats, *Neurobehav. Toxicol.*, 2, 85, 1980.
53. **Shigeta, S., Misawa, T., Aikawa, H., and Kondo, A.**, Repeated toluene exposure and Sidman avoidance response in rats, *Tokai J. Exp. Clin. Med.*, 4, 1, 1979.
54. **Ikeda, T. and Miyake, H.**, Decreased learning in rats following repeated exposure to toluene: preliminary report, *Toxicol. Lett.*, 1, 235, 1978.
55. **Benignus, V. A.**, Health effects of toluene: a review, *Neurotoxicology*, 2, 567, 1981.
56. **Hayden, J. W., Peterson, R. G., and Bruckner, J. V.**, Toxicology of toluene (methylbenzene): review of current literature, *Clin. Toxicol.*, 11, 549, 1977.
57. **Carlsson, A. and Lindquist, T.**, Exposure of animals and man to toluene, *Scand. J. Work Environ. Health*, 3, 135, 1977.
58. **Astrand, I., Ehrner-Samuel, H., Kilbom, A., and Ovrum, P.**, Toluene exposure. I. Concentration in alveolar air and blood and at rest and during exercise, *Work Environ. Health*, 9, 119, 1972.

59. **Garriott, J. C., Foerster, E., Juarez, L., DeLa Garza, F., Mendiola, I., and Curoe, J.**, Measurement of toluene in blood and breath in cases of solvent abuse, *Clin. Toxicol.*, 18, 471, 1981.
60. **Williams, R. T.**, *Detoxification Mechanisms*, John Wiley-Interscience, New York, 1959, 194.
61. **Pagnotto, L. D. and Liebermann, L. M.**, Urinary hippuric acid excretion as an index of toluene exposure, *Am. Ind. Hyg. Assoc. J.*, 28, 129, 1967.
62. **Wolf, M. A., Rowe, V. K., McCollister, D. D., Hollingsworth, R. L., and Oyen, F.**, Toxicological studies of certain alkylated benzenes and benzene, *Arch. Ind. Health*, 14, 387, 1956.
63. **Tossavainen, A.**, Styrene use and occupational exposure in the plastics industry, *Scand. J. Work Environ. Health*, 4(Suppl. 2), 7, 1978.
64. **Rogers, J. C. and Hooper, C. C.**, M.A.C. for styrene, *Ind. Med. Surg.*, 26, 32, 1957.
65. **Stewart, R. D., Dodd, H. C., Baretta, E. D., and Shaffer, A. W.**, Human exposure to styrene vapor, *Arch. Environ. Health*, 16, 656, 1968.
66. **Gamberale, F. and Hultengren, M.**, Exposure to styrene. II. Psychological functions, *Work Environ. Health*, 11, 86, 1974.
67. **Carpenter, C. P., Shaffer, C. B., Weil, C. S., and Smyth, H. F., Jr.**, Studies on the inhalation of 1:3-butadiene; with comparison of its narcotic effect with benzol, toluol, and styrene, and a note on the elimination of styrene by the human, *J. Ind. Hyg. Toxicol.*, 26, 69, 1944.
68. Styrene, polystyrene and styrene-butadiene copolymers, *IARC Monogr. on the Evaluation of the Carcinogenic Risk of Chemicals to Humans*, 231, 1979.
69. **Stepien, T.**, Two cases of lesions of the organ of vision with chemicals used in certain branches of industry and in agriculture, *Klin. Oczna*, 43, 169, 1973.
70. **Barsotti, M., Parmeggiani, L., and Sassi, C.**, Observations on occupational pathology in a polystyrene resin factory, *Med. Lav.*, 43, 418, 1952.
71. **Pratt-Johnson, J. A.**, Case report. Retrobulbar neuritis following exposure to vinyl benzene (styrene), *Can. Med. J.*, 90, 975, 1964.
72. **Araki, S., Abe, A., Ushio, K., and Fujino, M.**, A case of skin atrophy neurogenic muscular atrophy and anxiety reaction following long exposure to styrene, *Jpn. J. Ind. Health*, 13, 427, 1971.
73. **Axelson, D. and Gustavson, J.**, Some hygienic and clinical observations on styrene exposure, *Scand. J. Work Environ. Health*, 4(Suppl. 2), 215, 1978.
74. **Lilis, R., Lorimer, W. V., Diamond, S., and Selikoff, I. J.**, Neurotoxicity of styrene in production and polymerization workers, *Environ. Res.*, 15, 133, 1978.
75. **Lorimer, W. V., Lilis, R., Fischbein, A., Daum, S., Anderson, H., Wolff, M. S., and Selikoff, I. J.**, Health status of styrene-polystyrene polymerization workers, *Scand. J. Work Environ. Health*, 4(Suppl. 2), 220, 1978.
76. **Kohn, A. N.**, Ocular toxicity of styrene, *Am. J. Opthalmol.*, 85, 569, 1978.
77. **Cherry, N., Waldron, H. A., Wells, G. G., Wilkinson, R. T., Wilson, H. K., and Jones, S.**, An investigation of the acute behavioral effects of styrene on factory workers, *Br. J. Ind. Med.*, 37, 234, 1980.
78. **Götell, P., Axelson, O., and Lindelöf, B.**, Field studies on human styrene exposure, *Work Environ. Health*, 9, 76, 1972.
79. **Härkönen, H., Lindström, K., Seppäläinen, A. M., Asp, S., and Hernberg, S.**, Exposure-response relationship between styrene exposure and central nervous functions, *Scand. J. Work Environ. Health*, 4, 53, 1978.
80. **Härkönen, H.**, Relationship of symptoms to occupational styrene exposure and to the findings of electroencephalographic and psychological examinations, *Int. Arch. Occup. Environ. Health*, 40, 231, 1977.
81. **Seppäläinen, A. M. and Härkönen, H.**, Neurophysiological findings among workers occupationaly exposed to styrene, *Scand. J. Work Environ. Health*, 2, 140, 1976.
82. **Lindström, K., Härkönen, H., and Hernberg, S.**, Disturbances in psychological functions of workers occupationally exposed to styrene, *Scand. J. Work Environ. Health*, 2, 129, 1976.
83. **Rosén, I., Haeger-Aronsen, B., Rehnström, S., and Welinder, H.**, Neurophysiological observations after chemical styrene exposure, *Scand. J. Work Environ. Health*, 4(Suppl. 2), 184, 1978.
84. **Holmberg, P. C.**, Central nervous defects in two children of mothers exposed to chemicals in the reinforced plastics industry: chance or a causal relation?, *Scand. J. Work Environ. Health*, 3, 212, 1977.
85. **Hruba, E., Salcmanova, Z., and Schwartzova, K.**, Long term follow up of subjects exposed to possible styrene poisoning, *Cs. Neurol. Neurochir.*, 38, 116, 1975.
86. **Dolmierski, R., Kwiatkowski, S. R., and Nitka, J.**, Clinical and experimental research into the pathogenesis of toxic effect of styrene. VII. Appraisal of the nervous system in the workers exposed to styrene, *Bull. Instit. Marit. Trop. Med. Gdynia*, 27, 193, 1976.
87. **Järvisalo, J., Ed.**, Proceedings of the international symposium on styrene: occupational and toxicological aspects, Helsinki, *Scand. J. Work Environ. Health*, 4(Suppl. 2), 1, 1978.
88. **Härkönen, H.**, Styrene, its experimental and clinical toxicology: a review, *Scand. J. Work Environ. Health*, 4(Suppl. 2), 104, 1978.

89. **Spencer, H. C., Irish, D. D., Adams, E. M., and Rowe, V. K.**, The response of laboratory animals to monomeric styrene, *J. Ind. Hyg. Toxicol.*, 24, 295, 1942.
90. **Larsby, B., Tham, R., Odkvist, L. M., Hyden, D., Bunnfors, I., and Aschan, G.**, Exposure of rabbits to styrene. Electronystagmographic findings correlated to the styrene level in blood and cerebrospinal fluid, *Scand. J. Work Environ. Health*, 4, 60, 1978.
91. **Husain, R., Srivastava, S. P., Mushtag, M., and Seth, P. K.**, Effect of styrene on levels of serotonin, noradrenaline, dopamine and activity of acetylcholinesterase and monoamine oxidase in rat brain, *Toxicol. Lett.*, 7, 47, 1980.
92. **Agrawal, A. K., Srivastava, S. P., and Seth, P. K.**, Effect of styrene on dopamine receptors, *Bull. Environ. Contam. Toxicol.*, 29, 400, 1982.
93. **Savolainen, H. and Pfäffli, P.**, Effects of chronic styrene inhalation on rat brain protein metabolism, *Acta Neuropathol.*, 40, 237, 1977.
94. **Savolainen, H., Helojoki, M., and Tengén-Junnila, M.**, Behavioural and glial cell effects of inhalation exposure to styrene vapour with special reference to interactions of simultaneous peroral ethanol intake, *Acta Pharmacol. Toxicol.*, 46, 51, 1980.
95. **Seppäläinen, A. M. and Savolainen, H.**, Impaired nerve function in rats after prolonged exposure to vinyltoluene, *Arch. Toxicol.*, Suppl. 5, 100, 1982.
96. **Savolainen, H. and Pfäffli, P.**, Neurochemical effects of short-term inhalation exposure to vinyltoluene vapor, *Arch. Environ. Contam. Toxicol.*, 10, 511, 1981.

Chapter 7

PHENOL AND RELATED SUBSTANCES

John L. O'Donoghue

TABLE OF CONTENTS

I.	Phenol	142
	A. Synonyms	142
	B. Introduction	142
	C. Neurotoxicity	142
II.	Pentachlorophenol	142
	A. Synonyms	142
	B. Introduction	143
	C. Neurotoxicity	143
	D. Metabolism	144
III.	Hexachlorophene	144
	A. Synonyms	144
	B. Introduction	144
	C. Human Neurotoxicity	144
	D. Neurotoxicity in Animal Species and Environmental Release	145
	E. Experimental Neurotoxicity	145
	F. Recovery from Neurotoxicity	146
	G. Neuropathology	147
	H. Metabolism	148
	I. Mechanism of Action	148
	J. Other Effects	149
IV.	2,4-Dichlorophenoxyactic Acid	149
	A. Human Neurotoxicity	149
	B. Experimental Neurotoxicity	149

References ... 150

I. PHENOL

A. Synonyms   Monohydroxybenzene   Phenyl alcohol
                 Carbolic acid   Phenyl hydrate
                 Benzenol   Phenyl hydroxide

B. Introduction

Phenol (Figure 1) is a clear, colorless solid at room temperature. It is very soluble in water and is frequently used or handled as a concentrated aqueous solution. Production in the U.S. is on the order of 2 to 3 million pounds annually.[1] Primary uses include the manufacture of phenolic resins, bisphenol-A, alkyl phenols, caprolactam, and adipic acid. It is also used for bactericidal, fungicidal, and medical purposes.

C. Neurotoxicity

Chronic phenol toxiciy has been primarily of historical interest relating to Lister's introduction of phenol as an antiseptic in 1867. Phenol intoxication generally follows an acute course following either ingestion, skin exposure, or rarely inhalation. Affected individuals initially show central nervous system stimulation followed by severe, profound depression. Increased pulse rates, decreased blood pressure, cyanosis, dyspnea, stupor, convulsions, muscle tremors, darkened urine, and death are seen.[1-3] Direct damage occurs in the myocardium, liver, and kidneys.[1-3] Neundörfer and Wolpert[4] described a single case of residual neuropsychiatric effects, impotence, and depression in a man who recovered from an acute episode of phenol poisoning. Topical phenol tissue destruction is frequently painless due to local blockage of nerve impulses by phenol. Five percent phenol solutions applied to unabraded skin initially produces warmth and tingling followed by local anesthesia.[2] Thus even very severe tissue destruction may not produce pain. Exposure of peripheral nerves to phenol blocks nerve compound action potential within minutes.[5-7]

The neurolytic properties of phenol have been used therapeutically for the relief of pain and spasticity by local injection of phenol into the spinal and perineural spaces.[8] Morphological changes include myelin and axonal damage depending on the dose of phenol.[7]

Inadvertent intravascular injection of phenol in man may result in convulsions.[9] A number of phenolic compounds were tested for convulsive activity in rats.[10] In descending order of convulsive activity were catechol > resorcinol, p-cresol > o-cresol, phenol > m-cresol > hydroquinone > pyrogallol.

Following repeated phenol exposures 7 hr/day, 5 days/wk at 0.1 to 0.2 mg/ℓ to guinea pigs, signs of paralysis primarily of the hindquarters occurred after about 20 exposures over a 28-day period.[11] Rats and rabbits were not similarly affected. It was not clear from the report whether or not the paralysis was a terminal event. No neurohistiologic examinations were performed.

II. PENTACHLOROPHENOL

A. Synonyms   Penchlorol   Santophen 20
                 Penta   Dowicide 7
                 PCP   Cuprinol
                 Preventol P   Evrisan
                 Permacide   Santobrite
                 Chem-tol

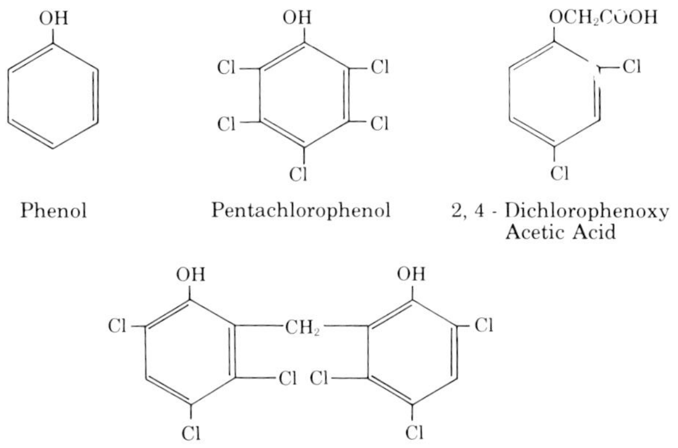

FIGURE 1. Phenol and related chemicals.

B. Introduction

Pentachlorophenol (PCP) is a crystalline solid which is usually dissolved in an organic solvent for use. It is very useful as a preservative of wood, wood products, starches, dextrins, paints, rubber, and glues, and to control termites and mollusks. Other uses include application as a general herbicide and defoliant. The toxicity of its salt, sodium pentochlorophenate, is the same as the parent compound.

C. Neurotoxicity

Although pentachlorophenol has been mentioned as capable of producing peripheral neuropathy, there is little data to support the statement.[12,13] Acute exposure to high doses of PCP commonly results in death. In a review of 51 cases of PCP poisoning, 30 resulted in death.[14] The clinical course in acute poisonings includes hyperpyrexia, tachypnea, tachycardia, hyperkinesis, muscle twitching, tremors, hyperglycemia, hyperperistalsis, weakness, and convulsions. Rigor mortis occurs rapidly after death.[15-19]

One individual working with a PCP wood preservative developed an acute case of PCP poisoning with sensory loss and motor paralysis of the hands and forearms.[18] Nerve damage was apparently due to absorption through the skin and direct toxicity to regional nerves similar to that seen with phenol.[2] Residual paralysis with sensory loss was present in one limb.

Imaizumi and Atsumi[20] found that survivors of acute PCP poisoning may have residual autonomic impairment, impaired circulation, and visual damage with arcuate scotoma.

Chronic or repeated exposure to PCP results in weight loss, eye and respiratory irritation, weakness, and asthma-like symptomalogy but no objective evidence of neurotoxicity.[14,19,21] Health surveys of workers using pentachlorophenol have not reported evidence of neurotoxicity.[22-24]

People engaged in the manufacture of pentachlorophenol or sodium pentachlorophenate have been reported to develop peripheral neuropathy, evidence of CNS disorders, porphyria, and chloracne due to the presence of dioxins in the manufacturing process.[25,26] Technical grades of PCP may contain 4 to 12% of other chlorinated phenols which may alter the toxicity of PCP.[15,27]

In the experimental laboratory, neurotoxicity due to PCP, or technical-grade PCP, has not been reported.[15,27,30] Rats consuming 200 ppm for 8 weeks, 500 ppm for 8 weeks, or 500 ppm for 8 months, and chickens consuming 1000 ppm for 8 weeks did not show evidence of clinical neurotoxicity or histopathological abnormalities in the brain.[27-30]

D. Metabolism

Cooper[31] has recently reviewed the metabolism of pentachlorophenol. In rats, PCP was eliminated by metabolism to tetrachlorohydroquinone and by excretion of PCP and its metabolites in the urine.[32] The highest concentrations of PCP were found in the liver and kidneys while the lowest were in the brain and fat.[32]

Administration of 20 mg/ℓ of technical-grade PCP in the drinking water of rats for 14 weeks resulted in the accumulation of PCP but not tetrachlorophenol in the brain.[23] Cerebral acid proteinase and superoxide dismutase activities were increased while glial glutathione levels were reduced.[33]

Effects of PCP on cerebral metabolism were investigated by studying PCP effects on isolated, perfused canine brain preparations.[34] PCP increased cerebral oxygen consumption 16% over control values, glucose consumption increased 50%, and lactate levels also increased. PCP appears to uncouple mitochondrial oxidative phosphorylation in the brain which is similar to its effects on other organs.[15]

III. HEXACHLOROPHENE

A. Synonyms

2,2′-Methylene bis(3,4,6-trichlorophenol)	Bilevon
Bis(2-hydroxy-3,5,6-trichlorophenyl)methane	Dermadex
2,2′-Dihydroxy-3,3′,5,5′,6,6′-hexachlorodiphenylmethane	Exofene
Trichlorohydroxyphenyl methane	Gamophene
Phisohex	Hexosan
G-11	Surgi-Cen
At-7	Surofene

B. Introduction

Hexachlorophene (HCP) is a crystalline solid soluble in organic solvents but relatively insoluble in water. Sodium or disodium salts were available until 1972, when HCP was restricted from over-the-counter products by the F.D.A. in the U.S. and by regulatory bodies in other countries. It had been used for about 20 years as an antibacterial ingredient in about 1500 products including shampoos, soaps, deodorants, and other health care and hygiene products. It was also commonly used in hospital nurseries to bathe infants, for presurgical scrubs, and in the treatment of burns. World-wide production prior to its ban was approximately 2500 tons.[35] Currently, hexachlorophene is sold as a prescription drug. Agricultural uses of HCP have included use as a pesticide, and fungicide on plants, and flukicide in animals. Commercial preperations include a 0.25% and 3% HCP emulsions and 0.25 to 1% talc powders.

C. Human Neurotoxicity

Several reviews are available on the toxicity of hexachlorophene.[35-43] The primary routes of human exposure to hexachlorophene are by application to the skin or by ingestion. Vaginal exposures have also been reported.[44] Ingestion of HCP has usually been unintentional involving mislabeled or unlabeled home containers[36,37,45-48] or rarely for treatment of fluke infestation.[49] Dermal exposure involves the once common practice of washing newborn babies with a 3% HCP emulsion[50-60] or washing the skin of burn patients.[61-63] When damaged, the skin is more permeable to HCP thus HCP absorption and exposure are greater.[63-66] In one instance, 204 children were poisoned by cutaneous exposure to a talc powder which unintentionally contained 6.3% hexachlorophene.[35]

Symptoms following ingestion of HCP are referable to both the gastrointestinal tract and the nervous system. Nausea, vomiting, cramps, diarrhea, fever, abnormal pupillary reflexes, twitching, papilledema, nystagmus, blindness, weakness, paralysis including facial paralysis, positive Babinski sign, convulsions, coma, or death may occur. Individuals with mild gastrointestinal signs have more severe nervous system signs.[37] Neurotoxicity and dermal irritation result from skin exposure to HCP.

Although the use of hexachlorophene has been curtailed, there is continuing controversy about its use in nurseries.[53,65,67-70] HCP use in premature infants or infants weighing less than 2 kg, infants with hyperbilirubinemia or hepatic disease, or on patients with burned or damaged skin carry the greatest risks of developing neurotoxicity.

D. Neurotoxicity in Domestic Animal Species and Environmental Release

Spontaneous neurotoxicity has also been reported in farm and pet animals usually due either to accidental ingestion of HCP[74] or due to the once common practice of bathing puppies with HCP soaps in the treatment of pustular dermatitis.[72,75] Environmental accumulation of HCP in stream organisms apparently resulted from the discharge of HCP soaps through waste treatment plants and untreated residential and institutional sources.[76] No neurotoxicity from environmental discharge has been recognized.

E. Experimental Neurotoxicity

HCP neurotoxicity has been experimentally reproduced in cats,[71] dogs,[72,73,77] sheep,[78-82] calves,[81] swine,[83] nonhuman primates,[44,84-86] amphibians,[87,88] mice,[84,89-91] rabbits,[38] and rats.[64,92-110]

The ability of hexachlorophene to damage the nervous system is similar across species. Differences among species relate primarily to the least effective neurotoxic dose and relative severity of lesions when different regions of the nervous system are compared, particularly involvement of the visual system. Comparisons between experimental species and human cases are confounded by the time courses involved. Nearly all human cases involve acute or subacute exposures whereas many experimental studies involve subchronic exposures. Within species there are differences in sensitivity to the neurotoxic effects of hexachlorophene which depend upon strain,[97,98] sex,[97,98] and most importantly, age.[72,83,84,89,90,97-100,104]

Lethal or near lethal oral doses of HCP (Table 1) result in diarrhea, central nervous system depression, increased respiratory rate, and hyperthermia.[97,98,112] Rigor mortis occurs soon after death as with other phenolic materials which uncouple oxidative phosphorylation.[97] Immature animals may develop convulsions, ataxia, weakness, or paralysis.[72,83,97]

Repeated oral or dermal application (Tables 2, 3, and 4) may result in growth retardation, tremor, weakness, posterior paralysis, and loss of visual acuity. The oral no-effect level in the rat is about 3.7 mg/kg.[97] Effects on the nervous system are age dependent.

Shuman et al.[104] found that 6- to 22-day-old rats were extremely susceptible to HCP and were damaged by as few as two baths with 3% HCP. Animals younger than 6 days were less severely affected and older rats were resistant. Ulsamer et al.[99] repeatedly gave nursing rat pups 10 mg/kg of HCP by gavage. Those which began receiving HCP at 10 to 13 days of age were most sensitive, dying after 2 to 3 doses. When dosing was begun at earlier ages, the pups developed tremors and functional limb deficits but recovered after 23 to 30 days of age even though HCP exposure continued. Rats older than 22 days of age when dosing was started remained clinically unaffected. Mice[89] and swine[83] also show recovery from clinical neurotoxicity as they age, even during continued dosing.

In rat pups, the concentration of HCP in the blood and brain are highest at 13 days of age, the period of greatest HCP sensitivity, and the period of rapid myelination (12 to 25 days of age) of the rat brain begins at this time.[99]

Peripheral nervous system myelin is also affected to a greater degree in mice younger than 16 days.[90,91] In mice older than 16 days, Persson et al.[91] did not produce the extensive myelin vacuolization seen in younger animals.

Table 1
HCP SINGLE DOSE TOXICITY

Species	Age	Route	Dose (mg/kg)	Effect	Ref.
Rat	Adult	p.o.	56	Female LD_{50}	100
			66	Male LD_{50}	100
			57.6	LD_{50}	97
		i.v.	7.5	Male LD_{50}	98
		i.p.	21.8	LD_{50}	97
	Weanling	p.o.	120	Female LD_{50}, posterior paralysis	98
			87	LD_{50}	97
		i.p.	40	LD_{50}	97
Swine	Newborn	p.o.	4	LD_{50}, CNS abnormalities and status spongiosus	83
Dog	2—3 wks	p.o.	3.75	Excitement, abnormal movement recovery	72
			7.5, 15, 30, 60	Convulsions, abnormal movements, coma, cerebral edema, death	

Note: i.p. = intraperitoneal; p.o. = gavage; i.v. = intravenous.

Table 2
TOXICITY IN ADULT RATS FED HCP

Diet concentration (ppm)	Approximate HCP dose (mg/kg/day)	Approximate duration (weeks)	Effect	Ref.
1000	—	2	Weight loss, paralysis, urinary and fecal incontinence, status spongiosus, lethal concentration	107, 110
800	—	2—30	Growth retardation, weak but can move	107
500	25	13	Weight loss, weakness after 2 weeks, paralysis after 3—5 weeks, status spongiosus, mortality low	92, 98, 103
100—200	8—20	13	Clinically unaffected, status spongiosus	98, 109
≤65	≤6.5	13	No adverse effect	97, 98, 109

F. Recovery from Neurotoxicity

Experimental animals develop a resistance to HCP with age and even animals showing paralysis may recover with increasing age while still receiving HCP.[89,99] Rats receiving 500 ppm of HCP in their diets for 10 weeks regained limb function almost entirely by 6 weeks after HCP discontinuation but minor microscopic lesions were still present 12 weeks after HCP discontinuation.[92] At 40 mg/kg HCP by gavage, posterior paralysis was absent after a 1-week recovery period and histologic brain vacuolization was detectable at 84 days recovery but not 168 days.[109] Effects of HCP on nerve conduction velocities recover within 1 week.[94] Behavioral deficiencies may be more slowly recoverable.[106]

Irreversible effects on the brain and the visual system are of considerable concern. Rat pups nursing on dams fed diets of 500 ppm HCP developed hydrocephalus *ex vacuo* and optic nerves contained degenerating axons and had undergone atrophy and gliosis.[103] Adult

Table 3
HCP TOXICITY: DIETARY ROUTE IMMATURE RATS

Diet concentration (ppm)	Approximate HCP dose (mg/kg/day)	Approximate duration (weeks)	Effect	Ref.
400	29	1	Weight loss, paralysis, status spongiosus, lethal concentration	97
200	15	16	Growth retardation, paralysis with recovery while exposure continued, status spongiosus	97
100	8	16	Growth retardation, clinically normal, status spongiosus	97
≤50	≤3.7	16	No adverse effects	97

Table 4
HCP TOXICITY FOLLOWING REPEATED DOSES BY GAVAGE TO RATS OF DIFFERING AGE

Age (days)	Dose (mg/kg/day)	Duration	Effect	Ref.
≥70	80	2—5 days	Lethal dose, status spongiosus	108
≥70	40	6 wk	Growth retardation, paralysis, status spongiosus	108, 109
≥70	20	1—6 wk	Clinically unaffected, status spongiosus	108, 109
≥22	10	8—87 days	Clinically unaffected, slight status spongiosus	98, 99, 108
10—13	10	2—3 days	Tremor, paralysis, status spongiosus, lethal dose	99
2—4	10	9—10 days	Tremor, paralysis, status spongiosus, recovery during dosing	99

animals are more susceptible to optic nerve atrophy and gliosis possibly due to greater constriction of the optic nerve by the adult ossified optic foramen.[81,100] Similar optic nerve atrophy has been reported in a woman chronically exposed to HCP. Irreversible visual deficits were reported in dogs receiving 3% and 10% HCP in an ointment applied daily.[77] From 3 to 5 weeks on, permanent mydriasis, loss of papillary reflexes, loss of visual faculty, deformation of the optic papilla, and retinal pigmentary change occurred.

G. Neuropathology

Hexachlorophene, like triethyltin, cuprizone, isonicotinic acid hydrazide, and 3,5-diiodo-4-chlorosalicylanilide, produces intramyelinic edema often referred to as status spongiosis because of the appearance of cystic fluid-filled vacuoles or spaces within myelinated portions of the central nervous system and peripheral nerves. Fluid accumulates within the myelin sheath by splitting of the sheath at the intraperiod line.

The fluid-filled vacuoles have little or no communication with the extracellular space since tracers such as horseradish peroxidase in the extracellular space enter the vacuoles only to a very limited extent.[88,113] Vacuolization appears to be a reversible lesion[88,90,92,109] which does not necessarily lead to demyelination or interfere with axonal conduction velocity[107] unless prolonged.

Axonal degeneration occurs following prolonged HCP intoxication.[90,103,107,113,114] Apparently, degenerative axonal changes are related to excess pressure. In the optic nerves, this may be brought about both by fluid accumulation and constriction of nerves by the optic foramen.[81,113] Powell et al.[114] suggested that increased endoneurial pressure measured within peripheral nerves of rats fed HCP led to axonal damage. Rats fed HCP for 11 days developed increased endoneurial pressure, but after recovery for 12 days showed few residual abnormalities. Rats fed for 21 days showed widespread axonal degeneration with residual damage after a 14 day recovery.

In addition to damage to the optic nerve, HCP directly damages the retina in dogs[77] and rats.[115] Photoreceptor cells in the retina undergo vacuolization in rats fed 900 ppm of HCP.[115] Dogs develop retinal damage after only 3 weeks of daily 3% or 10% ointment applied dermally.[77]

H. Metabolism

Hexachlorophene may be absorbed by oral, dermal, or vaginal routes. Oral and vaginal routes result in greater absorption than the dermal route.[116] The difference between the acute oral and i.v. rat LC_{50} suggests the HCP is poorly absorbed from the gut.[98] Absorption of HCP through the skin depends on the vehicle[98,135] used and the integrity of the skin.[64]

HCP is distributed into the blood, brain, liver, and adipose tissue.[99,116] Ulsamer et al.[99] found that blood and brain HCP levels were highest in 13-day-old rat pups which was the age group most sensitive to HCP. Adult rats and younger pups had lower HCP levels.[99] In pregnant rats, HCP is also distributed into the uterus, placenta, and embryo.[116] Infants from women bathed once with HCP have blood levels of 3 to 182 ppb HCP.[136]

Excretion after oral and vaginal doses is primarily in the feces.[116-118] In the rat,[117-118] small amounts of HCP are excreted in the urine but in the rabbit significant quantities are excreted by this route.[118] Milk from lactating rats contained HCP levels of 0.07 or 0.33 ppm when the animals were fed 20 to 100 ppm HCP.[98]

The half-life of HCP in infants bathed with HCP soap was 6.1 to 44.2 hr and appeared to follow first-order kinetics.[66] The liver is important in the metabolism of HCP[119] and infants with liver disease appear more susceptible to HCP toxicity.[66]

I. Mechanism of Action

Hexachlorophene, like other phenolic substances, uncouples oxidative phosphorylation in mitochondria resulting in hyperthermia and other physiological disturbances.[120,121,122] Although other myelinotoxic agents are also uncouplers of oxidative phosphorylation,[123] the relationship, if any, between uncouplers and myelin damage is not clear.

A number of biochemical changes occurs in neural tissue following HCP exposure. Brain wet weight increases as a result of increased water content. The brain of HCP rats may weigh 5 to 10% more than controls due to water accumulation.[99,102] Brain glucose levels increase slightly accompanying the water increase.[102,125,126]

Cammer et al.[124] have suggested that cerebral water accumulation is due to increased permeability of myelin lamellae. Myelin isolated from rats fed high levels of HCP (500 ppm) had normal lipid and protein composition except for a slight increase in the relative amount of low molecular weight basic protein.[124] In vitro, HCP produces an inhibition of protein and lipid synthesis with 1 hr, while in vivo, the incorporation of labeled precursors into brain ATP, phospholipids, sterols, protein, and RNA appeared unaffected by HCP.[101] Membrane associated ATPase and adenylatecyclase are inhibited in vitro in rat hepatocyte plasma membrane exposed to 1 μM to 1 mM HCP.[137]

Hanig et al.[127] briefly reported on the ability of hexachlorophene-related substances to produce elevation of cerebrospinal fluid (CSF) pressure similar to HCP. HCP analogues of the 2,2-methylene bischlorophenol class, including dichlorophene, and the 4-chlorophenol, tetrachlorophenol, and 4,6-dichlorophenol isomers did not increase CSF pressure. Nor did 2,4-dinitrophenol, isoniazid, cuprizone, bromobenzene, hexachlorobenzene, hexachlorocyclohexane, chlorpromazine, or DDT. Tribromosalicylanilide (200 mg/kg) may increase CSF pressure and tetraethyltin produced a more dramatic increase in pressure than HCP.

J. Other Effects

High oral[128] and vaginal[117] doses of HCP which cause maternal toxicity have been reported to result in teratogenicity in rats. Other studies at high oral doses[129] or at lower levels[130,131] or using hamsters[132] have not reported teratogenicity. Dietary exposure to 100 ppm HCP reduced rat pup survival and litter size.[98] Oral HCP administration to rats and sheep produced spermatogenic epithelium degeneration.[82,129] Topical HCP application which produced neurotoxicity resulted in long-lasting ejaculatory disturbances and an increased incidence and severity of prostatic cysts and fibrosis in rats.[133] Hypercoagulability of blood is reported to occur in dogs given HCP.[134]

IV. 2,4-DICHLOROPHENOXYACETIC ACID

A. Human Neurotoxicity

2,4-Dichlorophenoxyacetic acid (2,4-D) is a potent plant growth hormone, widely used as a herbicide. 2,4-D amine salts are more water soluble than the sodium salt. Few cases of 2,4-D poisoning with neurological signs have been reported.[138-146]

Berwick[139] described effects seen in a man following accidental ingestion of a herbicide which contained 24% 2,4-D. Fibrillation, twitching, and myotonia were followed by intercostal muscle paralysis. General skeletal muscle damage was indicated by increased serum enzymes and myoglobinuria. Sexual potency was lost for 4 months. Berwick[139] also mentioned a case by Todd[145] in which peripheral neuritis lasted two years. Myotonia and muscle weakness persisted in a man who recovered from severe acute toxicity following ingestion of a herbicide containing 10% 2,4-D and 20% 2-methyl-4-chlor-phenoxyacetic acid.[144]

Memory impairment, altered color vision, and polyneuritis followed recovery from ingestion of a herbicide containing 2,4-D and 2,4-dichlorophenoxypropionic acid.[140]

Goldstein et al.[143] considered skin absorption to be the main exposure in three cases involving severe sensory and motor deficits following use of 2,4-D containing sprays. Symptomatology included numbness, limb pain, loss of touch, pain, and temperature sensation, muscle stiffness, fasciculations, and paresis or paralysis. Exposures occurred over 2-day periods. Recovery in one case was incomplete years later.

A primary sensory neuropathy of the hands and feet developed in a farmer exposed by skin contact and possibly by inhalation to 98 to 99% pure 2,4-D.[138] Recovery was gradual. Foissac-Gegoux et al.[141] described unilateral facial sensory loss following 2,4-D exposure. Of 11 female field hands acutely poisoned by a herbicide containing 2,4-D, headache and vertigo were reported by 7, fatigue, numbness, pain in arms and legs, and irritability were reported by 5, and partial amnesia was reported by 2.[146]

Dudley and Thapar[147] suggested that areas of cerebral demyelination found in the brain of an elderly man with senile dementia who had committed suicide with 2,4-D were due to 2,4-D. Other lesions in the case were attributed to dementia. Singer et al.[148] studied a group of chemical workers exposed to 2,4-D, 2,4,5-trichlorophenoxyacetic acid, and chlorinated dioxin contaminants and found slowing of nerve conduction velocity when compared to a group of brake workers and a group from their laboratory staff.

B. Experimental Neurotoxicity

The most common neurological effect of acute 2,4-D intoxication in man appears to be a sensory or sensorimotor peripheral neuropathy but no animal studies are available which have specifically studied this phenomenon.

Acutely toxic doses of 2,4-D in animals result in transient myotonia[149-151] which occurs shortly after 2,4-D administration and persists for less than 24 hr.[152] Myotonia is characterized by temporary involuntary contraction or spasm of skeletal muscle when the animal attempts to move. Myotonia due to 2,4-D has been reported in dogs, monkeys, rats, mice, guinea pigs, and rabbits.[149-151] Lethal doses result in myotonia, weakness, central nervous system depression, and coma.[151] Repetitive myotonic doses of 2,4-D alter skeletal muscle ultrastructure and result in necrosis. Myotonia may also be produced by chemicals similar to 2,4-D including 2,4,5-trichlorophenoxyacetic acid[151] and 2-methyl-4-chlorophenoxyacetic acid (2,4,5-T).[154]

Calves given 2,4-D have been reported to have difficulty in swallowing, deafness, muscle weakness, and anorexia.[155] Pigs similarly treated show loss of coordination, muscle weakness, central nervous system depression, and gastrointestinal signs.[155]

Experimentally, 2,4-D affects cerebral electrical activity. Desi et al.[156] gave 200 mg/kg of 2,4-D i.p. either in single or repetitive daily doses to rats, cats, and dogs and found changes in the electroencephalogram pattern and in conditioned reflexes. Spinal cord demyelination was reported to have occurred in chronically treated rats although no effects referable to this change were reported.[156] 2,4-D protects mice genetically redisposed to audiogenic seizures.[157] The effective dose (ED_{50}) for seizure protection is 75 mg/kg or about half the dose producing myotonia.

REFERENCES

1. National Institute of Occupational Safety and Health, Criteria for a Recommended Standard. Occupational Exposure to Phenol, Department of Health, Education, and Welfare, Publ. No. 76-196, Washington, D.C., 1976.
2. **Harvey, S. C.,** Antiseptics and disinfectants; fungicides, ecoparasiticides, in *The Pharmacological Basis of Therapeutics*, 5th ed., Goodman, L. A. and Gilman, A., Eds., Macmillan, New York, 1975, 990.
3. **Deichmann, W. B. and Keplinger, M. L.,** Phenols and phenolic compounds, in *Patty's Industrial Hygiene and Toxicology*, Vol 2A, 3rd rev. ed., Clayton, G. D. and Clayton, F. E., Eds., John Wiley-Interscience, New York, 1981, 2567.
4. **Neundörfer, B. and Wolpert, E.,** Neuropsychiatric disorders following phenol poisoning, *Muench. Med. Wochenschr*, 118, 1177, 1976.
5. **Nathan, P. W. and Sears, T. A.,** Effects of phenol on nervous conduction, *J. Physiol. (London)*, 150, 565, 1960.
6. **Nathan, P. W., Sears, T. A., and Smith, M. C.,** Effects of phenol on the nerve roots of the cat: an electrophysiological and histological study, *J. Neurol. Sci.*, 2, 7, 1965.
7. **Schaumburg, H. H., Byck, R., and Weller, R. O.,** The effect of phenol on peripheral nerve. A histologcal and electrophysiologic study, *J. Neuropathol. Exp. Neurol.*, 29, 615, 1970.
8. **Wood, K. A.,** The use of phenol as a neurolytic agent: a review, *Pain*, 5, 205, 1978.
9. **Benzaw, H. T.,** Convulsions secondary to intravascular phenol: a hazard of celiac plexus block, *Anesth. Analg.*, 58, 150, 1979.
10. **Matsumoto, J., Kiyono, S., Nishi, H., Koike, J., and Ichihashi, I.,** The convulsive mechanism of phenol derivatives, *Med. J. Osaka*, 13, 313, 1963.
11. **Deichmann, W. B., Kitzmiller, K. V., and Witherup, S.,** Phenol studies. VII. Chronic phenol poisoning, with special reference to the effects upon experimental animals of the inhalation of phenol vapor, *Am. J. Clin. Pathol.*, 14, 273, 1944.
12. **Fullerton, P. M.,** Toxic chemicals and peripheral neuropathy: clinical and epidemiological features, *Proc. R. Soc. Med.*, 62, 201, 1969.
13. **McLeod, J. G.,** Peripheral neuropathy caused by drugs and toxic substances, *Aust. N. Z. J. Med.*, 3, 268, 1971.
14. **Anon.,** Pentachlorophenol poisoning in the home, *Calif. Health*, 27, 13, 1970.

15. **Deichmann, W. B. and Keplinger, M. L.,** Phenols and phenolic compounds, in *Patty's Industrial Hygiene and Toxicology,* Vol. 2A, 3rd ed. rev., Clayton, G. D. and Clayton, F. E., Eds., John Wiley-Interscience, New York, 1980, 2604.
16. **Haley, T. J.,** Human poisoning with pentachlorophenol and its treatment, *Ecotoxicol. Environ. Safety,* 1, 343, 1977.
17. **Bergner, H., Constantinidis, P., and Martin, J. H.,** Industrial pentachlorophenol poisoning in Winnipeg, *Can. Med. Assoc. J.,* 92, 448, 1965.
18. **Hernberg, S. and Pessi, Y.,** Peripheral nervous paralyses due to pentachlorophenolate, *Arch. Gewerbepath. Gewerbehyg.,* 21, 23, 1964.
19. **Anon.,** Hazards at home and at work, *Food Cosmet. Toxicol.,* 3, 845, 1965.
20. **Imaizumi, K. and Atsumi, K.,** Eye lesions due to agricultural chemicals, *Ganka,* 13, 717, 1971.
21. **Demidenko, N. M.,** Materials for establishing the maximum permissible concentration of pentachlorophenol in air, *Gig. Tr. Prot. Zabol.,* 13, 58, 1968.
22. **Bidstrup, P. L.,** Toxic chemicals and peripheral neuropathy, *Proc. R. Soc. Med.,* 62, 208, 1969.
23. **Ensberg, I. F., DeBruin, A., and Zielhuis, R. L.,** Health of workers exposed to a cocktail of pesticides, *Int. Arch. Arbeitsmed,* 32, 191, 1974.
24. **Klemmer, H. W., Wong, L., Sato, M. M., Reichert, E. L., Korsak, R. J., and Rashad, M. N.,** Clinical findings in workers exposed to pentachlorophenol, *Arch. Environ. Contam. Toxicol.,* 9, 715, 1980.
25. **Jiraśek, L., Kalenský, J., Kubec, K., Pazderova, J., and Lukáš, E.,** Choracne, porphyria cutanea tarda and other herbicide induced intoxications, *Hautarzt,* 27, 328, 1976.
26. **Vinogradova, V. K., Kalyaganov, P. I., Sudonina, L. T., and Yelizarov, G. P.,** Hygienic characteristics of working conditions and health status of workers engaged in the manufacture of sodium pentachlorophenolate, *Gig. Tr. Prof. Zabol.,* 17, 11, 1973.
27. **Kimbrough, R. D. and Linder, R. E.,** The effect of technical and purified pentachlorophenol on the rat liver, *Toxicol. Appl. Pharmacol.,* 46, 151, 1978.
28. **Debets, F. M., Strik, J. J., and Olie, K.,** Effects of pentachlorophenol on rat liver changes induced by hexachlorobenzene, with special reference to porphyria, and alterations in mixed function oxygenases, *Toxicology,* 15, 181, 1980.
29. **Knudsen, I., Verschuuren, H. G., Dentowkelaar, E. M., Kroes, R., and Helleman, P. F.,** Short-term toxicity of pentachlorophenol in rats, *Toxicology,* 2, 141, 1974.
30. **Stedman, T. M., Booth, W. H., Bush, P. B., Page, R. K., and Goetsch, D. D.,** Toxicity and bioaccumulation of pentachlorophenol in broiler chickens, *Poult. Sci.,* 59, 1018, 1980.
31. **Cooper, P.,** Getting rid of pentachlorophenol., *Food Cosmet. Toxicol.,* 17, 405, 1976.
32. **Braun, W. H., Young, J. D., Blau, G. E., and Gehring, P. J.,** The pharmacokinetics and metabolism of pentachlorophenol in rats, *Toxicol. Appl. Pharmacol.,* 41, 395, 1977.
33. **Savolainen, H. and Pekari, K.,** Neurochemical effects of peroral administration of technical pentachlorophenol, *Res. Commun. Chem. Pathol. Pharmacol.,* 23, 97, 1979.
34. **Drewes, L. R. and Remick, D. G.,** Uncoupling of brain respiratory metabolism by pentachlorophenol, *Fed. Proc. Fed. Am. Soc. Exp. Biol.,* 37, 1628, 1978.
35. **Martin-Bouyer, G., Leberton, R., Toga, M., Stolley, P. D., and Lockhart, J.,** Outbreak of accidental hexachlorophene poisoning in France, *Lancet,* 1, 91, 1982.
36. **Gump, W. S.,** Toxicological properties of hexachlorophene, *J. Soc. Cosmet. Chem.,* 20, 173, 1969.
37. **Kimbrough, R. D.,** Review of the toxicity of hexachlorophene, *Arch. Environ. Health,* 23, 119, 1971.
38. **Kimbrough, R. D.,** Review of recent evidence of toxic effects of hexachlorophene, *Pediatrics,* 51, 391, 1973.
39. **Kimbrough, R. D.,** Review of the toxicity of hexachlorophene, including its neurotoxicity, *J. Clin. Pharmacol.,* 13, 11, 1973.
40. **Kimbrough, R. D.,** Hexachlorophene: toxicity and use as an antibacterial agent, in *Essays in Toxicology,* Vol. 7, Hayes, W. J., Jr., Ed., Academic Press, New York, 1976, 99.
41. **Towfighi, J. and Gonatas, N. K.,** Hexachlorophene and the nervous system, in *Progress in Neuropathology,* Vol. 3, Zimmerman, H., Ed., Grune & Stratton, New York, 1976, 297.
42. **Towfighi, J.,** Hexachlorophene, in *Experimental and Clinical Neurotoxicology,* Spencer, P. S. and Schaumburg, H. H., Eds., Williams & Wilkins, Baltimore, 1980, 440.
43. **Kimbrough, R. D.,** Pharmacodynamics and neurotoxicity of hexachlorophene including ultrastructure of the brain lesion, *Clin. Toxicol.,* 9, 969, 1976.
44. **Lockhart, J. D.,** How toxic is hexachlorophene?, *Pediatrics,* 50, 229, 1972.
45. **Lustig, F. W.,** A fatal case of hexachlorophene (pHisoHex) poisoning, *Med. J. Aust.,* 50, 737, 1963.
46. **Pilapil, V. R.,** Hexachlorophene toxicity in an infant, *Am. J. Dis. Child,* 111, 333, 1966.
47. **Wear, J. B., Jr., Shanahan, R., and Ratliff, R. T.,** Toxicity of ingested hexachlorophene, *JAMA,* 181, 587, 1962.

48. **Martinez, A. J., Boehm, R., and Hadfield, M. G.**, Acute hexachlorophene encephalopathy: clinico-neuropathological correlation, *Acta Neuropathol.*, 28, 93, 1974.
49. **Liu, J., Wang, C., Yu, J., Wang, M., Chang, C., and Cheng, S.**, Hexachlorophene in the treatment of *Chlonorchiasis sinensis, Chinese Med. J.*, 82, 702, 1963.
50. **Gluck, L. and Wood, H.**, Effect of an antiseptic skin-care regimen in reducing staphylococcal colonization in new born infants, *N. Engl. J. Med.*, 265, 1177, 1961.
51. **Gezon, H. M., Thompson, D. J., Rogers, K. D., Hatch, T. F., and Taylor, P. M.**, Hexachlorophene bathing in early infancy, *N. Engl. J. Med.*, 270, 379, 1964.
52. **Herter, W. B.**, Hexachlorophene poisoning, *Kaiser Found. Med. Bull.*, 7, 228, 1959.
53. **Anderson, J. M., Cockburn, F., Forfar, J. O., Harkness, R. A., Kelly, R. W., and Kilshaw, B.**, Neonatal spongioform myelinopathy after restricted application of hexachlorophene skin disinfectant, *J. Clin. Pathol.*, 34, 25, 1981.
54. **Anderson, J. M., Kilshaw, B. H., and Harkness, R. A.**, Spongioform myelinopathy in premature infants, *Br. Med. J.*, 2, 175, 1975.
55. **Powell, H., Swarner, O., Gluck, L., and Lampert, P.**, Hexachlorophene myelinopathy in premature infants, *J. Pediatr.*, 82, 976, 1973.
56. **Gowdy, J. M. and Ulsamer, A. G.**, Hexachlorophene lesions in newborn infants, *Am. J. Dis. Child.*, 130, 247, 1976.
57. **Plueckhahn, V. D. and Collins, R. B.**, Hexachlorophene emulsions and antiseptic skin care of newborn infants, *Med. J. Aust.*, 1, 815, 1976.
58. **Goutières, F. and Aicardi, J.**, Accidental percutaneous hexachlorophene intoxication in children, *Br. Med. J.*, 2, 663, 1977.
59. **Shuman, R. M., Leech, R. W., and Alvord, E. C., Jr.**, Neurotoxicity of hexachlorophene in humans. I. A clinicopathological study of 248 children, *Pediatrics*, 54, 689, 1974.
60. **Shuman, R. M., Leech, R. W., and Alvord, E. C., Jr.**, Neurotoxicity of hexachlorophene in humans. II. A clinicopathological study of 46 premature infants, *Arch. Neurol.*, 32, 320, 1975.
61. **Larson, D. L.**, Studies show hexachlorophene causes burn syndrome, *Hospitals*, 42, 63, 1968.
62. **Mullick, F. G.**, Hexachlorophene toxicity. Human experience at the Armed Froces Institute of Pathology, *Pediatrics*, 51, 395, 1973.
63. **Chilcote, R., Curley, A., Loughlin, H. H., and Jupin, J. A.**, Hexachlorophene storage in a burn patient associated with encephalopathy, *Pediatrics*, 59, 457, 1977.
64. **Carroll, F. E., Salak, W. W., Howard, J. M., and Pairent, F. W.**, Absorption of antimicrobial agents across experimental wounds, *Surg., Gynecol., Obstet.*, 125, 974, 1967.
65. **Plueckhahn, V. D.**, Hexachlorophene and skin care of newborn infants, *Drugs*, 5, 98, 1973.
66. **Tyrala, E. E., Hillman, L. S., Hillman, R. E., and Dodson, W. E.**, Clinical pharmacology of hexachlorophene in newborn infants, *J. Pediatr.*, 91, 481, 1977.
67. **Bressler, R., Walson, P. D., and Fuginitti, V. A.**, Hexachlorophene in the newborn nursery. A risk benefit analysis and review, *Clin. Pediatr.*, 16, 342, 1977.
68. **Plueckhahn, V. D.**, Infant antiseptic skin care and hexachlorophene, *Med. J. Aust.*, 1, 93, 1973.
69. **Anon.**, Hexachlorophene as an antiseptic for infants, *Drug Ther., Bull.*, 10, 53, 1972.
70. **Mennuti, M. T.**, Drug and chemical risks to the fetus: occupational hazards for medical personnel, *Prog. Clin. Biol. Res.*, 36, 41, 1980.
71. **Hanig, J. P., Krop, S., Morrison, J. M., Jr., and Colson, S. H.**, Observations on hexachlorophene-induced paralysis in the cat and its antagonism by hypertonic urea, *Proc. Soc. Exp. Biol. Med.*, 152, 165, 1976.
72. **Edds, G. T. and Simpson, C. F.**, Hexaclorophene-phisohex toxicity in pups, *Am. J. Vet. Res.*, 35, 1005, 1974.
73. **Fletch, A. L., Walker, G. C., Percy, D. H., and Fletch, S. M.**, Hexachlorophene toxicity in the dog, *Univ. Toronto Med. J.*, 52, 95, 1975.
74. **Jack, E. S.**, Possible hexachlorophene poisoning in calves, *Vet. Res.*, 90, 198, 1972.
75. **Ward, B. C.**, Hexachlorophene toxicity in dogs, *Vet. Pathol.*, 12, 70, 1975.
76. **Sims, J. L. and Pfaender, F. K.**, Distribution and biomagnification of hexachlorophene in urban drainage areas, *Bull. Environ. Contam. Toxicol.*, 14, 214, 1975.
77. **Staben, P.**, The effect of hexachlorophene on the optic nerve and visual faculty in beagle dogs after prolonged dermal exposure, *Toxicol. Lett.*, 5, 77, 1980.
78. **Hall, G. A. and Reid, I. M.**, Hexachlorophene toxicity in sheep, *Lancet*, 2, 1251, 1972.
79. **Hall, G. A. and Reid, I. M.**, The effects of hexachlorophene on the nervous system of sheep, *J. Pathol.*, 114, 241, 1974.
80. **Udall, V. and Malone, J. C.**, Optic nerve atrophy after drug treatment, *Proc. Eur. Soc. Study Drug Toxicity*, 11, 2441, 1970.

81. **Udall, V.,** Drug induced blindness in some experimental animals and its relevance to toxicity, *Proc. R. Soc. Med.,* 65, 197, 1972.
82. **Thorpe, E.,** Some toxic effects of hexachlorophene in sheep, *J. Comp. Pathol.,* 79, 167, 1969.
83. **Robinson, G. R., Wagstaff, D. J., Colaianne, J. J., and Ulsamer, A. G.,** Experimental hexachlorophene intoxication in young swine, *Am. J. Vet. Res.,* 36, 1615, 1975.
84. **Tripier, M. F., Bérard, M., Toga, M., Martin-Bouyer, G., LeBreton, R., and Garat, J.,** Hexachlorophene and the central nervous system. Toxic effects in mice and baboons, *Acta Neuropathol. Suppl.,* 53, 65, 1981.
85. **Hart, E. R., Pittman, K., Drobeck, H. P., and Kurtzke, J. F.,** Recovery of infant monkeys from the effects of repeated bathing with an antibacterial scrub formulation containing hexachlorophene, *Toxicol. Appl. Pharmacol.,* 29, 117, 1974.
86. **Santolucito, J. A.,** The electroencephalogram and visual evoked potential of the squirrel monkey fed hexachlorophene, *Toxicol. Appl. Pharmacol.,* 22, 2a, 1972.
87. **Webster, H. deF., Reier, P. J., Kies, M. W., and O'Connell, M. F.,** A simple method for quantitative morphological studies of CNS demyelination: whole amounts of tadpole optic nerves examined by differential-interference microscopy, *Brain Res.,* 79, 132, 1974.
88. **Reier, P. J., Tabira, T., and Webster, H. deF.,** Hexachlorophene induced myelin lesions in the amphibian central nervous system. A freeze fracture study, *J. Neurol. Sci.,* 35, 257, 1978.
89. **Stenbäck, F.,** Hexachlorophene in mice. Effects after long term percutaneous applications, *Arch. Environ. Health,* 30, 32, 1975.
90. **Persson, L., Wingren, U., and Kristensson, K.,** Hexachlorophene induced lesions in the developing peripheral nervous system in mice, *Neuropathol. Appl. Neurobiol.,* 2, 167, 1976.
91. **Persson, L. A., Norlander, B., and Kristensson, K.,** Studies on hexachlorophene-induced myelin lesions in the trigeminal root transitional region in developing and adult mice, *Acta Neuropathol.,* 42, 115, 1978.
92. **Kimbrough, R. D. and Gaines, T. B.,** Hexachlorophene effects on the rat brain. Study of high doses by light and electron microscopy, *Arch. Environ. Health.,* 23, 114, 1971.
93. **Gellert, R. J., Wallace, C. A., Wiesmeier, E. M., and Shuman, R. M.,** Topical exposure of neonates to hexachlorophene: long-standing effects on mating behavior and prostatic development in rats, *Toxicol. Appl. Pharmacol.,* 43, 339, 1978.
94. **deJesus, P. and Pleasure, D. E.,** Hexachlorophene neuropathy, *Arch. Neurol.,* 29, 180, 1973.
95. **Lampert, P., O'Brien, J., and Garrett, R.,** Hexachlorophene encephalopathy, *Acta Neuropathol.,* 23, 326, 1973.
96. **Alder, S. and Zbinden, G.,** Use of pharmacological screening tests in subacute neurotoxicity studies of isoniazid, pyridoxine hydrochloride, and hexachlorophene, *Agents Actions,* 3, 233, 1973.
97. **Nakaue, H. S., Dost, F. N., and Buhler, D. R.,** Toxicity of hexachlorophene in the rat, *Toxicol. Appl. Pharmacol.,* 24, 239, 1973.
98. **Gaines, T. B., Kimbrough, R. D., and Linder, R. E.,** The oral and dermal toxicity of hexachlorophene in rats, *Toxicol. Appl. Pharmacol.,* 25, 332, 1973.
99. **Ulsamer, A. G., Yonder, P. D., Kimbrough, R. D., and Marzulli, F. N.,** Effect of hexachlorophene on developing rats: toxicity, tissue concentrations and biochemistry, *Food Cosmet. Toxicol.,* 13, 69, 1975.
100. **Towfighi, J., Gonatas, N. K., and McCree, L.,** Hexachlorophene induced changes in central and peripheral myelinated axons of developing and adult rats, *Lab. Invest.,* 31, 712, 1974.
101. **Pleasure, D., Towfighi, J., Silberberg, D., and Parris, J.,** Pathogenesis of hexachlorophene neuropathy. *In vivo* and *in vitro* studies, *Neurology,* 24, 1068, 1974.
102. **Lock, E. A.,** Increase in cerebral fluids in rats after treatment with hexachlorophene or triethyltin, *Biochem. Pharmacol.,* 25, 1455, 1976.
103. **Rose, A. L., Wisniewski, H. M., and Cammer, W.,** Neurotoxicity of hexachlorophene: new pathological and biochemical observations, *J. Neurol. Sci.,* 24, 425, 1975.
104. **Shuman, R. M., Leech, R. W., and Alvord, E. C., Jr.,** Neurotoxicity of topically applied hexachlorophene in the young rat, *Arch. Neurol.,* 32, 315, 1975.
105. **deJesus, P., Towfighi, J., and Snyder, D. R.,** Sural nerve conduction study in the rat: a new technique for studying experimental neuropathies, *Muscle Nerve,* 1, 162, 1978.
106. **Weiss, L. R., Williams, J. T., and Krop, S.,** Effect of hexachlorophene intoxication on learning in rats, *Toxicology,* 9, 331, 1978.
107. **Maxwell, I. C. and LeQuesne, P. M.,** Conduction velocity in hexachlorophene neuropathy: correlation between electrophysiological and histological findings, *J. Neurol. Sci.,* 43, 95, 1979.
108. **Kennedy, G. L., Jr. and Gordon, D. E.,** Histopathologic changes produced by hexachlorophene in the rat as a function of both magnitude and number of doses, *Bull. Environ. Contam. Toxicol.,* 16, 464, 1976.
109. **Kennedy, G. L., Jr., Dressler, I. A., Richter, W. R., Keplinger, M. L., and Calandra, J. C.,** Effects of hexachlorophene in the rat and their reversibility, *Toxicol. Appl. Pharmacol.,* 35, 137, 1976.

110. **Towfighi, J., Gonatas, N. K., and McCree, L.**, Hexachlorophene neuropathy in rats, *Lab. Invest.*, 29, 428, 1973.
111. **Slamovitz, T. L., Burde, R. M., and Klingele, T. G.**, Bilateral optic atrophy caused by chronic oral ingestion and topical application of hexachlorophene, *Am. J. Ophthalmol.*, 89, 676, 1980.
112. **Blockus, L. E., Chan, D. H. M., Goode, J. W., Keplinger, M. L., and Calandra, J. C.**, A possible mechanism of action of hexachlorophene intoxication, *Toxicol. Appl. Pharmacol.*, 22, 277, 1972.
113. **Towfighi, J. and Gonatas, N.**, The distribution of peroxidases in the sciatic nerves of normal and hexachlorophene intoxicated developing rats, *J. Neurocytol.*, 6, 39, 1977.
114. **Powell, H. C., Myers, R. R., Zweifach, B. W., and Lampert, P. W.**, Endoneurial pressure in hexachlorophene neuropathy, *Acta Neuropathol.*, 41, 139, 1978.
115. **Towfighi, J., Gonatas, N. K., and McCree, L.**, Hexachlorophene retinopathy in rats, *Lab. Invest.*, 32, 330, 1975.
116. **Kimmel, C. A., Moore, W., Jr., Hysell, D. K., and Stara, J. F.**, Teratogenicity of hexachlorophene in rats. Comparison of uptake following various routes of administration, *Arch. Environ. Health.*, 28, 43, 1974.
117. **Gandolfi, A. J., Dost, F. N., and Buhler, D. R.**, Absorption, excretion and metabolism of hexachlorophene in the rat and rabbit, *Fed. Proc. Fed. Am. Soc. Exp. Biol.*, 31, 605, 1972.
118. **Wit, J. G. and VanGenderen, H.**, Some aspects of the fate of hexachlorophene (2,2'methylene bis[3,4,6-trichlorophenol]) in rabbits, rats and dairy cattle, *Acta Physiol. Pharmacol. Neerl.*, 11, 123, 1962.
119. **Gandolfi, A. J. and Buhler, D. R.**, Biliary metabolites and enterohepatic circulation of hexachlorophene in rats, *Xenobiotica*, 4, 693, 1974.
120. **Caldwell, R. S., Nakaue, H. S., and Buhler, D. R.**, Biochemical lesion in rat liver mitochondria induced by hexachlorophene, *Biochem. Pharmacol.*, 21, 2425, 1972.
121. **Nakaue, H. S., Caldwell, R. S., and Buhler, D. R.**, Bisphenols: uncouplers of phosphorylating respiration, *Biochem. Pharmacol.*, 21, 2273, 1972.
122. **Cammer, W. and Moore, C. L.**, Effect of hexachlorophene on the respiration of brain and liver mitochondria, *Biochem. Biophys. Res. Commun.*, 46, 1887, 1972.
123. **Cammer, W.**, Uncoupling of oxidative phophorylation *in vitro* by the neurotoxic fragrance compound acetyl ethyl tetramethyl tetralin and its putative metabolite, *Biochem. Pharmacol.*, 29, 1531, 1980.
124. **Cammer, W., Rose, A. L., and Norton, W. T.**, Biochemical and pathological studies of myelin in hexachlorophene intoxication, *Brain Res.*, 98, 547, 1975.
125. **DeLucia, R., Medeiros, L. O., Medeiros, L. R., Aizenstein, M. L., Valle, L. B. S., and Oliveira-Filho, R. M.**, Effect of hexachlorophene on the metabolism of glucose and glutamate in rat brain, *Gen. Pharmacol.*, 9, 321, 1978.
126. **Harris, R. L., Turkus, J., and Veech, R. L.**, Effect of hexachlorophene upon intermediary metabolites, redox, and phosphorylation states in rat brain, *Pediatrics*, 54, 118, 1974.
127. **Hanig, J. P., Yoder, P., and Krop, S.**, Toxicity and effects upon rat cerebrospinal fluid pressure of hexachlorophene analogs, *Toxicol. Appl. Pharmacol.*, 37, 186, 1976.
128. **Oakley, G. P. and Shepard, T. H.**, Possible teratogenicity of hexachlorophene in rats, *Teratology*, 5, 264, 1972.
129. **Thorpe, E.**, Some pathological effects of hexachlorophene in the rat, *J. Comp. Pathol.*, 77, 137, 1967.
130. **Kennedy, G. L., Jr., Smith, S. H., Keplinger, M. L., and Calandra, J. C.**, Effect of hexachlorophene on reproduction in rats, *J. Agric. Food Chem.*, 23, 866, 1975.
131. **Kennedy, G. L., Jr., Smith, S. H., Plank, J. B., Keplinger, M. L., and Calandra, J. C.**, Reproductive and peri- and postnatal studies with hexachlorophene, *Food Cosmet. Toxicol.*, 14, 421, 1976.
132. **Alleva, F. R.**, Failure of neonatal injection of hexachlorophene to affect reproduction in hamster, *Toxicology*, 1, 357, 1973.
133. **Gellert, R. J., Wallace, C. A., Wiesmeier, E. M., and Schuman, R. M.**, Topical exposure of neonates to hexachlorophene: long-standing effects on mating behavior and prostatic development in rats, *Toxicol. Appl. Pharmacol.*, 43, 339, 1978.
134. **Bhargava, A. S., Staben, P., Nieuweboer, B., and Günzel, P.**, Effect of hexachlorophene on the coagulation process in beagle dogs, *Arzneimittelforsch*, 26, 2183, 1976.
135. **Taylor, T., Chasseaud, L. F., Down, W. H., and Medd, R. K.**, The percutaneous absorption of hexachlorophene by piglets, *Food Cosmet. Toxicol.*, 10, 857, 1972.
136. **Curley, A., Hawk, R. E., Kimbrough, R. D., Nathanson, G., and Finberg, L.**, Dermal absorption of hexachlorophene in infants, *Lancet*, 2, 296, 1971.
137. **Mavier, P., Stengel, D., and Hanoune, J.**, Inhibition of adenylate cyclase and ATPase activities from rat liver plasma membrane by hexachlorophene, *Biochem. Pharmacol.*, 25, 305, 1976.
138. **Berkley, M. C. and McGee, K. R.**, Neuropathy following exposure to a dimethylamine, salt of 2,4-D, *Arch. Int. Med.*, 111, 351, 1963.

139. **Berwick, P.,** 2,4-Dichlorophenoxyacetic acid poisoning in man: some interesting clinical and laboratory findings, *JAMA,* 214, 1114, 1970.
140. **Brandt, M. R., Jr.,** Herbatox poisoning; a brief review and report of a new case, *Ugeskr. Laeg.,* 133, 500, 1971.
141. **Foissac-Gegoux, P., Lelievre, A., Basin, B., and Warot, P.,** Polynévrite apres usage d'un deshervant: l'acid 2,4-D, *Lille Méd.,* 7, 1049, 1962.
142. **Fullerton, P. M.,** Toxic chemicals and peripheral neuropathy: clinical and epidemiological features, *Proc. R. Soc. Med.,* 62, 201, 1969.
143. **Goldstein, N. P., Jones, P. H., and Brown, J. R.,** Peripheral neuropathy after exposure to an ester of dichlorophenoxyacetic acid, *JAMA,* 171, 1306, 1959.
144. **Park, J., Darrien, I., and Prescott, L. F.,** Pharmacokinetic studies and severe intoxication with 2,4-D and mecoprop, *Proc. Eur. Soc. Toxicol.,* 18, 154, 1977.
145. **Todd, R. L.,** A case of 2,4-D intoxication, *J. Iowa Med. Sci.,* 52, 6635, 1962.
146. **Bezuglyi, V. P., Fokina, K. V., Komarova, L. I., Sivitskaia, I. I., Ilina, V. I., and Gorskaia, N. Z.,** Clinical manifestations of long-term sequels of acute poisoning with 2,4-dichlorophenoxyacetic acid, *Gig. Tr. Prof. Zabol.,* 3, 24, 1976.
147. **Dudley, A. W., Jr. and Thapar, N. T.,** Fatal human ingestion of 2,4-D, a common herbicide, *Arch. Pathol.,* 94, 270, 1972.
148. **Singer, R., Moses, M., Valciukas, J., Lilis, R., and Selikoff, I. J.,** Nerve conduction velocity studies of workers employed in the manufacture of phenoxy herbicides, *Environ. Res.,* 29, 297, 1982.
149. **Hill, E. V. and Carlisle, H.,** Toxicity of 2,4-dichlorophenoxyacetic acid for experimental animals, *J. Ind. Hyg. Toxicol.,* 29, 85, 1947.
150. **Kwieciński, H.,** Myotonia induced by chemical agents, *Crit. Rev. Toxicol.,* 8, 279, 1981.
151. **Dalgaard-Mikkelsen, S. and Poulsen, E.,** Toxicology of herbicides, *Pharmacol. Rev.,* 14, 225, 1962.
152. **Iyer, V., Whiting, M., and Fenichel, G.,** Neural influence on experimental myotonia, *Neurology,* 27, 73, 1977.
153. **Elo, H. and Ylitalo, P.,** Substantial increase in the levels of chlorophenoxyacetic acids in the central nervous system of rats as a result of severe intoxication, *Acta Pharmacol. Toxicol.,* 41, 280, 1977.
154. **Danon, J. M., Karpati, G., and Carpenter, S.,** Subacute skeletal myopathy induced by 2,4-dichlorophenoxyacetate in rats and guinea pigs, *Muscle Nerve,* 1, 89, 1978.
155. **Erwe, K.,** Studies on the analytical chemistry and toxicology of phenoxy herbicides, *Sven. Farm. Tidskr.,* 70, 837, 1966.
156. **Desi, I., Sos, J., Olasz, J., Sule, F., and Markus, V.,** Nervous system effects of a chemical herbicide, *Arch. Environ. Health,* 4, 95, 1962.
157. **Sudak, F., Essman, W. B., and Hamburgh, M.,** Antioconvulsive effects of 2,4-dichlorophenoxy-acetic acid in mice susceptible to audiogenic seizures, *Exp. Neurol.,* 6, 30, 1962.

Chapter 8

CYCLIC HALOGENATED HYDROCARBONS AND RELATED SUBSTANCES

John L. O'Donoghue

TABLE OF CONTENTS

I.	Polychlorinated Biphenyls	158
	A. Introduction	158
	B. Human Neurotoxicity	158
	C. Experimental Neurotoxicity	158
II.	Polybrominated Biphenyls	159
	A. Introduction	159
	B. Human Neurotoxicity	159
	C. Experimental Neurotoxicity	159
III.	2,3,7,8-Tetrachlorodibenzo-*p*-Dioxin	160
	A. Synonyms	160
	B. Introduction	160
	C. Human Neurotoxicity	160
	D. Experimental Neurotoxicity	161
IV.	Chlordecone	161
	A. Introduction	161
	B. Human Neurotoxicity	162
	C. Experimental Neurotoxicity	162
	D. Neuropathology	163
	E. Toxicokinetics	163
	F. Mechanism of Action	163
	G. Other Effects	164
V.	Neurotoxicity of Other Pesticides	164
References		166

I. POLYCHLORINATED BIPHENYLS

A. Introduction

Polychlorinated biphenyls (PCBs) compose a group of 209 possible isomers of unsubstituted biphenyl. Commercial materials are liquids composed of mixtures of PCBs and have been sold under a variety of trade names. Chemically, PCBs are very inert and are soluble in organic solvents but not water. Chlorodibenzofurans and chloronaphthalenes may be present in some commercial samples. Prior to restrictions placed on PCBs in 1972, they were found in heat transfer fluids, hydraulic fluids, lubricants, plasticizers, coatings, inks, carbonless copying paper, adhesives, pesticides, electric wire insulation, immersion oil for microscopy, water permeability barriers, and other products. Current uses are in capacitors and transformers.

Exposure to PCBs occur in their manufacture, through widespread environmental contamination due to their use and improper disposal, and in certain instances, by contamination of food products during processing.

B. Human Neurotoxicity

Several reviews on the toxicity of PCBs have been published.[1-6] In general, PCBs have a low order of acute toxicity but because of their high fat solubility and poor metabolism, their chronic toxicity is much higher.[4] A major target organ for PCBs is the liver.[3-7]

Evidence for human neurotoxicity due to PCBs grew out of an epidemic of chloracne-like skin disease which occurred during 1968 in Japan. More than a thousand people were affected by the disease called "Yusho" or oil disease which was believed to be caused by the leakage of Kanechlor® 400, which was mainly tetrachlorobiphenyl, from heat transfer pipes used in the processing of rice bran oil.[8-10] The contaminated oil contained 2000 to 3000 ppm of Kanechlor® 400[9] and was unusual because it contained about 5 mg/kg of polychlorinated dibenzofurans or about 250× more than usually found in PCBs.[3]

Presenting symptoms and signs were acniform eruptions, dark brownish nails, and cheese-like discharges from the meibomian glands as well as some complaints of numbness and neuralgic limbs. Murai and Kuroiwa[10] reported on the neurological status of 21 "Yusho" cases which had been selected by dermatological findings and not because of neurologic complaints. Forty-eight percent of the cases had one or more neuropathic symptoms of numbness, pain, hypoesthesia, and areflexia. A similar number of cases also had reduced nerve conduction velocity in one or more peripheral nerve. One case had decreased motor nerve conduction velocity. The clinical and electrophysiological data suggest a sensory neuropathy.[10] In follow-up examinations 12 years after the "Yusho" epidemic, sensory abnormalities were still prevalent,[11,12] although there has been a gradual diminution in general clinical symptoms.[13] Chen et al.[14] have reported sensory and nerve conduction changes in two children exposed to PCBs.

C. Experimental Neurotoxicity

Although there have been many studies performed on PCBs, there are relatively few which have examined the nervous system in a controlled manner. Tilson et al.[15] found that when female mice were exposed to 32 mg of 3,4,3',4'-tetrachlorobiphenyl (TCB) on days 10 to 16 of gestation, their offspring developed permanent neurobehavioral deficits observable in adulthood. Less affected mice were deficient in traversing a rod and in avoidance recognition. More seriously affected offspring also showed deficits in forelimb strength and visual placement and developed a spinning syndrome characterized by circling, bobbing, and hyperactivity. Spinal and cranial nerve roots had histologic malformations consisting of abnormal nerve fibers and glia.[16] It has been suggested that the underlying deficit is in dopaminergic synaptogenesis.[16,17] Aroclor 1242® also decreased norepinephrine and dopamine levels in killifish brain.[18] Aroclor 1248® fed at a concentration of 2.5 ppm in the diet of female monkeys is reported to result in locomotor hyperactivity and learning errors in their offspring which correlated with PCB body burden.

II. POLYBROMINATED BIPHENYLS

A. Introduction

Polybrominated biphenyls (PBBs) are a potentially large group of materials but only hexabromobiphenyl, octabromobiphenyl, and decabromobiphenyl isomers have been commercialized. Commercial production and use of biphenyls in the U.S. ceased in 1977 but imported finished products may contain them.[21] PBBs have been used as fire retardants in acrylonitrile-butadiene-styrene plastics, coatings, lacquers, and polyurethane foam.

PBBs like PCBs have relatively low acute toxicity but because they accumulate in the body and because they are poorly metabolized, they present a considerable risk for chronic toxicity. Reviews of PBB toxicity are available.[21-24]

B. Human Neurotoxicity

From their commercial introduction in 1970 until their accidental introduction into the Michigan food chain in 1973, very little was known about the toxicity of PBBs. In 1973, an estimated 500 to 1000 lb of Firemaster FF-1® was unintentionally mixed into animal feeds in replacement of magnesium oxide. Firemaster FF-1® is a pulverized form of Firemaster PB-6® with 2% calcium trisilicate added. Firemaster PB-6® is a mixture of about 56% 2,4,5,2',4',5'-hexabromobiphenyls, 27% 2,3,4,5,2'4',5'-heptabromobiphenyls, 11% pentabromobiphenyls, other biphenyls, and low concentrations of brominated naphthalenes.[22-28]

Results of this contamination were the destruction of approximately 1.5 million chickens, 29,800 cattle, 5920 hogs, 1470 sheep, and widespread contamination of the food chain.[21,29]

The neurologic consequences to farm families exposed to PBBs in this incident are controversial. Comparative studies between Michigan farm populations exposed to PBBs and Wisconsin farm populations unexposed to PBBs found a higher prevalence of skin, neurological, and musculoskeletal symptoms in Michigan farmers, which the authors attributed at least partly to PBBs.[30] The results of neurobehavioral performance tests suggested that there was an association in older Michigan farmers between higher serum PBB levels and neurobehavioral dysfunction;[31] but serum PBB levels in Michigan men and women exhibiting the most prominent neurologic symptoms were not higher than Wisconsin residents.[32]

Weil and colleagues[33] examined 33 children born to families on quarantined Michigan farms between 1973 and 1975. Examinations conducted in 1977 showed little differences when the children were compared to non-PBB exposed children except that there were some indications of a relationship between PBB levels and performance on developmental tests.

Stross et al.[34] did not find a correlation between the characteristic symptoms of 23 farmers exposed to PBBs and either PBB levels in serum, bile or fat, electromyograms, or nerve conduction changes. Twenty-eight PBB chemical workers with body fat PBB levels of 12.8 ppm were not neurologically impaired.[34] Reactive depression due to the many problems which affected farmers involved in the PBB incident was postulated as a cause of the symptoms.

C. Experimental Neurotoxicity

Tilson et al.[35] examined performance on a battery of neurobehavioral tests by rats and mice given 0.03 to 30 mg/kg of Firemaster FF-1® or 0.168 to 16.8 mg/kg of 2,4,5,2',4',5'-hexabromobiphenyl, the major component of Firemaster FF-1®. Dosing with these PBBs for a period of 30 days depressed motor reflexes, impaired forelimb grip strength, decreased motor activity, and decreased body weight gain. During a 30-day recovery period, rats tended to get worse but mice improved. The mechanism by which PBBs induced these effects was unclear but may involve systemic toxicity especially since body weight gain was depressed. The purified hexabromobiphenyl was less active in this study than Firemaster FF-1®, suggesting that other materials in the Firemaster were at least partially responsible for the behavioral deficits. In assays for hyperkeratotic activity, Needham et al.[36] showed that the major components of Firemaster FF-1® did not demonstrate hyperkeratotic activity but the more polar fractions did.

Operant conditioning was used to test whether or not Firemaster FF-1® at doses of 1, 3, or 6 mg/kg of body weight for 20 days would affect learning or performance of a simple discriminant task.[37] Neither of these parameters were changed by PBB administration but increases (1 mg/kg group) and decreases (6 mg/kg group) in responses were observed suggesting that PBBs may cause hyperactivity or depression depending on the administered dose. Calcium binding to synaptic membranes and calcium uptake by synaptosomes was reduced in the brains of rats receiving 1 mg/kg PBB but was not affected by either 3 or 6 mg/kg PBB.[38]

III. 2,3,7,8-TETRACHLORODIBENZO-*p*-DIOXIN

A. Synonyms   2,3,6,7-Tetrachlorodibenzo-*p*-dioxin   Dioxan
             2,3,7,8-Tetrachlorodibenzo-1,4-dioxin   TCDD
             Dioxin   TCDBD

B. Introduction

2,3,7,8-Tetrachlorodibenzo-*p*-dioxin (TCDD) is one of the most toxic materials known and is the most acutely toxic chlorinated dibenzodioxin. Single dose LD_{50} values are 22.5 µg/kg for male rats and 44.7 µg/kg for female rats.[39] TCDD is not a commercial product although it does have flame retardant and pesticidal properties.[40] It occurs as a by-product of 2,4,5-trichlorophenol synthesis when the normal reaction temperature of 180°C is exceeded. TCDD may be found in products derived from 2,4,5-trichlorophenol such as the herbicide 2,4,5-trichlorophenoxy-acetic acid.

Other chlorodibenzo-*p*-dioxins may occur in pentachlorophenol and other chlorophenol products although newer manufacturing processes have greatly reduced their formation.[41]

Exposure to TCDD and other chlorodibenzo-*p*-dioxins has occurred in the production of chlorophenols following reactor explosions, by exposure to the herbicide 2,4,5-trichlorophenoxyacetic acid (2,4,5-T), and by the improper disposal of waste oils.[40,42-45] Reviews of chloro-dibenzo-*p*-dioxin toxicity are available.[40,42-44]

C. Human Neurotoxicity

Data on TCDD neurotoxicity are primarily case reports of individuals or groups accidentally exposed. The lack of normal comparison groups in some of these reports makes evaluation of TCDD neural effects imprecise. Systemic toxicity including gastrointestinal symptoms, chloracne, porphyria cutanea tarda, hyperpigmentation, hirsutism, and increased skin fragility overshadow neurological complaints.

Following an uncontrolled reaction in a 2,4,5,-T plant in West Virginia, 228 workers were exposed to TCDD.[44] In addition to chloracne, hyperpigmentation, and cold intolerance, neurological symptoms (muscle pain, fatigue, and nervousness) were present.

Polyneuritis and sensory impairment in workers followed an accident in a trichlorophenol plant.[45,46] Thiess and Goldmann found one case of asymmetric leg paralysis and one case of hearing loss in a follow-up of the 53 exposed people.[40]

Psychopathological changes, sleep disturbances, muscle weakness, orthostatic hypotension, appetite and weight loss, abdominal pain and liver damage occurred in workers manufacturing 2,4,5,-T.[47] Telegina and Bikbulutova reported that 18 of 128 people producing 2,4,5-T developed a neurasthenic syndrome.[48] Jirśek and colleagues[49-52] followed the condition of 55 people exposed to TCDD in the manufacture of 2,4,5,-T and pentachlorophenol for periods up to 10 years. Seventeen workers had symptoms referable to the nervous system. These included weakness and pain in the legs, somnolence, insomia, and psychic abnormalities referred to as neurasthenia with vegetative symptoms. A variety of metabolic disorders of protein, lipid, and porphyrin metabolism also were identified.

Poland et al.[53] reexamined 73 men working in a 2,4,5-T factory where TCDD intoxication had previously occurred.[55] Two men had unexplained leg fatigue but no weakness on neurological examination.[53] The severity of chloracne correlated with hyperpigmentation, hirsutism, eye irritation, and a high score on the manic scale of the Minnesota Multiphasic Personality Inventory. Control populations were not studied.

Oliver[54] described cases involving three scientists working with TCDD. Two men developed chloracne after synthesizing TCDD in laboratory hoods. Two and one half years later, one of the men developed headaches, excessive fatigue, loss of vigor and drive, decreased concentration, loss of libido, hair changes, and uncharacteristic anger. The symptoms subsided in 6 months. A third man working with dioxin developed similar signs, but not chloracne, about 2 years after working with TCDD. All three men had hypercholesterolemia. The changes in blood cholesterol and the dermatological effects suggest that TCDD may have also been responsible for the neurological symptoms.

An accident at a trichlorophenol factory in Meda, Italy led to a contamination of a large, densely populated area with dioxin. Filippini et al.[56] examined 308 people living in the contaminated area by clinical and electrophysiologic means. No cases of acute polyneuropathy were identified but there was some evidence of neurological effect when the exposed population was compared to a "normal" population. The "normal" population consisted of an unexposed population with no evidence or history of disease predisposing to peripheral neuropathy. Twenty-six people had symptoms or clinical and electrophysiological signs of peripheral neuropathy while sixteen people had only electrophysiological abnormalities.

Exposure of the environment to TCDD-containing herbicides has produced controversy about the cause of neurological complaints among Vietnam veterans. Agent Orange was a herbicide containing 2,4-dichlorophenoxyacetic acid and 2,4,5-T. Extensive use of this herbicide was made in Vietnam resulting in possible widespread TCDD dispersal. Persistent peripheral sensory complaints and psychiatric problems have been associated with Agent Orange exposure.[57]

In another episode where waste oil containing TCDD was used to control dusts in horse arenas, a large number of people and animals were intoxicated with TCDD but neurological complaints were absent.[58,59]

D. Experimental Neurotoxicology

Although there are a number of studies of experimental TCDD toxicity,[39,42,43,60,61] the nervous system has not been specifically studied.

IV. CHLORDECONE

A. Introduction

Chlordecone is an organochlorine insecticide which is (structurally) related to Mirex, differing only in the substitution of a keto group in chlordecone for two chlorine atoms. Chlordecone is also known as 1,1a,3,3a,4,5,5a,5b,6-decachlorooctahydro-1,3,4-metheno-2H-cyclobuta[c,d]pentalen-2-one, decachloroketone, compound 1189, merex, and by its

abnormalities[78,84] and was directly proportional to brain chlordecone levels[84] in rodents given chlordecone.

Inhibition of synaptosomal membrane ATPase has been postulated to interfere with the uptake of central nervous system catacholamines as does ouabain, a potent inhibitor of catacholamine uptake.[101] Adrenal gland catacholamine was reduced 54% by giving rats diets containing 200 ppm of chlordecone for 8 days.[77] Epinephrine was selectively reduced 63% and norepinephrine increased 28% by this treatment. In vivo and in vitro, mouse brain synaptosomal uptake of dopamine, neorepinephrine,[105] and γ-aminobutyric acid[106] was inhibited by chlordecone. In vivo brain concentrations of Ca^{++} were depressed in an age dependent manner,[107] and in vitro, mitochondrial oxidative phosphorylation and associated Ca^{++} transport were inhibited. At high concentrations of chlordecone, synaptosomal membranes were destabilized.[108,109] Further evidence of chlordecone effects on synaptosomal membrane processes include reduced binding of ligands for specific neurotransmitter receptor sites and increased nonreceptor proteins in specific regions of the brain.[110]

G. Other Effects

While neurotoxicity initiated the investigations of chlordecone, other significant toxic effects have been studied extensively. These include neoplastic[113] and nonneoplastic hepatotoxicity,[67,73,74,111] enhancement of carbon tetrachloride hepatotoxicity,[112] testicular dysfunction,[67,74] and female reproductive toxicity.[74,85-87,94]

V. NEUROTOXICITY OF OTHER PESTICIDES

A number of pesticides, particularly organochlorine insecticides, have specific effects on the nervous system producing acute neurologic signs especially tremor and convulsions which are generally reversible. Table 1 lists several of the more common pesticides. A review by Taylor and Calabrese[114] covers the neurotoxicity of these materials.

Table 1
NEUROTOXICITY OF SOME PESTICIDES IN MAN AND LABORATORY ANIMALS

Pesticide	Effects in man	Experimental effects	Ref.
Dichlorodiphenyl trichloroethane (DDT)	Tremor, convulsions, gait disturbances, paresthesias, peripheral neuropathy, EEG changes	Tremors, convulsions, gait disturbances, excitement, weakness, EEG changes	114—116
Endrin	Incoordination, tremor, convulsions, EEG changes	Convulsions, EEG changes	114, 116
Methoxychlor	CNS excitation	Tremor, convulsions, apprehension, nervousness	114
Aldrin	Convulsions, EEG changes	Tremors, convulsions, excitement, EEG changes	114, 115
Dieldrin	Tremor, convulsions, vertigo, muscle twitching, myoclonic jerks, EEG changes	Convulsions, EEG changes	114, 115
Endosulfan	Agitation, convulsions, EEG changes	Tremors, convulsions, muscle contractions	114, 117
Kelthane	None reported	Weakness, coma, no EEG changes.	115, 116
Chlordane	Tremor, convulsions, ataxia, incoordination	Tremor, convulsions, hyperexcitable, EEG changes	114, 115
Heptachlor	Tremors, irritability	Tremors, convulsions, EEG changes	114—116
Hexachlorocyclohexane or Benzenehexachloride and its γ-isomer, Lindane	Convulsions, irritability, lack of coordination, EEG changes	Excitability, CNS stimulation, convulsions	114—116
Toxaphene	Convulsions, muscle fasiculations	Tremor, convulsions, excitability, tetanic muscle contractions	114, 115
Mirex	None reported	Hyperexcitability, weakness	114
Chlordimeform		Abnormal behavior	118
Cycloheximide		Ataxia, abnormal gait, periodic falling, and wing droop in mallards	119
N,N-diethyl-m-toluamide	Difficulty walking, slurred speech, convulsions	CNS depression, terminal convulsions, tremors	114, 120, 121
Pyrethroids	None reported	Tremor, sinuous writhing, convulsions, abnormal behavior, EEG changes	114, 122, 123

REFERENCES

1. Conference on PCB's, *Environ. Health Perspec.*, 1, 1, 1972.
2. **Hutzinger, O., Safe, S., and Zitko, V.,** *The Chemistry of PCB's*, CRC Press, Boca Raton, Fla., 1974.
3. Polychlorinated biphenyls and polybrominated biphenyls, *IARC Monogr. on the Evalution of the Carcinogenic Risk of Chemicals to Humans*, 18, 43, 1978.
4. **Kimbrough, R. D.,** The toxicity of polychlorinated polycylic compounds and related chemicals, *Crit. Rev. Toxicol.*, 2, 445, 1974.
5. **Peakall, D. B. and Risebrough, R. W.,** PCB's and their environmental effects, *Crit. Rev. Environ. Control*, 4, 469, 1973.
6. U.S. Environmental Protection Agency, PCB's in the United States. Industrial Use and Environmental Distribution, PB-252-012, National Technical Information Service, Springfield, Va., 1976.
7. **Norback, D. H. and Allen, J. R.,** Chlorinated aromatic hydrocarbon induced modifications of the hepatic endoplasmic reticulum: concentric membrane arrays, *Environ. Health Perspect.*, 1, 137, 1972.
8. **Kuratsune, M.,** An abstract of results of laboratory examinations of patients with Yusho and of animal experiments, *Environ. Health Perspect.*, 1, 129, 1972.
9. **Kuratsune, M., Yoshimura, T., Matsuzaka, J., and Yamaguchi, A.,** Epidemiologic study on Yusho, a poisoning caused by ingestion of rice oil contaminated with a commercial brand of polychlorinated biphenyls, *Environ. Health Perspect.*, 1, 119, 1972.
10. **Murai, Y. and Kuroiwa, Y.,** Peripheral neuropathy in chlorobiphenyl poisoning, *Neurology*, 21, 1173, 1971.
11. **Shibasaki, H.,** Neurological studies of patients with rice oil disease 12 years after the onset, *Fukuoka Acta Med.*, 72, 230, 1981.
12. **Toshihiko, H., Tominaga, H., Tanaka, K., Ito, H., and Kaji, M.,** Study on vibratory sensation of patients with Yusho (polychlorinated biphenyl poisoning), *Fukuoka Acta Med.*, 72, 214, 1981.
13. **Yoshihura, S. and Yoshimura, H.,** Study of Yusho: progress in this decade, *Eisei Kagaku*, 27, 144, 1981.
14. **Chen, R. C., Chang, Y. C., Chang, K. J., Lu, F. J., and Tung, T. C.,** Peripheral neuropathy caused by chronic polychlorinated biphenyls poisoning, *J. Taiwan Med. Assoc.*, 80, 47, 1981.
15. **Tilson, H. A., Davis, G. J., McLachlan, J. A., and Lucifer, G. W.,** The effects of polychlorinated biphenyls given prenatally on the neurobehavioral development of mice, *Environ. Res.*, 18, 466, 1979.
16. **Chow, S. M., Miike, T., Payne, W. M., and Davis, G. J.,** Neuropathology of «spinning syndrome» induced by prenatal intoxication with a PCB in mice, *Ann. N.Y. Acad. Sci.*, 320, 373, 1979.
17. **Agrawal, A. K., Tilson, H. A., and Bondy, S. C.,** 3,4,3',4'-Tetrachlorobiphenyl given to mice prenatally produces long-term decreases in striatal dopamine and receptor binding sites in the caudate nucleus, *Toxicol. Lett.*, 7, 417, 1981.
18. **Fingerman, S. W. and Russell, L. C.,** Effects of the polychlorinated biphenyl Aroclor 1242 on locomotor activity and on the neurotransmitters dopamine and norepinephrine in the brain of the gulf killifish, *Fundulus grandis*, *Bull. Environ. Contam. Toxicol.*, 25, 682, 1980.
19. **Bowman, R. E., Heironimus, M. P., and Allen, J. R.,** Correlation of PCB body burden with behavioral toxicology in monkeys, *Pharmacol. Biochem. Behav.*, 9, 49, 1978.
20. **Bowman, R. E., Heironimus, M. P., and Barsotti, D. A.,** Locomotor hyperactivity in PCB-exposed Rhesus monkeys, *Neurotoxicology*, 2, 251, 1981.
21. **DiCarlo, F. J., Seifter, J., and DeCarlo, V. J.,** Assessment of the hazards of polybrominated biphenyls, *Environ. Health Perspect.*, 23, 351, 1978.
22. **Damstra, T., Jurgelski, W., Jr., Posner, H. S., Vouk, V. B., Bernheim, N. J., Guthrie, J., Luster, M., and Falk, H. L.,** Toxicity of polybrominated biphenyls (PBBs) in domestic and laboratory animals, *Environ. Health Perspect.*, 44, 175, 1982.
23. Polybrominated biphenyls, *IARC Monogr. on the Evaluation of the Carcinogenic Risk of Chemicals to Humans*, 18, 107, 1978.
24. Workshop on scientific aspects of polybrominated biphenyls, *Environ. Health Perspect.*, 23, 1978.
25. **Hass, J. R., McConnell, E. E., and Haran, D. J.,** Chemical and toxicologic evaluation of Firemaster BP-6, *J. Agric. Food Chem.*, 26, 94, 1978.
26. **Moore, R. W. and Aust, S. D.,** Purification and characterization of polybrominated congeners, *Biochem. Biophys. Res. Commun.*, 84, 936, 1978.
27. **Robl, M. G., Jenkins, D. H., Wingender, R. J., Gordon, D. E., and Keplinger, M. L.,** Toxicity and residue studies in dairy animals with Firemaster FF-1 (polybrominated biphenyls), *Environ. Health Perspect.*, 23, 91, 1978.
28. **O'Keefe, P. W.,** Trace contaminants in a polybrominated biphenyl fire retardant and a search for these compounds in environmental samples, *Bull. Environ. Contam. Toxicol.*, 22, 420, 1979.
29. **Carter, L. J.,** Michigan's PBB incident: chemical mix-up leads to disaster, *Science*, 192, 240, 1976.

30. **Anderson, H. A., Lilis, R., Selikoff, I. J., Rosenman, K. D., Valciukas, J. A., and Freedman, S.,** Unanticipated prevalence of symptoms among dairy farmers in Michigan and Wisconsin, *Environ. Health Perspect.*, 23, 217, 1978.
31. **Valciukas, J. A., Lilis, R., Wolff, M. S., and Anderson, H. A.,** Comparative neurobehavioral study of a polybrominated biphenyl-exposed population in Michigan and a nonexposed group in Wisconsin, *Environ. Health Perspect.*, 23, 199, 1978.
32. **Lilis, R., Anderson, H. A., Wolff, M. S., and Petrocci, M.,** The neurotoxicity of polybrominated biphenyls: results of a medical field survey, *Ann. N.Y. Acad. Sci.*, 320, 337, 1979.
33. **Weil, W. B., Spencer, M., Benjamin, D., and Seagull, E.,** The effect of polybrominated biphenyl on infants and young children, *J. Pediatr.*, 98, 47, 1981.
34. **Stross, J. K., Smokler, I. A., Isbister, J., and Wilcox, K. R.,** The human health effects of exposure to polybrominated biphenyls, *Toxicol. Appl. Pharmacol.*, 58, 145, 1981.
35. **Tilson, H. A., Cabe, P. A., and Mitchell, C. L.,** Behavioral and neurological toxicity of polybrominated biphenyls in rats and mice, *Environ. Health Perspect.*, 23, 257, 1978.
36. **Needham, L. L., Hill, R. H., Jr., Orti, D. L., Patterson, D. G., Kimbrough, R. D., Groce, D. F., and Liddle, J. A.,** Investigation of hyperkeratoxic activity of polybrominated biphenyls in Firemaster FF-1, *J. Toxicol. Environ. Health*, 9, 877, 1982.
37. **Geller, I., Hartman, R. J., Garcia, C., and Seifter, J.,** Effects of polybrominated biphenyl on a discrimination task in rats, *Neurobehav. Toxicol.*, 1, 263, 1979.
38. **Gause, E. M., Ross, D. H., Hamilton, M. G., Leul, B. Z., Seifter, J., and Geller, I.,** Correlation of systemic and biochemical effects of PBB with behavioral effects, *Neurobehav. Toxicol.*, 1, 269, 1979.
39. **Schwetz, B. A., Norris, J. M., Sparschu, G. L., Rowe, V. K., Gehring, P. J., Emerson, J. L., and Gerbig, C. G.,** Toxicity of chlorinated dibenzo-p-dioxins, *Environ. Health Perspect.*, 5, 87, 1973.
40. International Agency for Research on Cancer, Chlorinated dibenzodioxins, in *IARC Monogr. on the Evaluation of the Carcinogenic Risk of Chemicals to Man*, 15, 1977, 41.
41. **Johnson, R. L., Gehring, P. J., Kociba, R. J., and Schwetz, B. A.,** Chlorinated dibenzodioxins and pentachlorophenol, *Environ. Health Perspect.*, 5, 171, 1973.
42. **Kimbrough, R. D.,** The toxicity of polychlorinated polycylic compounds and related chemicals, *Crit. Rev. Toxicol.*, 2, 445, 1974.
43. National Institute of Environmental Health Sciences Conference on Chlorinated Dibenzodioxins and Dibenzofurans, *Environ. Health Perspect.*, 5, 19, 1973.
44. **Firestone, D.,** The 2,3,7,8-tetrachlorodibenzo-p-dioxin problem: a review, *Ecol. Bull.*, 27, 39, 1978.
45. **Goldmann, P. J.,** Severe acute chloracne caused by trichlorophenol decomposition products, *Arbeitsmed. Sozialmed. Arbeitshyg.*, 7, 12, 1972.
46. **Goldmann, P. J.,** Severe acute chloracne, one mass intoxication caused by 2,3,6,7-tetrachlorodibenzodioxin, *Der Hautarzt.*, 24, 149, 1973.
47. **Bauer, H., Schulz, K. H., and Spiegelberg, U.,** Berufliche Vergiftungen bei der herstellung von chlorphenol-berbindungen, *Arch. Gewerbepath. Gewerbehyg.*, 18, 538, 1961.
48. **Telegina, K. A. and Bikhulatova, L. I.,** Affection of the follicular apparatus of the skin in workers occupied in production of butylether of 2,4,5-trichlorophenoxyacetic acid, *Vestn. Dermatol. Venerol.*, 44, 35, 1970.
49. **Jirasek, L., Kalenský, J., and Kubeck, K.,** Acne chlorina and porphyria cutanea tarda during the manufacture of herbicides, *Cs. Derm.*, 48, 306, 1973.
50. **Jirasek, L., Kalenský, J., Kubeck, K., Pazderova, J., and Lukáš, E.,** Acne chlorina, porphyria cutanea tarda and other manifestations of general intoxication during the manufacture of herbicides. II, *Cs. Derm.*, 49, 145, 1974.
51. **Jirasek, L., Kalenský, J., Kubeck, K., Pazaerova, J., and Lukáš, E.,** Chloracne, porphyria cutanea tarda and other herbicide induced intoxications, *Hautarzt*, 27, 328, 1976.
52. **Pazderova-Vejlupkova, J., Lukáš, E., Nemcova, M., Pickova, J., and Jirasek, L.,** Chronic poisoning by 2,3,7,8-tetrachlorodibezo-p-dioxin, *Prac. Lek.*, 32, 204, 1980.
53. **Poland, A. P., Smith, D., Metter, G., and Possick, P.,** A health survey of workers in a 2,4-D and 2,4,5-T plant with special attention to chloracne, porphyria cutanea tarda, and psychologic parameters, *Arch. Environ. Health*, 22, 316, 1971.
54. **Oliver, R. M.,** Toxic effects of 2,3,7,8-tetrachlorodibenzo-1,4-dioxin in laboratory workers, *Br. J. Ind. Med.*, 32, 49, 1975.
55. **Bleiberg, J., Wallen, M., Brodkin, R., and Applebaum, I. L.,** Industrially acquired porphyria, *Arch. Dermatol.*, 89, 793, 1964.
56. **Filippini, G., Bordo, B., Crenna, P., Massetto, N., Musicco, M., and Boeri, R.,** Relationship between clinical and electrophysiological findings and indicators of heavy exposure to 2,3,7,8-tetrachlorodibenzo dioxin, *Scand. J. Work Environ. Health*, 7, 257, 1981.
57. **Bogen, G.,** Symptoms in Vietnam veterans exposed to Agent Orange, *JAMA.*, 242, 2391, 1979.

58. **Kimbrough, R. D., Carter, C. D., Liddle, J. A., Cline, R. E., and Phillips, P. E.,** Epidemiology and pathology of a tetrachlorodibenzodioxin poisoning episode, *Arch. Environ. Health,* 32, 77, 1977.
59. **Carter, C. D., Kimbrough, R. D., Liddle, J. A., Cline, R. E., Zack, M. M., Jr., Barthel, W. F., Koehler, R. E., and Phillips, P. E.,** Tetrachlorodibenzodioxin: an accidental poisoning episode in horse arenas, *Science,* 188, 738, 1975.
60. **Harris, M. W., Moore, J. A., Vos, J. G., and Gupta, B. N.,** General biological effects of TCDD in laboratory animals, *Environ. Health Perspect.,* 5, 101, 1973.
61. **Gupta, B. N., Vos, J. G., Moore, J. A., Ziwkl, J. G., and Bullock, B. C.,** Pathologic effects of 2,3,7,8-tetrachlorodibenzo-*p*-dioxin in laboratory animals, *Environ. Health Perspect.,* 5, 125, 1973.
62. **Allen, J. R., Barsotti, D. A., Van Miller, J. P., Abrahamson, L. J., and Lalich, J. J.,** Morphological changes in monkeys consuming a diet containing low levels of 2,3,7,8-tetrachlorodibenzo-*p*-dioxin, *Food Cosmet. Toxicol.,* 15, 401, 1977.
63. **Lofgren, C. S., Bartlett, F. J., Stringer, C. E., Jr., and Banks, W. A.,** Imported fire ant and toxic bait studies: further tests with granulated mirex-soybean oil bait, *J. Econ. Entomol.,* 57, 695, 1964.
64. **Sandrock, K., Bjeniek, D., Klein, W., and Kirte, E.,** Bertäge zur ökologischen chemie. LXXXVI. Isolie rung und structurauf — klaurung von Kelevan-^{14}C-metabolism und bilanz in karoffelin und boden, *Chemosphere,* 5, 199, 1974.
65. **Taylor, J. R., Selhorst, J. B., and Calabrese, V. P.,** Chlordecone, in *Experimental and Clinical Neurotoxicology,* Spencer, P. S. and Schaumburg, H. H., Eds., Williams & Wilkins, Baltimore, 1980, 407.
66. **Huff, J. E. and Gerstner, H. B.,** Kepone: a literature survey, *J. Environ. Pathol. Toxicol.,* 1, 377, 1978.
67. **Cannon, S. B., Veazey, J. M., Jr., Jackson, R. S., Burse, V. W., Hayes, C., Straub, W. E., Landrigan, P. J., and Liddle, J. A.,** Epidemic Kepone poisoning in chemical workers, *Am. J. Epidemiol.,* 107, 529, 1978.
68. **Carlson, D. A., Konyha, K. D., Wheeler, W. B., Marshall, G. P., and Zaylskie, R. G.,** Mirex in the environment: its degradation to Kepone and related compounds, *Science,* 194, 939, 1976.
69. **Taylor, J. R., Martinez, A. J., Houff, S. A., Harbison, J. W., Selhorst, J. B., and Jackson, R. S.,** Neurologic disorder induced by Kepone: preliminary report, *Neurology,* 26, 358, 1976.
70. **Taylor, J. R., Selhorst, J. B., Houff, S. A., and Martinez, A. J.,** Chlordecone intoxication in man. I. Clinical observations, *Neurology,* 28, 626, 1978.
71. **Guzelian, P. S.,** Comparative toxicology of chlordecone (Kepone) in humans and experimental animals, *Annu. Rev. Pharmacol. Toxicol.,* 22, 89, 1982.
72. **Sanborn, G. E., Selhorst, J. B., Calabrese, V. P., and Taylor, J. R.,** Pseudotumor cerebri and insecticide intoxication, *Neurology,* 29, 1222, 1979.
73. **Guzelian, P. S.,** Therapeutic approaches for chlordecone poisoning in humans, *J. Toxicol. Environ. Health,* 8, 757, 1981.
74. **Cannon, S. B. and Kimbrough, R. D.,** Short-term chlordecone toxicity in rats including effects on reproduction, pathological organ changes, and their reversibility, *Toxicol. Appl. Pharmacol.,* 47, 469, 1979.
75. **Larson, P. S., Egle, J. L., Jr., Hennigar, G. R., Lane, R. W., and Borzelleca, J. F.,** Acute, subchronic, and chronic toxicity of chlordecone, *Toxicol. Appl. Pharmacol.,* 48, 29, 1979.
76. **Egle, J. L., Jr., Guzelian, P. S., and Borzelleca, J. F.,** Time course of the acute toxic effects of sublethal doses of chlordecone (Kepone), *Toxicol. Appl. Pharmacol.,* 48, 533, 1979.
77. **Baggett, J. M., Thureson-Klein, A., and Klein, R. L.,** Effects of chlordecone on the adrenal medulla of the rat, *Toxicol. Appl. Pharmacol.,* 52, 313, 1980.
78. **Jordan, J. E., Gice, T., Mishra, S. K., and Desaiah, D.,** Acute chlordecone toxicity in rats: a relationship between tremor and ATPase activities, *Neurotoxicology,* 2, 355, 1981.
79. **Mehendal, H. M.,** Onset and recovery from chlordecone- and mirex-induced hepatobiliary dysfunction, *Toxicol. Appl. Pharmacol.,* 58, 132, 1981.
80. **Curtis, L. R. and Mehendale, H. M.,** The effect of Kepone pretreatment on biliary excretion of xenobiotics in the male rat, *Toxicol. Appl. Pharmacol.,* 47, 295, 1979.
81. **Huang, T. P., Ho, I. K., Mehendale, H. M., and Hume, A. S.,** Neurotoxicity induced by oral administration of Kepone (chlordecone) in the mouse, *Fed. Proc. Fed. Am. Soc. Exp. Biol.,* 38, 845, 1979.
82. **Huang, T. P., Ho, I. K., and Mehendale, H. M.,** Assessment of neurotoxicity induced by oral administration of chlordecone (Kepone®) in the mouse, *Neurotoxicology,* 2, 113, 1981.
83. **Huang, T. P., Ho, I. K., and Mehendale, H. M.,** Correlation between neurotoxicity and chlordecone (Kepone) levels in brain and plasma in the mouse, *Neurotoxicology,* 2, 373, 1981.
84. **Ho, I. K., Fujimori, K., Huang, T. P., and Chang-Tusi, H.,** Neurochemical evaluation of chlordecone toxicity in the mouse, *J. Toxicol. Environ. Health,* 8, 701, 1981.
85. **McFarland, L. A. and Lacy, P. B.,** Physiologic and endocrinologic effects of the insecticide Kepone in the Japanese quail, *Toxicol. Appl. Pharmacol.,* 15, 441, 1969.
86. **Eroschenko, V. P. and Place, T. A.,** Prolonged effects of Kepone on strength and thickness of egg shells from Japanese quail fed different calcium level diets, *Environ. Pollut.,* 13, 255, 1977.

87. **Eroschenko, V. P.,** Estrogenic activity of the insecticide chlordecone in the reproductive tract of birds and mammals, *J. Toxicol. Environ. Health,* 8, 731, 1981.
88. **Sherman, M. and Ross, E.,** Acute and subacute toxicity of insecticides to chicks, *Toxicol. Appl. Pharmacol.,* 3, 521, 1961.
89. **Couch, J. A., Winstead, J. T., and Goodman, L. R.,** Kepone induced scoliosis and its histological consequences in fish, *Science,* 197, 585, 1977.
90. **Hansen, D. J., Goodman, L. R., and Wilson, A. J., Jr.,** Kepone chronic effects on embryo, fry, juvenile and adult sheepshead minnows, *Cyprinodon variegatus, Chesapeake Sci.,* 18, 227, 1977.
91. **Dietz, D. D. and McMillan, D. E.,** Comparative effects of Mirex and Kepone on schedule-controlled behavior in the rat. I. multiple fixed-ratio 12 fixed-interval 2-min schedule, *Neurotoxicology,* 1, 369, 1979.
92. **Dietz, D. D. and McMillian, D. E.,** Comparative effects of Mirex and Kepone on schedule-controlled behavior in the rat. II. Spaced-responding, fixed-ratio, and unsignalled avoidance schedules, *Neurotoxicology,* 1, 387, 1979.
93. **Chernoff, N. and Rogers, E. H.,** Fetal toxicity of Kepone in rats and mice, *Toxicol. Appl. Pharmacol.,* 38, 189, 1976.
94. **Rosenstein, L., Brice, A., Rogers, N., and Lawrence, S.,** Neurotoxicity of Kepone in perinatal rats following in utero exposure, *Toxicol. Appl. Pharmacol.,* 41, 142, 1977.
95. **Martinez, A. J., Taylor, J. R., Dyck, P. J., Houff, S. A., and Isaacs, E.,** Chlordecone intoxication in man. II. Ultrastructure of peripheral nerves and skeletal muscle, *Neurology,* 28, 631, 1978.
96. **Martinez, A. J., Taylor, J. R., Isaacs, E., Dyck, P. J., and Houff, S. A.,** Kepone poisoning: ultrastructure of nerves and skeletal muscles, *J. Neuropathol. Exp. Neurol.,* 35, 323, 1976.
97. **Shah, P. V., Monroe, R. J., and Guthrie, F. E.,** Comparative rates of dermal penetration of insecticides in mice, *Toxicol. Appl. Pharmacol.,* 59, 414, 1981.
98. **Egle, J. L., Jr., Fernandez, S. B., Guzelian, P. S., and Borzelleca, J. F.,** Distribution and excretion of chlordecone (Kepone) in the rat, *Drug Metab. Dispose.,* 6, 91, 1978.
99. **Boylan, J. J., Egle, J. L., and Guselian, P. S.,** Cholestyramine: use as a new therapeutic approach for chlordecone (Kepone) poisoning, *Science,* 199, 893, 1978.
100. **Fariss, M. W., Blanke, R. V., Sandy, J. J., and Guzelian, P. S.,** Demonstration of major metabolic pathways for chlordecone (Kepone) in humans, *Drug Metabl. Dispos.,* 8, 434, 1980.
101. **Desaiah, D.,** Interaction of chlordecone with biological membranes, *J. Toxicol. Environ. Health,* 8, 719, 1981.
102. **Desaiah, D., Mehendale, H. M., and Ho, I. K.,** Kepone inhibition of mouse brain synaptosomal ATPase activities, *Toxicol. Appl. Pharmacol.,* 45, 268, 1978.
103. **Desaiah, D., Gilliland, T., Ho, I. K., and Mehendale, H. M.,** Inhibition of mouse brain synaptosomal ATPase and oubain binding by chlordecone, *Toxicol. Lett.,* 6, 4, 1980.
104. **Bansal, S. K. and Desaiah, D.,** Effects of chlordecone and its structural analogs on *p*-nitrophenyl phosphatase, *Toxicol. Lett.,* 12, 83, 1982.
105. **Chang-Tsui, Y.-Y. and Ho, I. K.,** Effect of Kepone® (chlordecone) on synaptosomal catecholamine uptake in the mouse, *Neurotoxicology,* 1, 643, 1980.
106. **Chang-Tsui, Y.-Y. and Ho, I. K.,** Effects of Kepone (chlordecone) on synaptosomal gamma amino butyric acid uptake in the mouse, *Neurotoxicology,* 1, 357, 1979.
107. **Hoskins, B. and Ho, I. K.,** Chlordecone-induced alterations in content and subcellular distribution of calcium in mouse brain, *J. Toxicol. Environ. Health,* 9, 535, 1982.
108. **End, D. W., Carchman, R. A., Ameen, R., and Dewey, W. L.,** Inhibition of rat brain mitochondrial calcium transport by chlordecone, *Toxicol. Appl. Pharmacol.,* 51, 189, 1979.
109. **End, D. W., Carchman, R. A., and Dewey, W. L.,** Neurochemical correlates of chlordecone neurotoxicity, *J. Toxicol. Environ. Health,* 8, 707, 1981.
110. **Seth, P. K., Agrawal, A. K., and Bondy, S. C.,** Biochemical changes in the brain consequent to dietary exposure of developing and mature rats to chlordecone (Kepone), *Toxicol. Appl. Pharmacol.,* 59, 262, 1981.
111. **Mehendale, H. M.,** Chlordecone-induced hepatic dysfunction, *J. Toxicol. Environ. Health,* 8, 743, 1981.
112. **Davis, M. E. and Mehendale, H. M.,** Functional and biochemical correlates of chlordecone exposure and its enhancement of CCl4 hepatotoxicity, *Toxicology,* 15, 91, 1980.
113. **Reuber, M. D.,** Carcinomas of the liver in rats ingesting kepone, *Neoplasma,* 26, 231, 1979.
114. **Taylor, J. R. and Calabrese, V. P.,** Organochlorine and other insecticides, in *Handbook of Clinical Neurology,* Vol. 36, Vinken, P. J. and Bruyn, G. W., Eds., North Holland, Amsterdam, 1979, chap. 16.
115. **Deichmann, W. B.,** Halogenated cyclic hydrocarbons, Part 2, in *Patty's Industrial Hygiene and Toxicology,* Vol. 2B, 3rd rev. ed., Clayton, G. D. and Clayton, F. E., Eds., John Wiley-Interscience, New York, 1981, chap. 49.
116. **Joy, R. M.,** Convulsive properties of chlorinated hydrocarbon insecticides in the cat central nervous system, *Toxicol. Appl. Pharmacol.,* 35, 95, 1976.

I. ACRYLAMIDE

A. Synonyms
2-Propenamide
Acrylic amide
Akrylamid

B. Introduction

Monomeric acrylamide ($CH_2=CHCONH_2$) is a white, crystalline solid which is highly soluble in water. Monomer production has been increasing rapidly since its introduction in 1954 and is approaching 10^8 pounds annually.[1] It is used as a chemical intermediate, but its greatest use is in the production of polymers. Polyacrylamides are very useful as flocculation agents to clarify waste and drinking water and as stabilizers in paper. Acrylamide pumped into the soil stabilizes soil in tunneling and drilling operations. Commercial grout mixtures may contain N-substituted acrylamides and catalysts, such as dimethylaminopropionitrile, which are also neurotoxic.

C. Human Neurotoxicity

Cases of human neurotoxicity have all been associated with exposure to the acrylamide monomer and not polyacrylamide, although it may contain monomer. All cases of neurotoxicity have involved industrial populations working with the monomer except for a family intoxicated by drinking and bathing in well water contaminated during soil injection.[1-20] Absorption of acrylamide occurs by dermal, intestinal, and pulmonary routes. Generally, industrial poisonings have been thought to be due to skin absorption although poisonings have occurred even with protective clothing. Whether these cases are due to breaks in work practice or to inhalation of monomer is not clear. Women, men, and children are susceptible to acrylamide. An exact count of the number of reported cases is not available but numbers are in the range of 50 to 60 people. Human neurotoxicity is due to repeated exposures to relatively low amounts of monomer with a latency period of approximately 2 months before symptoms appear.

Symptoms include easy fatigability, difficulty climbing stairs, weakness in the hands and feet, dizziness, tremor, and ataxia. Loss of deep tendon reflexes, numbness in the feet, abnormal distal sensation such as tingling or cold, and increased sensitivity to touch occur early. Secondary muscle atrophy, particularly in the hands and feet, and weight loss are later effects. A sensory and motor peripheral neuropathy is the usual manifestation of acrylamide, but in some cases urinary retention occurs, indicating an effect on the autonomic nervous system. In other individuals, central nervous system effects are apparent. These include abnormal sleepiness, poor memory, confusion, hallucinations, slurred speech, hyperactive reflexes, abnormal behavior, positive Romberg sign, abnormal EEG patterns, and changes in the visual fields.[2,6,10] Taken together, these findings indicate that acrylamide has a much wider effect on the nervous system than simply peripheral neuropathy, and at least in some circumstances autonomic and central nervous system damage results. Recovery from toxicity is described as good but may take months to a couple of years.[2-10]

D. Experimental Neurotoxicity

Neurotoxicity in laboratory animals was recognized prior to large scale production of acrylamide.[11-14] Dogs, cats, rats, mice, monkeys, and chickens are susceptible.[11-15] This may indicate that the effect(s) of acrylamide is on a process(es) or structure(s) which is phylogenetically common or fundamental to the nervous system. Single doses of acrylamide produce severe acute dysfunction of the nervous system but not central-peripheral neurotoxicity as is seen in man.[11,13] Single doses of 0.075 g/kg to 5 g/kg, followed by a latency period of 12 hr to 0.25 hr depending on the dose, caused cats to show ataxia, tremors, weakness, signs of mass sympathetic discharge, behavior suggestive of hallucinations, and tonic-clonic seizures.[11] Repeated seizures were frequently lethal. Rats receiving doses near the LD_{50} showed tremors and then either completely recovered or died.[14]

Repeated doses result in a different clinical pattern. Fullerton and Barnes[14] showed that doses of 100 mg/kg of acrylamide could be at least 2 weeks apart and still produce peripheral neuropathy. In laboratory animals, ataxia, abnormal posture, weakness, particularly in the hindlimbs, paralysis, and enlarged urinary bladders are prominent signs. In severe cases, the forelimbs and cranial musculature (hoarseness and dysphagia) can be affected. Among the different species, the monkey is more susceptible than the cat, and chickens have a variable response.[12,15,16] The no-effect level is between 0.2 to 1 mg/kg/day for 93 days in the rats and 0.3 to 1 mg/kg/day for 1 year in cats.[17,18]

Older rats appear to be more sensitive to acrylamide than young animals. Fullerton and Barnes[14] found that 52-week-old rats given three doses of 100 mg and 26 week-old rats given four doses had a severe neuropathy where 5- and 8-week-old rats were only mildly affected. Fetal rats appear resistant to acrylamide even at doses which result in maternal neurotoxicity.[19] The only reported effect of acrylamide on fetal rats is a change in the affinity and the number of dopamine binding sites in the brain.[20]

E. Neuropathology

Morphological changes due to acrylamide are primarily based on animal studies since there are few pathology reports of human cases. Early studies on acrylamide focused on physiological effects on the central nervous system since no lesions were identified and because clinical signs of ataxia especially in cats were thought to have a central origin.[11,13] Fullerton and Barnes[14] first described peripheral nerve axonal degeneration as a major effect of acrylamide and noted its similarity to the "dying-back" neuropathies. Fullerton and Barnes,[14] and Hopkins[21] found that there was a correlation between the length of the axon and the severity of axonal damage, the distal end of the longest axons being more affected than the proximal ends. Additional work has shown that the most affected axons may not be the longest but rather long axons which have the greatest diameter.[22-28] Sensory axons appear to be much more sensitive to acrylamide than motor axons, particularly the sensory axons supplying Pacinian corpuscles and muscle spindle fibers.[23,28] It is the greater sensitivity of these specialized receptors which appears to account for the early loss of vibration sensitivity and tendon reflexes and electrophysiological changes in these organs. Based on early sensory abnormalities, recommendations for sensory screening of workers have been made (Volume I, Chapter 1).

Axonal changes occurring in acrylamide-treated animals have been described as focal axonal swelling with increased numbers of neurofilaments, abnormal accumulations of organelles, and invaginations of adjacent myelin sheath into the axons.[22,28-32] Later effects include myelin retraction and degeneration and axonal fragmentation. From the limited number of reports in man, peripheral axonal pathology appears to be similar to the experimental situation.[3,5]

While axonal pathology is most severe in large myelinated axons, the whole spectrum of axons is probably affected, including unmyelinated axons.[33,34] Effects on unmyelinated axons account for urinary bladder distension or urinary retention observed in animals and some human cases, abnormalities in penile retraction in animals, sympathetic anomalies in the hands and feet of workers, and megaesophagus in experimental dogs.[14,33,34]

While initial concern for acrylamide neurotoxicity on the central nervous system shifted to the peripheral effects, it is again shifting back to central effects with the observation that axons in the spinal cord and brain are affected at the same time peripheral axons are damaged.[22,23,29] Damage to central axons may account for behavioral, visual, and memory changes reported in a portion of the clinical cases.[9] The role of neuronal damage is also being reevaluated because of reports detailing cerebellar Purkinje cell and sensory neuron alteration or degeneration.[35-37]

F. Toxicokinetics

Monomeric acrylamide is readily absorbed by all routes of exposure. The route of administration appears to have no effect on the onset of neurotoxicity. Kuperman[11] showed that acrylamide-induced ataxia occurred in cats at the same total dose (102 mg/kg) and after equivalent latencies no matter which route was used.

Distribution of acrylamide throughout total body water occurs within minutes of an i.v. dose and blood concentrations rapidly decline with a half-life of <2 hr.[38] Approximately 60% of a dose is excreted by the kidneys as two metabolites and a small amount of acrylamide.[39,40] Urinary excretion following multiple doses is nearly constant day to day.[40] Following a single i.v. dose, rats excreted ~6% of the dose by exhalation as carbon dioxide within 8 hr.[39]

C^{14}-labeled acrylamide can be identified in a major organs, presumably protein-bound.[39] In the liver, acrylamide is conjugated with glutathione and excreted in the bile.[38] Binding to sulfhydryls, both protein and nonprotein, occurs equally both with acrylamide and N-(hydroxymethyl) acrylamide, an analogue which is less neurotoxic, suggesting that sulfhydryl binding per se does not correlate with neurotoxicity.[39] None the less, sulfhydryl binding is an important route of detoxification since diethyl maleate, which depletes glutathione levels, reduces the time of onset of acrylamide neuropathy.[41]

Tissue binding may play a very important role in the chronic neuropathy induced by acrylamide. Hashimoto and Aldridge[39] found that radiolabeled acrylamide persists in brain and spinal cord for 14 days correlating well with Fullertons and Barnes[14] report that demonstrated cumulative doses of acrylamide could be spaced as far apart as 14 days and still result in neuropathy. Neuropathy was also associated with a plateauing of red blood cell acrylamide at 400 μg Eq/g.[40] This occurred after 9 days of exposure of rats to water providing 30 mg/kg/day acrylamide.

Hepatic metabolism may result in activation or detoxification of acrylamide but outside of conjugation with glutathione, little data are available. Repeated injections of 30 mg/kg of acrylamide for 2 weeks did not affect hepatic glucuronyl transferase, total hepatic protein, microsomal protein, or hepatic/body weight ratios.[42] Dixit et al.[43] reported that 50 mg/kg acrylamide for 5 days inhibited hepatic glutathione-S-transferase activity in an age-dependent manner.

Effects of modifiers of hepatic function are sometimes contradictory. Agrawal et al.[44] found that SKF-525A and methylmercuric chloride, blockers of hepatic mixed function oxidases, blocked acrylamide enhancement of striatal dopamine receptor activity. SKF-525A has also been reported to increase the acute toxicity of acrylamide.[16]

Pretreatment with phenobarbital or DDT, enhancers of hepatic mixed function oxidase, is reported not to affect the onset or recovery from acrylamide neuropathy in one study and to delay the onset of a decrement in rotarod performance in another study.[15,16] The reason for this difference is unclear.

G. Mechanism(s) of Action

A great deal of work has been performed trying to identify the mechanism(s) of action of acrylamide. To date, no mechanism has received general acceptance. Interference with axonal transport mechanisms has been reported (see Droz and Chretien[45] for review) with acrylamide intoxication. Whether alteration of axonal flow is a cause or an effect of acrylamide neurotoxicity and what underlying reactions result in alterations of axonal flow are receiving extensive study. Three main mechanisms have received much attention, i.e., interference with vitamin activity, protein alterations, and interference with axonal glycolysis.

Kaplan et al.[16] noted a resemblance between the structure of nicotinamide and a potential dimer of acrylamide and suggested that the greater sensitivity of cats to acrylamide might be based on their inability to convert tryptophan to nicotinamide. Rats deficient in pyridoxine or thiamin were no more susceptible to acrylamide than rats fed normal diets.[16] Supplementation with thiamin disulfide, pyridoxine, and cobalamin reduced the severity of neuropathic lesions in rats and thiamin propyl disulfide alone delayed the onset of ataxia in rabbits receiving acrylamide.[46,47] It was not clear whether the effects of thiamin disulfide were due to its vitamin activity or due to the presence of sulfur moieties to which acrylamide may bind nonspecifically. Supplementation with vitamin A and E did not affect the onset of neuropathy or speed recovery.[15]

Acrylamide binding to proteins has been studied closely. Binding to neural protein and protein in other major organs is long lasting.[39,48] Protein synthesis, determined by in vivo incorporation of valine and leucine, was decreased prior to onset of neuropathy.[49] A similar change was found with methylene bisacrylamide which is much less neurotoxic than acrylamide, implying that interference in protein metabolism does not necessarily lead to neuropathy.[49]

Interference with glycolysis or energy metabolism has been proposed as a mechanism by which several axonotoxic chemicals might exert their effects.[50] In vitro studies using tissues from rats previously treated with acrylamide failed to show inhibition of oxygen uptake by brain slices, changes in pyruvate and lactate levels, or effects on oxidative phosphorylation in rat liver mitochondria.[39] These results may not bear directly on whether acrylamide interferes with energy metabolism since the tissues selected are not those most affected by acrylamide, metabolic activation of acrylamide may be necessary, although no direct proof of this exists, and in vitro responses of acrylamide on other areas such as protein metabolism do not reproduce in vivo effects. In vitro and in vivo (cats and rats) studies of acrylamide have shown inhibition of neuron specific enolase (NSE) and glyceraldehyde-3-phosphate dehydrogenase activities. NSE is inhibited both in the central and peripheral nervous systems of acrylamide-treated cats, particularly on the distal region of peripheral nerves which are the most severely damaged by acrylamide.[51] Of particular interest would be whether or not acrylamide inhibition of glycolytic enzymes occurs prior to neuropathy and whether nonneurotoxic analogues of acrylamide produce similar effects.

H. Other Effects

Few effects other than neurotoxicity have been reported, although acrylamide appears to affect the general metabolism causing reduction in body weight gain and decreased feed efficiency conversion to body mass in rats. Testicular germinal cell atrophy occurs in chronically exposed rats.[13,17] In man, dermatitis affecting the palms of the hands and soles of the feet is commonly reported prior to clinical neuropathy.

II. RELATED SUBSTANCES

Table 1 is a list of chemicals which have been examined for acrylamide-like effects. N-hydroxymethylacrylamide, N-methylacrylamide, and N,N-dimethylacrylamide produce a typical acrylamide neuropathy but are approximately 30%, 13%, and 4% as potential as acrylamide, respectively.[15] N-N-Dimethylacrylamide produces tremors, spasticity, and difficulty walking in cats but not a typical acrylamide syndrome.[52] N-Isopropylacrylamide at a cumulative dose of 840 mg/kg over 18 days results in paralysis of the hindlimbs and head tremors in cats.[52] N,N'-Methylenebisacrylamide has been used as a nonneurotoxic analogue of acrylamide because it does not result in clinical signs of neurotoxicity, but Jennekens, as reported by Schotman et al.,[49] found minor changes in the peripheral motor nerve endings in rats which had received a cumulative dose of 3000 mg/kg.

Table 1
SUBSTANCES TESTED FOR ACRYLAMIDE-LIKE EFFECTS

Name	Formula	Clinical neurotoxicity	Ref.
Acrylamide	$CH_2=CHCONH_2$	+	11—15, 58—60
N-Methylacrylamide	$CH_2=CHCONHCH_3$	+	15
N-Hydroxymethylacrylamide (methylolacrylamide)	$CH_2=CHCONHCH_2OH$	+	15, 52, 53, 57
N-Isopropylacrylamide	$CH_2=CHCONHCH(CH_3)_2$	+	4, 57
N,N-Dimethylacrylamide	$CH_2=CHCON(CH_3)_3$	+	52
		−	57
N,N-Diethylacrylamide	$CH_2=CHCON(C_2H_5)_2$	+	15
		−	57
Diacetone acrylamide	$CH_2=CHCONHC(CH_3)_2CH_2COCH_3$	−	57
N-t-Butylacrylamide	$CH_2=CHCONCH(CH_3)_3$	−	52, 57
N-t-Octylacrylamide	$CH_2=CHCON(CH_2)_4C(CH_3)_3$	−	52, 57
N-Isobutyoxymethylacrylamide	$CH_2=CHCONHCH_2OCH_2CH(CH_3)_2$	−	57
N,N-Pentamethyleneacrylamide	$CH_2=CHCON\langle C_5H_{10}\rangle$	−	15
N,N'-Methylene-bis-acrylamide	$(CH_2=CHCONH)_2CH_2$	+[a]	49
		−	15, 52, 57
N-N-Bis acrylamidoacetic acid	$(CH_2=CHCONH)_2CHCOOH$	−	15
3,3-Iminodipropionamide	$NH(CH_2CH_2CONH_2)_2$	−	15
S-β-Propionamide glutathione	$Glutathion-SCH_2CH_2CONH_2$	−	15
2-Methylacrylamide	$CH_2=C(CH_3)CONH_2$	−	52, 53
		+	44, 45
Allylacetamide	$CH_2=CHCH_2CONH_2$	−	53
Iodoacetamide	ICH_2CONH_2	−	57
Crotonamide	$CH_3CH=CHCONH_2$	−	53, 57
Senecioic acid amide	$(CH_3)_2C=CHCONH_2$	−	53
Sodium acrylate	$CH_2=CHCOONa$	−	53
Methyl methacrylate	$CH_2=C(CH_3)COOCH_3$	−	15
Acrylonitrile	$CH_2=CHCN$	−	53
Ethylcrotonate	$CH_3CHCOOC_2H_5$	−	15

[a] Jennekens[30,49] reported minor histological lesions in motor endings of rats but no clinical neurotoxicity.

In addition to being neurotoxic themselves, N-hydroxymethylacrylamide, N-methylacrylamide, and N,N-diethylacrylamide decrease the time to onset of acrylamide neurotoxicity when combined exposure occurs.[15] Neurotoxicity is not due to their metabolism to acrylamide.[15]

Any modification of acrylamide's structure results in a less potent product. Addition to the amide group decreased potency and, if the addition is sufficiently long, results in loss of activity. Increasing the spacing between the vinyl bond and the amide group (i.e., allyl acetamide) or substitution of methyl or other groups adjacent to the vinyl bond (i.e., crotonamide) results in loss of activity suggesting that other unsaturated amides are not likely to have biological activity similar to acrylamide. During chronic toxicity studies on a food additive, 2-(2-furyl)-3-(5-nitro-2-furyl) acrylamide, male rats given diets containing 0.4% of this compound developed ataxia and muscle atrophy after about 4 months.[54] A study of workers exposed to diacetone acrylamide, a lubricating oil additive, did not find sufficient evidence to link it to complaints of nervous system dysfunction.[55]

REFERENCES

1. National Institute of Occupational Safety and Health, Criteria Document: Recommendations for an Occupational Exposure Standard for Acrylamide, Department of Health Education, and Welfare, Publ. No. 77—112, Washington, D.C., 1976.
2. **Garland, T. A. and Patterson, M. W. H.,** Six cases of acrylamide poisoning, *Br. Med. J.,* 4, 134, 1967.
3. **Davenport, J. G., Farrell, D. F., and Sumi, M. K.,** "Giant axonal neuropathy" cased by industrial chemicals: neurofilamentous axonal masses in man, *Neurology,* 26, 919, 1976.
4. **Mapp, C., Mazzotla, M., Bartolucci, G. B., and Fabbri, L.,** Nervous system disease caused by acrylamide: first cases in Italy, *Med. Lav.,* 68, 1, 1968.
5. **Fullerton, P. M.,** Electrophysiological and histological observations on peripheral nerves in acrylamide poisoning in man, *J. Neurol. Neurosurg. Psychiatry,* 32, 186, 1969.
6. **Takahashi, M., Ohara, T., and Hashimoto, K.,** Electrophysiological study of nerve injuries in workers handling acrylamide, *Int. Acrh. Arbeitsmed.,* 28, 1, 1971.
7. **Kesson, D. M., Baird, A. W., and Lawson, D. H.,** Acrylamide poisoning, *Postgrad. Med. J.,* 53, 16, 1977.
8. **Spencer, P. S. and Schaumburg, H. H.,** A review of acrylamide neurotoxicity. I. Properties, uses and human exposure, *Can. J. Neurol. Sci.,* 1, 143, 1974.
9. **Schaumburg, H. H. and Spencer, P. S.,** Clinical and experimental studies of distal axonopathy — a frequent form of brain and nerve damage produced by environmental chemical hazards, *Ann. N. Y. Acad. Sci.,* 329, 14, 1979.
10. **Igisu, H., Goto, I., Kawamura, Y., Kato, M., and Izumi, K.,** Acrylamide encephaloneuropathy due to well water pollution, *J. Neurol. Neurosurg. Psychiatry,* 38, 581, 1975.
11. **Kuperman, A. S.,** Effects of acrylamide on the central nervous system of the cat, *J. Pharmacol. Exp. Ther.,* 123, 180, 1958.
12. **Leswing, R. J. and Ribelin, W. E.,** Physiologic and pathologic changes in acrylamide neuropathy, *Arch. Environ. Health,* 18, 23, 1969.
13. **McCollister, D. D., Oyen, F., and Rowe, V. K.,** Toxicology of acrylamide, *Toxicol. Appl. Pharmacol.,* 6, 172, 1964.
14. **Fullerton, P. M. and Barnes, J. M.,** Peripheral neuropathy in rats produced by acrylamide, *Br. J. Ind. Med.,* 23, 210, 1966.
15. **Edwards, P. M.,** Neurotoxicity of acrylamide and its analogues and effects of these analogues and other agents on acrylamide neuropathy, *Br. J. Ind. Med.,* 32, 31, 1975.
16. **Kaplan, M. L., Murphy, S. D., and Gilles, F. H.,** Modification of acrylamide neuropathy in rats by selected factors, *Toxicol. Appl. Pharmacol.,* 24, 564, 1973.
17. **Burek, J. D., Albee, R. R., Beyer, J. E., Bell, T. J., Carreon, R. M., Morden, D. C., Wade, C. E., Hermann, E. A., and Gorzinski, S. J.,** Subchronic toxicity of acrylamide administered to rats in the drinking water followed by up to 144 days of recovery, *J. Environ. Pathol. Toxicol.,* 4, 157, 1980.
18. Anonymous, TSCA Chemical Assessment Series. Assessment of Testing Needs: Acrylamide. Support Document for Decision Not to Require Testing of Health Effects, Toxic Substances Control Act, Section 4, Report No. EPA-569/11-80-016, U.S. Environmental Protection Agency, Washington, D.C., 1980.
19. **Edwards, P. M.,** The insensitivity of the developing rat fetus to the toxic effects of acrylamide, *Chem.-Biol. Interact.,* 12, 13, 1976.
20. **Agrawal, A. K. and Squibb, R. E.,** Effects of acrylamide given during gestation on dopamine receptor binding in rat pups, *Toxicol. Lett.,* 7, 233, 1980.
21. **Hopkins, A. P.,** Effects of acrylamide on the peripheral nervous system of the baboon, *J. Neurol Neurosurg. Psychiatry,* 33, 805, 1970.
22. **Spencer, P. S. and Schaumburg, H. H.,** Ultrastructural studies on the dying-back process. IV. Differential vulnerability of PNS and CNS fibers in experimental central-peripheral distal axonopathies, *J. Neuropathol. Exp. Neurol.,* 36, 300, 1977.
23. **Spencer, P. S. and Schaumberg, H. H.,** A review of acrylamide neurotoxicity. II. Experimental animal neurotoxicity and pathologic mechanisms, *Can. J. Neurol. Sci.,* 1, 151, 1974.
24. **Lowndes, H. E., Baker, T., Cho, E. S., and Jortner, B. S.,** Position sensitivity of de-efferented muscle spindles in experimental acrylamide neuropathy, *J. Pharmacol. Exp. Ther.,* 205, 40, 1978.
25. **Spencer, P. S. and Schaumburg, H. H.,** Nervous system degeneration produced by acrylamide monomer, *Environ. Health Perspect.,* 11, 129, 1975.
26. **Sumner, A. J. and Ashbury, A. K.,** Acrylamide neuropathy: selective vulnerability of sensory fibers, *Trans. Am. Neurol. Assoc.,* 99, 79, 1974.
27. **Schaumburg, H. H., Wisniewski, H. M., and Spencer, P. S.,** Ultrastructural studies of the dying-back process. I. Peripheral nerve terminal degeneration in systemic acrylamide intoxication, *J. Neuropathol. Exp. Neurol.,* 33, 260, 1974.

Chapter 10

MISCELLANEOUS ORGANIC NITROGEN AND AROMATIC COMPOUNDS

John L. O'Donoghue

TABLE OF CONTENTS

I. Nitrobenzene 182
   A. Introduction 182
   B. Human Neurotoxicity 182
   C. Experimental Neurotoxicity 182
   D. Related Substances 183

II. p-Bromophenylacetylurea 183
    A. Synonyms 183
    B. Introduction 183
    C. Neurotoxicity 183
    D. Neuropathology 183
    E. Metabolism 183
    F. Mechanism of Action 184
    G. Other Effects 184

III. N-3-Pyridylmethyl-N'-p-Nitrophenylurea 184
     A. Synonyms 184
     B. Introduction 184
     C. Human Neurotoxicity 184
     D. Neurotoxicity in Animal Species 185
     E. Neuropathology 185
     F. Metabolism 185
     G. Mechanism of Action 186
     H. Other Effects 186

IV. 2-Acetoxy-4'-Chloro-3,5-Diiodobenzanilide 186

V. 2'-Chloro-2,4-Dinitro-5',6-Di(Trifluoromethyl)-Diphenyl Amine 186

VI. Paraquat 186

VII. Biscyclohexanone Oxaldihydrazone 187
     A. Introduction 187
     B. Experimental Neurotoxicity and Neuropathology 187
     C. Related Substances 188

VIII. 2-t-Butylazo-2-Hydroxy-5-Methyl Hexane 188
      A. Synonyms 188
      B. Introduction 188
      C. Acute toxicity 189
      D. Neurotoxicity 189
      E. Experimental Neurotoxicity 189

IX. Acetyl Ethyl Tetramethyl Tetralin 189
    A.   Synonyms 189
    B.   Introduction 190
    C.   Neurotoxicity 190
    D.   Neuropathology 191
    E.   Mechanism of Action 191
    F.   Metabolism of AETT 192
    G.   Studies on Related Substances 192
    H.   Other Effects of AETT 192

X.   Other Neurotoxic Organic Nitrogen Chemicals 192

References 193

I. NITROBENZENE

A. Introduction

The primary use for nitrobenzene ($C_6H_5NO_2$) is in the manufacture of aniline and aniline-based dyes. It has a bitter almond-like odor and because of this has been used in soaps, perfumes, and as an odor mask under the name oil or essence of mirbane. Other uses include as a paint solvent, shoe polish ingredient, and an intermediate in chemical synthesis.

B. Human Neurotoxicity

Methemoglobin formation and hepatotoxicity are the primary concerns with nitrobenzene intoxication. Signs and symptoms possibly referable to the nervous system have been reviewed by Beauchamp et al.[1] These include severe headache, nausea, vertigo, confusion, and coma suggestive of toxicity to the central nervous system and general weakness, paresthesias, hyperalgesia, and polyneuritis suggestive of peripheral nervous system toxicity.[1-6] These changes for the most part have been transitory or recoverable. Pathological changes in nitrobenzene-poisoned individuals have been nonspecific and include edema, hemorrhage, and softening of the brain.[1]

C. Experimental Neurotoxicity

Signs of acute poisoning in experimental animals include nystagmus, opisthotonus, loss of righting reflex, tremors, paralysis, and coma.[1,7] Two types of neuropathological changes have been described.[1] The first is cerebellar Purkinje cell degeneration described in dogs, pigeons, and chickens.[1] Cerebellar Purkinje cells are susceptible to anoxia and changes in them are frequently produced artifactually during tissue collection. The second change is focal malacia or vacuolization occurring in the cerebellar peduncles and medulla oblongata of rats[8,9] and the medulla of rabbits.[10] The malacic foci were accompanied by reactive gliosis.[9] In rats, bilaterally symmetrical malacic foci were produced by a single dose of 450 mg/kg nitrobenzene.[9]

Methemoglobinemia, testicular atrophy, and hepatic necrosis were seen in rats given one dose of nitrobenzene. Degeneration of the adrenal cortex resulted from repeated doses given to rabbits.[8]

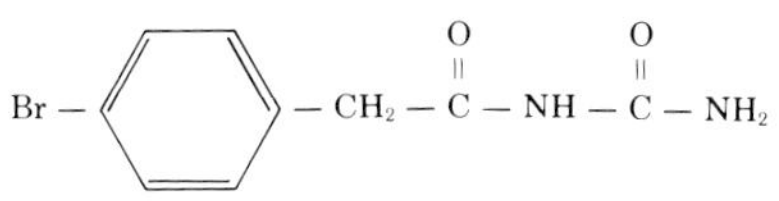

$$Br-C_6H_4-CH_2-\overset{O}{\underset{\|}{C}}-NH-\overset{O}{\underset{\|}{C}}-NH_2$$

FIGURE 1. *p*-Bromophenylacetylurea.

D. Related Substances

p-Nitrobenzamide ($NO_2C_6H_4CONH_2$) when incorporated into the diets of five male rats at a concentration of 1.0% resulted in the development of abnormal gaits in all animals by 14 to 15 days.[11] Abnormal gaits were seen as hyperextension of the hindlimbs and overstepping with the hind paws. Tails were held unusually high. Bilaterally symmetrical areas of malacia in the cerebellum and brainstem involved the deep cerebellar nuclei, ventral cochlear nucleus, lateral and superior vestibular nuclei, and dorsolateral to the superior olivary nucleus. The lesions were very similar if not identical to those described with nitrobenzene.

Other toxic effects included anemia, Heinz body formation, testicular germinal cell atrophy, and degeneration of the adrenal zona fasiculata. Methemoglobinemia was not present in animals given 300 mg/kg by gavage; a dose which is approximately half the rat oral LD_{50} (673 mg/kg). Diets of 0.1% *p*-nitrobenzamide produced none of the pathological effects of the 1.0% diets.

II. *p*-BROMOPHENYLACETYLUREA

A. Synonyms

N-(Aminocarbonyl)-4-bromobenzene-acetamide

B. Introduction

p-Bromophenylacetylurea is an analogue of the anticonvulsant phenylacetylurea or phenacemide. No industrial use of this substance has been reported, but it is of interest because of its relationship to phenylureas generally.

C. Neurotoxicity

Diezel and Quadbeck[12] first reported on the neurotoxicity of *p*-bromophenylacetylurea when they found that it was the only one of a group of halogenated analogues of phenylacetylurea which was neurotoxic. Single oral doses of 200 mg/kg or two doses on successive days had no effect on rats for 7 to 10 days. After this period, the rate developed progressive weakness and ataxia. The weakness gradually appears in the distal hindlimbs, progresses to paralysis, and may subsequently involve the forelimbs.[12-19] Recovery begins 3 weeks after dosing. Hens dosed with *p*-bromophenylacetylurea do not develop neuropathy.[13]

D. Neuropathology

p-Bromophenylacetylurea produces a typical "dying-back" axonal pathology involving both myelinated axons of the peripheral and central nervous systems. In the peripheral nervous system, the distal ends of the axons of the hindlimb were first affected. Spinal lesions were most severe in the distal ends of the dorsal columns but also affected the spinocerebellar and tectospinal tracts.[14-16,20,21]

E. Metabolism

The neurotoxic properties of this material appear to be highly specific since the *p*-chlor- and *m*-bromo-analogues are inactive.[12,13] A considerable amount of *p*-bromophenylacetylurea or an active metabolite must be excreted in the urine because Cavanagh[14] reported that untreated rats kept in the same cage with treated animals were paralyzed, suggesting absorption of excreted material.

The liver may play a role in detoxification of *p*-bromophenylacetylurea since conditions which would be expected to depress liver function, such as partial hepatectomy and carbon tetrachloride exposure, enhanced neurotoxicity.[22]

F. Mechanism of Action

No inhibition of acetylcholine esterase or benzoylcholine esterase due to *p*-bromophenylacetylurea was detected.[22] Cavanagh and Chen[16,23] have postulated that protein synthesis may be impaired. They have demonstrated that incorporation of C^{14}-labeled glycine into proteins was reduced in vivo in spinal ganglia from rats dosed twice with 200 mg/kg of *p*-bromophenylacetylurea and that in vitro incorporation of C^{14}-labeled leucine was also impaired. They have demonstrated that incorporation of C^{14}-labeled glycine into proteins was reduced in vivo in spinal ganglia from rats dosed twice with 200 mg/kg of *p*-bromophenylacetylurea and that in vitro incorporation of C^{14}-labeled leucine was also impaired. Amino acid incorporation was reduced prior to and following onset of paralysis.

G. Other Effects

Damage to renal proximal tubule epithelium and atrophy of testicular germinal epithelium have been seen following i.p. injection of *p*-bromophenylacetylurea.[16]

III. *N*-3-PYRIDYLMETHYL-*N'*-*p*-NITROPHENYLUREA

A. Synonyms
N-(4-Nitrophenyl)-*N'*-(3-pyridinylmethyl)urea
1-Nitrophenyl-3-(3-pyridylmethyl)urea
Vacor®
Pyriminil
Pyriminyl
Pyrinuron
RH 787

B. Introduction

This substance was marketed as a new single-dose rodenticide for warfarin-resistant rats under the name Vacor® rat killer. It was available as a 10% tracking powder or a 2% bait until 1979 when the manufacturer suspended U.S. sales.[24] Availability of the rodenticide from remaining stocks and in other countries is possible. The rat is the most sensitive species to single doses of *N*-3-pyridylmethyl-*N'*-*p*-nitrophenylurea (PNU) (oral LD_{50} = 12 mg/kg) followed by cats (62 to 200 mg/kg).[25]

Exposure to people is through accidental or suicidal ingestions of the rodenticide.

C. Human Neurotoxicity

Human poisoning due to PNU may have involved several hundred cases as Lewitt[26] has suggested. The clinical syndrome in the reported cases has been very similar with differences based on the severity of clinical abnormalities and on the dose needed to produce effects.[26-43] Cases have involved both men and women from 19 to 50 years of age and children from 25 months to 7 years of age. The dose of PNU consumed ranged from 0.39 to 7.02 g for 17 cases. Forty-one percent of the cases consumed two packets of the rodenticide (~1.56 g PNU). Seventy-six percent of the cases consumed two or more packets of the rodenticide.

Although a primary effect of PNU is destruction of pancreatic β-islet cells, a distinctive

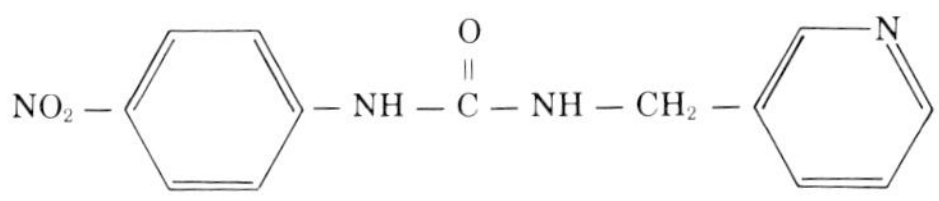

FIGURE 2. *N*-3-Pyridymethyl-*N'*-*p*-nitrophenylurea.

neuropathy, most severely affecting the autonomic nervous system, consistently develops in people who have consumed the rodenticide.[26-43] A mild sensorimotor peripheral neuropathy and some evidence of CNS damage also occurs. β-Dimethylaminopropionitrile (Chapter 2) and PNU similarly produce signs of autonomic damage which is more severe than sensory or motor nerve damage.

Severe autonomic neuropathy is an early and potentially life-threatening effect if cardiac arrhythmias or severe postural hypotension develop. Postural hypotension, in particular, occurred in nearly all cases of PNU toxicity and generally improved little if at all with time, although recovery is possible.[27] Autonomic dysfunction also included abnormal pupillary responses, impotence, decreased sweating, urinary retention, dysphagia, and gastrointestinal hypomotility.[26,29,43] Damage to central hypothalamic centers may also occur since hypothermia and anorexia have been reported.[26]

A mild sensorimotor neuropathy commonly occurs within a day of PNU ingestion but in some cases is delayed a few days.[26] The general pattern is distal sensory impairment, absence of muscle-stretch reflexes, and muscle weakness. Sensory deficits generally are more severe than motor deficits. Severe sensory neuropathies with loss of deep-tendon reflexes, pain, and proprioception develop in some cases.[38,42]

Encephalopathic effects include confusion, decreased attention, coma, cerebellar ataxia, tremor, hyperactivity, nystagmus, seizures, and increased cerebropsinal fluid protein.[26,33,38] Central nervous system effects are complicated by the severe diabetic state and secondary complications which follow PNU ingestion.

D. Neurotoxicity in Animal Species

Neurotoxicity due to PNU has also been reported in horses.[44] A dose of 250 to 500 mg of PNU ingested by a 3-year-old quarter horse resulted in severe muscle fasciculations, dilated pupils, and profuse sweating. Three months later, the horse had recovered. Three other horses were similarly affected and also developed intense abdominal pain, hindlimb weakness, ataxia, and persistent inappetence. Devi and Krishnamoorthy[45] reported that a single dose of 1 mg produced signs of paralysis in rats.

E. Neuropathology

Dorsal root ganglia, paravertebral sympathetic ganglia, and central and peripheral sensory pathways have shown damage on autopsy examination of a few cases. Complete neuropathological analyses have not been reported. Papasozomenos[38] reported a case in which sympathetic and dorsal root ganglia were degenerating or absent 19 days after PNU ingestion. The dorsal roots and their central pathway were severely degenerate while the sural nerve showed swollen axons and myelin thinning. Isolated skeletal muscle fibers were also degenerating or regenerating. Degenerating myelin has been reported by others in the dorsal roots or peripheral nerves.[26,36] A single case of cerebral edema has been reported.[26] The limited evidence available suggests that autonomic and peripheral sensory changes may be due to PNU destruction of autonomic and sensory neuronal cell bodies with secondary axonal and myelin changes.

F. Metabolism

Analysis of liver from a human poisoning case demonstrated the presence of PNU and *p*-nitroaniline.[46] It was suggested that *p*-nitroaniline was formed by amide hydrolysis of PNU.

G. Mechanism of Action

PNU appears to kill rats by interference with or antagonism to nicotinamide.[47] Prompt administration of nicotinamide prevents or lessens PNU toxicity if it is given within a few hours of PNU.[27,33,47] The relationship between PNU and nervous system nicotinamide has not been studied. The possible similarity between PNU neurotoxicity and neurotoxicity due to an analogue of nicotinamide, 6-aminonicotinamide, has been suggested, but 6-aminonicotinamide toxicity in animals produces severe damage to the motor system rather than the sensory system.[48]

A possible mechanism for PNU-induced postural hypotension was reported by LeWitt,[26] who found that baseline norepinephrine level was abnormally low and that levels remained low after the induction of postural hypotension in a poisoned patient. It was suggested that PNU produced a selective sympathectomy similar to that produced by 6-hydroxydopamine.[26]

PNU produced a noncompetitive inhibition of rat serum acetylcholinesterase in vitro.[45] In vivo, poisoned rats had increased levels of both serum and red blood cell acetylcholine indicating that PNU inhibited hydrolysis of acetylcholine.

H. Other Effects

In addition to insulin-dependent diabetes and neuropathy, PNU commonly produces evidence of abdominal pain and may produce ischemic myocardial damage.[26] Many other effects may result secondary to diabetes mellitus.

IV. 2-ACETOXY-4′-CHLORO-3,5-DIIODOBENZANILIDE

Kurtz et al.[49] reported that feeding diets containing 0.4% or approximately 350 mg/kg of 2-acetoxy-4′-chloro-3,5-diiodobenzanilide (Tremerad, Clioxanide) for 3 weeks to rats resulted in a reversible vacuolization or *status spongiosis* of both central and peripheral myelin sheaths.

A structurally similar material, 3,4′,5-tribromosalicylanilide, fed at levels of up to 500 ppm for 11 weeks did not produce similar effects.[50]

V. 2′-CHLORO-2,4-DINITRO-5′,6-DI (TRIFLUOROMETHYL)-DIPHENYL AMINE

2′-Chloro-2,4-dinitro-5′,6-di(trifluoromethyl)-diphenyl amine (CDTD) has been studied for use as a plant miticide.[51] It is among the most effective uncouplers of mitochondrial oxidative phosphorylation reported.[52]

When given to rats in single oral doses of 75 to 200 mg/kg, four repeated oral doses of 25 mg/kg, or fed in the diet at 200 ppm for 4 weeks, it produces hindlimb weakness.[52] The basis for this weakness is rapid rise in cerebrospinal fluid pressure resulting from the accumulation of fluid within the myelin sheath.[51] Morphologically, the fluid results in splitting of the intraperiod line of the myelin sheath resulting in vacuolization or status spongiosus of the sheath which appears to be identical to effects of hexachlorophene (Chapter 7).[51] The most severely affected areas are the optic nerves.

VI. PARAQUAT

Paraquat (1,1′-dimethyl-4,4′-bipyridinium or methyl viologen) and its bromide, chloride, or iodide salts are quaternary nitrogen compounds which have wide use as herbicides.[53]

Primary toxic effects of paraquat are on the lungs[53] and neurologic complications are uncommon.

Ten human fatalities following paraquat poisoning have involved cerebral vascular damage and hemorrhage. Mukada et al.[54] described a case of widespread hemorrhagic leukoencephalopathy in the brain and spinal cord. Demyelination associated with hemorrhage and degeneration of Purkinje and cerebellar granule cells were also present. Grant et al.[55] described eight patients with cerebral edema and hemorrhage, microglial and astrocytic changes, and meningeal inflammation. Yamashita and Ikuta[56] described neuronal degeneration and myelin damage associated with minor cerebrospinal hemorrhage in a man dying 12 hr after paraquat dichloride ingestion. A single case of cerebral hemorrhage was reported following fatal ingestion of diquat, a related herbicide.[57]

Lipofusion-like accumulations in neurons and less frequently other types, neuronal chromatolysis, and gray matter amyloid bodies were described in two patients who survived fatal doses of paraquat for approximately 1 week.[58]

Peripheral neuropathy following paraquat ingestion has been reported twice. Weidenbach[59] described a 34-year-old man who attempted suicide by ingesting paraquat after having drunk alcohol. Nine weeks later, transient unilateral disturbance of feeling, paresis, and reflex weakening were reported. No relationship to paraquat ingestion was definitely made. A single case of polyradiculitis has been reported in a middle-aged man by Nakamura et al.[60]

Comparable neurologic effects due to paraquat have not been reported either spontaneously or experimentally in animals. Acute effects of paraquat in animals include hyperexcitability, incoordination, and convulsions.[53]

VII. BISCYCLOHEXANONE OXALDIHYDRAZONE

A. Introduction

Biscyclohexanone oxaldihydrazone, also known as bis(cyclohexylidene hydrazide)oxalic acid, bis(cyclohexylidene hydrazide)ethanedioic acid, and cuprizone, is a white crystalline powder used as a chelating agent in the detection of copper in serum and paper pulp.[61] No neurotoxic effects due to cuprizone have been reported in humans.

B. Experimental Neurotoxicity and Neuropathology

Diets containing 0.1 to 0.75% cuprizone produce intramyelinic edema and demyelination in the brains of weanling male mice.[62-72] The majority of studies have used diets containing 0.5% cuprizone. At this level cuprizone is extremely toxic, resulting in decreased body weight gain, weakness, paresis of the hindlimbs, doming of the skull due to hydrocephalus, convulsions, and high mortality after 3 weeks on the diet.[62,66,68,71]

Neuropathologic lesions consist of dilatation of the lateral ventricles and third ventricle resulting in hydrocephalus or hydrancephaly, edema, status spongiosis, or vacuolization of myelin sheaths, demyelination, and glial reactions.[62-71]

Hydrocephalus or dilation of the lateral and third ventricles results from stenosis of the aqueduct of Sylvius.[71] Aqueductal stenosis is preceded by edema and ependymal cell degeneration and loss and is accompanied by aqueductule formation. It has been suggested that aqueductal stenosis is a sequelum of pressure exerted on the aqueduct by edematous mesencephalic tissue.[71] Hydrocephalus does not occur in older mice or other species.[71,72]

Spongiosus and demyelination vary in severity in different regions of the brain. The most consistent and severe lesions are located in the superior cerebellar penducles and the brain stem, largely in the tegmentum of the pons and midbrain.[62,63,65,66,68] Less severe changes occur in the internal capsule, anterior commissure, thalamic nuclei, and the cerebellar and cerebral white matter.[65,66] In severe cases, vacuoles may be located in the cerebral cortex.

Spongiosus or vacuolization of the myelin sheath appears to be the result of degeneration of oligodendroglia caused by cuprizone, and not a direct effect of cuprizone on myelin. The first changes following cuprizone ingestion occur in oligondendroglia after 2 weeks of exposure[67,69,73] Mitochondria hypertrophy, swell and degenerate, the cytoplasm of the oligondendroglia becomes dense, and nuclei become pyknotic.[67,69,73] After 4 weeks of exposure, oligodendroglia are severely depleted. Demyelination begins in the outer tongue of oligodendroglia cytoplasm.[67,69,73] This is followed by splitting of the intraperiod line producing the light microscopic appearance of spongiosus or vacuolization, breakdown of myelin sheaths, and phagocytosis of the resulting debris. Demyelination of the superior cerebellar peduncle is nearly complete by 5 weeks of exposure.[67]

Even with severe demyelination, axonal changes are mild or absent. Glial abnormalities include the appearance of Alzheimer's type II glia, astrocyte organelle alterations, and increase in astrocytic processes.[62,63,66,69]

Remyelination occurs rapidly and nearly completely in both young and old mice following return of animals to stock diets.[65,68,69,71-77]

Cuprizone toxicity is most severe in weanling male mice. Female mice may not be affected[73] and adult mice are less sensitive.[68,70,77] Certain strains of mice may also have varying sensitivity.[67] Rats, guinea pigs, and Syrian hamsters do not develop clinical neurologic deficits on cuprizone diets but they may develop mild status spongiosus.[78,79]

A variety of substances have been tried as antidotes to cuprizone but most have had no effect on neural lesions. Glutamic acid, glutamine, arginine, γ-aminobutyric acid, pyridoxine, folic acid, riboflavin, thiamin, vitamin B_{12}, vitamin A, multivitamins, diuretics, glucose, and copper have been tested as antidotes.[64,72] Copper supplementation reduces hydrocephalus but does not affect spongiosus.[72] Pyridoxine and γ-aminobutyric acid given together reduced mortality, the incidence and severity of status spongious, and the severity of hydrocephalus.[64]

The status spongiosus induced by cuprizone has been compared to that induced by hexachlorophene,[80] triethyltin,[80] cycloleucine,[81] intracisternal injection of ethidium bromide,[82] and isoniazid.[83] Similarities in pathogenesis of status spongiosus appear most commonly between cuprizone, ethidium bromide, and isoniazid.

C. Related Substances

A number of other hydrazides or derivatives also have neurotoxic effects. Hydrazine and dimethylhydrazine commonly result in convulsions in experimental animals.[84] Phenylisopropyl hydrazine and phenylisobutylhydrazine produce degenerative changes in the inferior olivary nuclei and pyriform lobe of dogs[85] as have other substituted hydrazines.[86] Monochloroacetaldehyde 2,4-dinitrophenyl hydrazone, a fungicide, was reported to produce abnormal posture in chickens fed lethal levels and convulsions, ataxia, rigidity, and inability to stand in dogs fed 1000 to 1500 ppm repeatedly.[87]

VIII. 2-t-BUTYLAZO-2-HYDROXY-5-METHYL HEXANE

A. Synonyms Lucel® 7
2-(1,1-Dimethylethyl)azo-5-methyl-2-hexanol

B. Introduction

2-t-Butylazo-2-hydroxy-5-methyl hexane belongs to a family of chemicals known as azo foaming agents.[88] This particular azo foaming agent is used for thermoset resins including unsaturated polyesters, vinyl esters, urethanes, phenolics, and acrylics.[89] The commercial product has a minimum purity of 90% and may contain triethylamine as a stabilizer.[89] It is unstable if not refrigerated, and may decompose releasing methyl isoamyl ketone, isobutane, or isobutylene.[90]

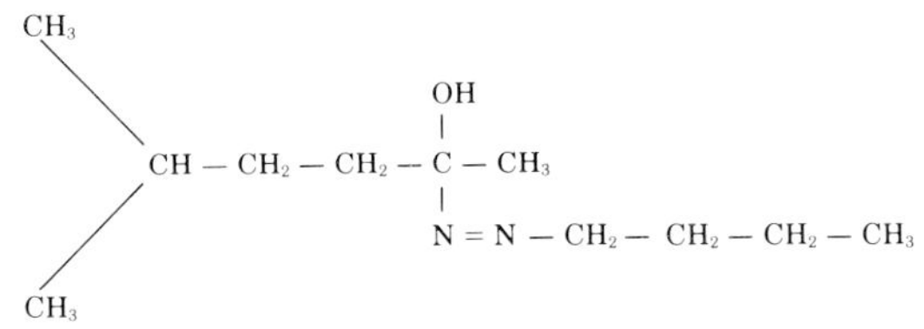

FIGURE 3. 2-*t*-Butylazo-2-hydroxy-5-methyl hexane.

C. Acute Toxicity

Acute toxicity studies using rats have shown the LC_{50} to be 7.0 mg/ℓ and the oral LD^{50} to be 919 mg/kg of body weight.[2] The dermal LD^{50} was 500 mg/kg of body weight in rabbits.[2] Application of the test material to the eye in ocular irritation studies showed that the test material was not an irritant if the eye was washed within 5 min of compound application, but it was a possible irritant if washing was delayed for 24 hr. An Ames test for mutagenicity was negative.[2] The acute toxicity test results indicate that 2-*t*-butylazo-2-hydroxy-5-methyl hexane may be toxic following oral, dermal, or inhalation exposure.[2]

D. Neurotoxicity

Four cases of neurotoxicity were reported in a plant manufacturing reinforced plastic bath tubs.[91] The affected people were 25 to 33 years of age and were all spray applicators who had direct contact with 2-*t*-butylazo-2-hydroxy-5-methyl hexane for 2 to 4 weeks before the onset of symptoms. Clinical abnormalities included a mixed sensory-motor neuropathy, memory loss, decreased attention span, and loss of color and peripheral vision. Nerve conduction and electromyography studies, approximately 1 year later, supported the presence of a residual distal peripheral neuropathy in the four cases. 2-*t*-Butylazo-2-hydroxy-5-methyl hexane had been marketed from approximately mid-1978 to late 1980.

No other cases of neurotoxicity with this chemical have been reported. Support for the possible neurotoxicity of this chemical is based on the association between the introduction of the foaming agent and the onset of neurotoxicity, a period of 1 to 3 months. It has also been suggested that there is a structural relationship to *n*-hexane, but this seems very unlikely. The clinical symptomatology, especially the prominence of memory loss, decreased attention span, and abnormalities in vision differ from those expected following *n*-hexane exposure (Chapter 4).

E. Experimental Neurotoxicity

Rats were given 250, 375, or 500 mg/kg of 2-*t*-butylazo-2-hydroxy-5-methyl hexane topically in ethanol 5 days/week for up to 5 weeks.[92] Clinical abnormalities included weight loss, hindlimb weakness, quadriparesis, corneal opacity, nasal discharge, and incontinence. Neuropathological examination showed a central-peripheral distal axonopathy which differed from that produced by *n*-hexane and methyl-*n*-butyl ketone.[92]

IX. ACETYL ETHYL TETRAMETHYL TETRALIN

A. Synonyms

AETT	1-(3-Ethyl-5,6,7,8-tetrahydro-5,5,8,8-tetramethyl-2-naphthalenyl)ethanone
Musk 36 A	3′-Ethyl-5′,6′,7′,8′-tetrahydro-5′,5′,8′,8′-tetramethyl-2′-acetonaphthane
Versalide®	6-Aceytl-1,1,4,4,-tetramethyl-7-ethyl-1,2,3,4-tetralin
Polycyclic musk	1,1,4,4-Tetramethyl-6-ethyl-7-acetyl-1,2,3,4-tetrahydronaphthalene
Musk tetraline	7-Acetyl-6-ethyl-1,1,4,4-tetramethyl-1,2,3,4-tetrahydronaphthalene

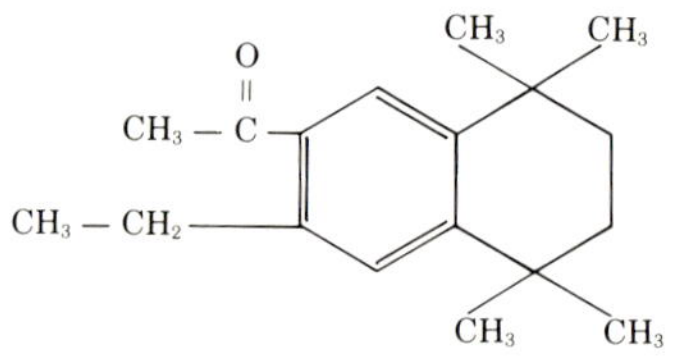

FIGURE 4. Acetyl ethyl tetramethyl tetralin.

B. Introduction

Acetyl ethyl tetramethyl tetralin (AETT) was present in many cosmetics, soaps, and deodorants from the 1950s to 1977 when it was voluntarily removed from the market by the fragrance and cosmetic industries. AETT has been listed as a temporary food flavor for use in Europe.[92] Until 1977, approximately 100,000 lb/year were used in the USA.[93] The highest use concentration was in perfumes where 0.3% AETT was usual, but the level of AETT could have been as high as 1.2%.[93]

C. Neurotoxicity

No human neurotoxicity has been reported due to AETT even though exposure from cosmetics was common and could have been as high as 0.3 mg/kg/day.[94] During exposures of humans to 4% AETT in petrolatum during a skin sensitization test, no evidence of neurotoxicity was reported.[93] Opdyke[93] has recently reviewed much of the data on AETT toxocity.

Neurotoxicity was first suspected with AETT following routine percutaneous toxicity studies when a perfume diluted to 20% (w/v) with corn oil was applied to the shaved backs of rats 5 days/week for 4 weeks resulting in blue discoloration of the skin and internal organs, particularly the brain and spinal cord. The dose of AETT given in the perfume amounted to approximately 38 mg/kg/day. Following a series of studies on the perfume, AETT was identified as the agent responsible for the blue color.

Male and female rats exposed daily to 50 or 100 mg/kg of AETT for 30 days exhibited blue skin discoloration, abnormal gait, tremors, and ataxia. Application of AETT in a variety of vehicles at concentrations of 0.1 to 1% (0.6 to 18 mg/kg), 5 days/week for 26 weeks produced blue skin and hyperexcitability at all dose levels. Blue discoloration of viscera and the CNS was visible in animals killed after 13 and 26 weeks of exposure. Animals allowed a 4- or 12-week recovery period still showed discoloration. Blue discoloration was not observed in animals receiving doses ≤0.9 mg/kg of AETT. Females appeared to be more affected than males.[93]

Groups of 20 female rats were administered 50, 100, 200, or 400 mg/kg of AETT as a 10% (w/v) ethanol solution 7 days/week for up to 14 weeks. A recovery period of approximately 20 weeks was allowed for some animals. Doses of 200 and 400 mg/kg were lethal in 14 or 3 days, respectively, and 2 deaths were observed at 100 mg/kg. All surviving animals at 50 and 100 mg/kg showed abnormal neurological function including hyperexcitability, incoordination, tremors, ataxia, and an abnormal hunched posture. Autopsy examination up to approximately 20 weeks post exposure showed blue discoloration of viscera and the CNS. Repeated dermal exposures have shown the no-effect level in rats to be 3 mg/kg after 13 weeks of exposure.[93]

AETT apparently does not affect the developing murine nervous system. No embryotoxicity, teratogenicity, or tissue discoloration was observed in fetuses from rats treated on days 6 to 16 of gestation with 5 or 30 mg/kg of AETT dissolved in ethanol. Maternal tissues were discolored blue.[93]

D. Neuropathology

AETT produces an unusual, complex neuropathological syndrome in the rat involving both central and peripheral neurons and myelin sheaths. As well as discoloration of the skin, all internal organs including the brain, spinal cord, dorsal root ganglia, and to a lesser extent, the peripheral nerves, spinal roots, and optic nerves were discolored blue to green-grey. The discoloration may not have a histological counterpart since colored viscera appeared relatively normal.[95-99]

On the other hand, both the CNS and PNS showed pigmentary deposition, widespread demyelination, and scattered axonal degeneration.[96-98] Two types of pigment deposition were observed. The first was the more common, consisting of membrane-bound granules containing lamellar or curvilinear structures with the histochemical and electron microscopic characteristics of ceroid lipopigment. The abnormal accumulation of these granules has been compared to human ceroid lipofuscinoses. Abnormal ceroid-like granules were present in neurons throughout the nervous system, oligodendrocytes, Schwann cells, and astrocytes.

The second type of granule was larger, was not bound by a membrane, and occurred extracellularly near degenerating neurons as well as within neurons. The granules contained dense granular material, clear areas with cytoplasmic debris, and rarely lamellar structures. The second type of granule appears in a few neurons later in the disease than the ceroid granules and may represent an aggregation and further degeneration of the ceroid granules. These structures have been compared to a type of Lafora body or Bunina body found in human neurological diseases.[96-99]

Later in the course of the neuropathy, pathological changes occur in central and peripheral myelin producing a type of lesion referred to as *status spongiosus* or widespread symmetrical vacuolation of central white matter. The myelin lamellae split at the intraperiod line and become separated by fluid giving the microscopic appearance of vacuoles in the myelin sheath. The vacuolation occurs within the internode resulting in segmental myelin damage which is followed by the appearance of phagocytes which strip the damaged myelin from the axon. Remyelination of the demyelinated axons then occurs, hence the Schwann cell and oligodendrocyte are able to produce myelin even while mature myelin is being damaged by AETT. The demyelination caused by AETT has been compared to that caused by hexachlorophene, triethyltin, isoniazid, and cuprizone.[96-99]

E. Mechanism of Action

It is presumed that AETT alters the neuron, Schwann cells, and oligodendrocytes since pigmentary granules appear in these sites. Demyelination due to AETT exposure may be due either to direct damage to myelin or damage to the myelin-forming cells, i.e., Schwann cells and oligodendrocytes.[96-99]

Sterman and Spencer[100] performed a series of experiments to examine the effect of AETT on Schwann cells under conditions of focal demyelination, axonal transection, and during axonal regeneration and remyelination. Exposure of rats to 50 mg/kg/day of AETT added to their diets in ethanol resulted in increased numbers of Schmidt-Lanterman incisures, thought to be important for maintenance of myelin, after 6 weeks of exposure. By 10 weeks, splitting or bubbling of myelin was observed, followed by segmental demyelination. When AETT was administered and the environment was changed by induction of focal demyelination by creating a perineurial window, tibial nerve transection to produce axonal degeneration or sural nerve crushing to induce axonal regeneration and remyelination, Schwann cell function was not altered by AETT. This suggests that some Schwann functions are not seriously impaired by AETT and that AETT may not primarily affect the Schwann cell.

Cammer[101] studied the effect of AETT and a possible oxidative metabolite, diacetyl tetramethyl tetralin, on the ability of rat liver mitochondria to inhibit mitochondrial oxidation. Both AETT and its suggested metabolite inhibited mitochondrial respiration by uncoupling oxidative phosphorylation as do other agents such as hexachlorophene (Chapter 7) which produce myelin vacuolization. The relationship between inhibition of isolated mitochondria and myelin vacuolization is unclear but a common mechanism could be involved. Production of energy deficiency in myelin-supporting cells could lead to alterations in membrane permeability and the influx of water into myelin resulting in vacuolization.[101]

F. Metabolism of AETT

Early studies on the comparative toxicity of AETT revealed that New Zealand white rabbits exposed dermally to doses of AETT comparable to doses received by rats did not develop organ discoloration.[93] Intraperitoneal doses of 1000 mg/kg of AETT in rabbits did not result in the effects noted in rats. Acute i.p. toxicity studies showed the relative toxicity to be rat > mouse = hamster > guinea pig > rabbit. Only rabbits in this series did not exhibit blue discoloration of internal organs.[93]

In vitro studies with whole rat organs immersed in a commercial AETT-ethanol solution produced blue discoloration of the ethanol while immersion in very pure samples of AETT did not. Injection of the purified AETT into rats again produced blue discoloration of tissues. This suggested AETT was metabolically activated in vivo.[93]

A possible metabolite of AETT, o-diacetyl tetramethyl tetralin (ODTT), was synthesized and tested.[93] ODTT reproduced AETT discoloration when administered to rats. Unlike AETT, the injection site turned blue with ODTT suggesting the ODTT itself may be responsible for tissue discoloration. Immersion of rat brain and rabbit brain (which was negative with AETT) in an ODTT ethanol solution resulted in blue discoloration of the solution.[93] I.p. injection of 1000 mg/kg of ODTT did not produce typical AETT lesions in rabbits but 15 mg/kg of ODTT i.v. produced paralysis and blue organ discoloration including neural tissues.[93] This suggests that the rabbit may be able to detoxify AETT and ODTT. Rhesus monkeys also apparently metabolize AETT differently than rats since doses of 50 mg/kg/day of AETT orally for 16 days produced typical lesions in rats but not monkeys.[93] Differences in metabolism may account for the absence of reports of toxicity in humans.

Studies using carbon-14 labeled AETT administered to rats showed wide distribution of AETT with accumulation in lipid storage areas.[93] AETT rapidly entered all tissues and was slowly redistributed allowing time for the build-up of consecutive doses. AETT appeared to be excreted by the liver and then reabsorbed resulting in an enterohepatic cycle.

G. Studies on Related Substances

While there are similarities among chemicals producing the odor of musk, molecular properties important to neurotoxicity have not been identified.[102] Musk ambrette (2,6-dinitro-3-methoxy-4-*tert*-butyl toluene) produced distal axonal damage in central and peripheral axons and spinal root demyelination which appears to differ from AETT neurotoxicity.[103] Other musk odorants including galoxide, celestolide, tonalid, and phantolid did not produce neurotoxicity.[95,104,105] Another tetralin, 2-(*N,N*-dipropyl)amino-5,6-dihydroxytetralin, has been reported to produce hyperactivity and abnormal biting behavior in guinea pigs.[106]

H. Other Effects of AETT

AETT has been reported to produce other effects in rats including poor body weight gain or body weight loss, decreased feed efficiency, increased serum biliribuin, increased blood urea nitrogen (females), and depressed red and white blood cell counts.[95,104]

X. OTHER NEUROTOXIC ORGANIC NITROGEN CHEMICALS

4-Aminopyridine is used as an avicide for crop protection, has been reported to poisonings of both man[107] and animals,[108] and has been studied for use in treatment to improve neuromuscular function. It enhances the release of acetylcholine at neuromuscular junctions by altering potassium conductance across axonal membranes during membrane depolarization.[109]

1-Methyl-4-phenyl-1,2,5,6-tetrahydropyridine is used as a chemical intermediate. Langston et al.[110] have reported that its presence in illicit drugs led to degeneration in the substantia nigra and severe chronic parkinsonism.

Levine and Sowinski[111] reported that a series of piperazines and piperidines which were all tertiary amines produced hydropic degeneration of the choroid plexus in rats. An unrelated aliphatic tertiary amine, 3,3'-methyliminobis-(N-methylpropylamine), caused edema and necrosis in the hypothalamus and medulla oblongata.[112] This work suggests that tertiary amines may have particular affinity for the central nervous system.

REFERENCES

1. **Beauchamp, R. O., Irons, R. D., Rickert, D. E., Couch, D. B., and Hamm, T. E., Jr.,** A critical review of the literature on nitrobenzene toxicity, *Crit. Rev. Toxicol.,* 11, 33, 1982.
2. **Hamilton, A.,** Industrial poisoning by compounds of the aromatic series, *J. Ind. Hyg.,* 1, 200, 1919.
3. **Stifel, R. E.,** Methemoglobinemia due to poisoning by shoe dye, *JAMA,* 72, 395, 1919.
4. **Carter, F. W.,** An unusual case of poisoning, with some notes on non-alkaloid organic substances, *Med. J. Aust.,* 2, 558, 1936.
5. **Ikeda, M. and Kita, A.,** Excretion of p-nitrophenol and p-aminophenol in the urine of a patient exposed to nitrobenzene, *Br. J. Ind. Med.,* 21, 210, 1964.
6. **Larens, L., Bierme, R., Jorda, M. F., Cathala, B., and Fabre, H.,** Acute, toxic methemoglobinemia from accidental ingestion of nitrobenzene, *Eur. J. Toxicol.,* 7, 12, 1974.
7. **Smith, R. P., Alkaitis, A. A., and Shafer, P. R.,** Chemically induced methemoglobinemias in the mouse, *Biochem. Pharmacol.,* 16, 317, 1967.
8. **Bond, J. A., Chism, J. P., Rickert, D. E., and Popp, J. A.,** Induction of hepatic and testicular lesions in Fischer-344 rats by single oral doses of nitrobenzene *Fund. Appl. Toxicol.,* 1, 389, 1981.
9. **Gross, E. A., Bond, J. A., Lyght, O., and Morgan, K. T.,** Characterization of nitrobenzene-induced cerebellar malacia in F-344 rats, 5th CIIT Conference on Toxicology: Toxicity of Nitroaromatic Compounds, Raleigh, N.C., January 28 and 29, 1982.
10. **Matsumara, H. and Yoshida, T.,** Experimental studies of nitrobenzol poisoning, *Kyusho, J. Med. Sci.,* 10, 259, 1959.
11. Eastman Kodak Company, unpublished data, Rochester, N.Y., 1983.
12. **Diezel, P. B. and Quadbeck, G.,** Nervenschadigung durch *p*-bromophenylacetylharnstoff, *Arch. Exp. Path. Pharmak.,* 238, 534, 1960.
13. **Barnes, J. M.,** Toxic substances and the nervous system, *Sci. Basis Med. Annu. Rev.,* 183, 1969.
14. **Cavanagh, J. B.,** Toxic substances and the nervous system, *Br. Med. Bull.,* 25, 268, 1969.
15. **Cavanagh, J. B., Chen, F. C.-K., Kyu, M. H., and Ridley, A.,** The experimental neuropathy in rats caused by *p*-bromophenylacetylurea, *J. Neurol. Neurosurg. Psychiatry,* 41, 471, 1968.
16. **Cavanagh, J. B.,** Peripheral neuropathy caused by chemical agents, *Crit. Rev. Toxicol.,* 2, 365, 1973.
17. **Jakobsen, J. and Brimijoin, S.,** Axonal transport of enzymes and labelled proteins in experimental axonopathy induced by *p*-bromophenylacetylurea, *Brain Res.,* 229, 103, 1981.
18. **Jakobsen, J., Lambert, E. H., Carlson, G., and Brimijoin, S.,** Clinical and electrophysiological characteristics of the experimental neuropathy caused by *p*-bromophenylacetylurea, *Exp. Neurol.,* 75, 158, 1982.
19. **Troncoso, J. C., Griffin, J. W., Price, D. L., and Hess-Kozlow, H.,** Pathology of the peripheral neuropathy induced by *p*-bromophenylacetylurea, *Lab. Invest.,* 46, 215, 1982.
20. **Ohnishi, A. and Ikeda, M.,** Morphometric evaluation of primary sensory neurons in experimental *p*-bromophenylacetylurea intoxication, *Acta Neuropathol.,* 52, 111, 1980.
21. **Ohnishi, A. and Ikeda, M.,** Experimental dying back phenomenon — morphometric evaluation of primary sensory neuron in para-bromophenylacetylurea neuropathy, *Rinsho Shinkeigaku,* 19, 44, 1979.
22. **Chen, F. C.-K. and Cavanagh, J. B.,** Factors affecting neurotoxicity by *p*-bromophenylacetylurea in rats, *Br. J. Exp. Pathol.,* 52, 315, 1971.

82. **Yasima, K. and Susuki, K.,** Ultrastructural changes of oligodendroglia and myelin sheaths induced by ethidium bromide, *Neuropathol. Appl. Neurobiol.,* 5, 49, 1979.
83. **Blakemore, W. F.,** Isoniazid, in *Experimental and Clinical Neurotoxicology,* Spencer, P. S. and Schaumburg, H. H., Eds., Williams & Wilkins, Baltimore, 1980, chap. 33.
84. **Reinhardt, C. F. and Brittelli, M. R.,** Heterocyclic and miscellaneous nitrogen compounds, in *Patty's Industrial Hygiene and Toxicology,* Vol. 2A, 3rd rev. ed., Clayton, G. D. and Clayton, F. E., Eds., John Wiley-Interscience, New York, 2798.
85. **Highman, B. and Maling, H. M.,** Neuropathologic lesions in dogs after prolonged administration of phenyl isopropylhydrazine (JB835), *J. Pharmacol. Exp. Ther.,* 137, 344, 1971.
86. **Palmer, A. C. and Noel, P. R.,** Neuropathological effects of prolonged administration of some hydrazine monoamine inhibitors in dogs, *J. Pathol. Bacteriol.,* 86, 463, 1963.
87. **Ambrose, A. M., Borzelleca, J. F., Larson, P. S., Smith, B. R., Jr., and Hennigar, G. R., Jr.,** Toxicologic studies on monochloracetaldehyde 2,4-dinitrophenyl hydrazone, a foliar fungicide, *Toxicol. Appl. Pharmacol.,* 8, 472, 1966.
88. **Harpell, G. A., Gallagher, R. B., and Novits, M. F.,** Use of azo foaming agents to produce reinforced elastomeric foams, *Rubber Chem. Technol.,* 50, 678, 1977.
89. **Noller, D. C.,** 2-t-Butylazo-2-hydroxy-5-methyl hexane, Material Safety Data Sheet, Lucidol Division, Pennwalt Corp., Buffalo, N.Y., 1978.
90. **Anon.,** Toxic occupational neuropathy — Texas, Morbidity and Mortality Weekly Rep. No. 29, U.S. Department of Health, Education and Welfare, 1980, 503.
91. **Spencer, P. S., Bischoff, M. C., and Beaubernard, C.,** Experimental neurotoxicity of Lucel®-7 (2-t-butylazo-2-hydroxy-5-methyl-hexane), *Toxicologist,* 3, 54, 1983.
92. Council of Europe, Natural Flavouring Substance, Their Sources, and Added Artificial Flavouring Substances. Partial agreement in the Social and Public Health Field, List 2, No. 2220, 1974, 328.
93. **Opkyke, D. L. J.,** Monographs on fragrance raw materials. AETT, *Food Cosmet. Toxicol.,* 17, 357, 1979.
94. **Eiermann, H. J.,** Regulatory issues concerning acetylethyl tetramethyltetralin and 6-methylcoumarin, *Contact Dermatitis,* 6, 120, 1980.
95. **Arthur, B. H., Johnson, W. D., Griffing, W. J., and Emmerson, J. L.,** Toxicity of topically applied acetyl ethyl tetramethyl tetralin (AETT) in rats, *Toxicol. Appl. Pharmacol.,* 48, A202, 1979.
96. **Spencer, P. S., Foster, G. V., Sterman, A. B., and Horoupian, D.,** Acetyl ethyl tetramethyl tetralin, in *Experimental and Clinical Neurotoxicology,* Spencer, P. S. and Schaumburg, H. H., Eds., Williams & Wilkins, Baltimore, 1980, chap. 20.
97. **Spencer, P. S., Sterman, A. B., Bischoff, M., Horoupian, D., and Foster, G. V.,** Experimental myelin disease and ceroid accumulation produced by the fragrance compound acetyl ethyl tetramethyl tetralin, *Trans. Am. Neurol. Assoc.,* 103, 185, 1978.
98. **Spencer, P. S., Sterman, A. B., Horoupian, D. S., and Foulds, M. M.,** Neurotoxic fragrance produces ceroid and myelin disease, *Science,* 204, 633, 1979.
99. **Spencer, P. S., Sterman, A. B., Horoupian, D., Bischoff, M., and Foster, G.,** Neurotoxic changes in rats exposed to the fragrance compound acetyl ethyl tetramethyl tetralin, *Neurotoxicology,* 2, 221, 1979.
100. **Sterman, A. B. and Spencer, P. S.,** The pathogenesis of primary internodal demyelination produced by acetyl ethyl tetramethyl tetralin: evidence for preserved Schwann cell somal function, *J. Neuropathol. Exp. Neurol.,* 40, 112, 1981.
101. **Cammer, W.,** Uncoupling of oxidative phosphorylation (*in vitro*) by the neurotoxic fragrance compound acetyl ethyl tetramethyl tetralin and its putative metabolite, *Biochem. Pharmacol.,* 29, 1531, 1980.
102. **Brugger, W. E. and Jurs, P. C.,** Extraction of important molecular features of musk compounds using pattern recognition techniques, *J. Agric. Food Chem.,* 25, 1158, 1977.
103. **Spencer, P. S., Bischoff, M. C., Moreno, O. M., and Opdyke, D. L.,** Neurotoxic properties of musk ambrette, *Toxicology,* 3, 54, 1983.
104. **Stults, F. H., Ford, D. M., and Moran, E. J.,** Subchronic dermal toxicity of acetyl ethyl tetramethyl tetralin (AETT), galaxolide, and celestolid in rats: emphasis on neurobehavioral effects, *Toxicol. Appl. Pharmacol.,* (Abstr.), A34, 1980.
105. **Gressel, Y., Troy, W. R., and Foster, G. V.,** Safety evaluation of four bicyclic musk fragrance-chemicals relative to the neurotoxin, acetyl ethyl tetramethyl tetralin (AETT), *Dev. Toxicol. Environ. Sci.,* 8, 53, 1980.
106. **Costall, B., DeSouza, C. X., and Naylor, R. J.,** Topographical analysis of the actions of 2-(N,N-dipropyl)amino-5,6-dihydroxy-tetralin to cause biting behaviour and locomotor hyperactivity from the striatum of the guinea-pig, *Neuropharmacology,* 19, 623, 1980.
107. **Spyker, D. A., Lynch, C., Shabonowitz, J., and Sinn, J. A.,** Poisoning with 4-aminopyridine: 3 cases, *Clin. Toxicol.,* 16, 487, 1980.
108. **Nicholson, S. S. and Prejean, C. J.,** Suspected 4-aminopyridine toxicosis in cattle, *J. Am. Vet. Med. Assoc.,* 178, 1277, 1981.

109. **Glover, W. E.,** The aminopyridines, *Gen. Pharmacol.*, 13, 259, 1992.
110. **Langston, J. W., Ballard, P., Tetrud, J. W., and Irwin, I.,** Chronic parkinsonism in humans due to a product of meperidine-analog synthesis, *Science*, 219, 979, 1983.
111. **Levine, S. and Sowinski, R.,** T-lymphocyte depletion and lesions of chroid plexus and kidney induced by tertiary amines in rats, *Toxicol. Appl. Pharmacol.*, 40, 147, 1977.
112. **Levine, S. and Sowinski, M. S.,** Hypothalamic and medullary lesions caused by an aliphatic triamine unrelated to gold thioglucose, *J. Neuropathol. Exp. Neurol.*, 41, 54, 1982.

INDEX

A

Acetaldehyde syndrome, 45—48
Acetone, 81
2-Acetoxy-4'-chloro-3,5-diiodobenzanilide, 184
Acetoxymethylmethylnitrosamine, 47
Acetylcholine, defect in release of, 45
7-Acetyl-6-ethyl-1,1,4,4-tetramethyl-1,2,3,4-tetrahydronaphthalene, see Acetyl ethyl tetramethyl tetralin
Acetyl ethyl tetramethyl tetralin (AETT), 187—190
   mechanism of action, 189—190
   metabolism, 190
   neuropathology, 189
   neurotoxicity, 188
3-Acetylpyridine, 33
6-Acetyl-1,1,4,4-tetramethyl-7-ethyl-1,2,3,4-tetralin, see Acetyl ethyl tetramethyl tetralin
Acidosis, metabolic, 84
Acids, see also specific compounds, 114—118
Acro-osteolysis, 110—111
Acrylamide, 64, 169—174
   mechanism of action, 172—173
   neuropathology, 171—172
   neurotoxicity
     in experimental animals, 170—171
     in humans, 170
   no-effect level, 171
   toxicokinetics, 172
Acrylic amide, see Acrylamide
Acrylonitrile, 27, 31—32, 174
   metabolism, 32
   neuropathology, 32
   neurotoxicity
     in experimental animals, 31—32
     in humans, 31
Adhesive, 40, 62, 128, 156
Adrenal cortex, degeneration of, 88, 180—181
Aerosol, 107
Aethylis chloridum, see Ethyl chloride
AETT, see Acetyl ethyl tetramethyl tetralin
Agent Orange, 159
Akinesia, 28
Akrylamid, see Acrylamide
β-Alanine, 30
Alcoholic beverages, 80—82
Alcoholism, 45—46
Alcohols, see also specific compounds, 80—85
   halogenated, 99—118
Aldrin, 163
Aliphatic hydrocarbons, halogenated, see also specific compounds, 99—118
Alkanes, see also specific compounds, 62—82
Alkyl aryl phenylphosphorothioates, 7
Alkyl esters, 6—7
Allylacetamide, 174
Allyl chloride, 113—114
   neurotoxicity
     in experimental animals, 114
     in humans, 114
Amines, 191
γ-Aminobutyric acid, 118, 162
N-(Aminocarbonyl)-4-bromobenzene-acetamide, see p-Bromophenylacetylurea
1-Amino-3-chloro-2-propanol, 118
Aminodipropionitrile, see β,β'-Iminodipropionitrile
2-Amino-hexanoic acid, 74, 76
6-Amino-1-hexanol, 81
6-Aminonicotinamide, 118, 184
β-Aminopropionitrile, 27, 30
4-Aminopyridine, 190
Ammonium bromide, 106
n-Amylbromide, see 1-Bromopentane
Amyotrophic lateral sclerosis, 29
Analgesic, 111—112
Anemia, 181
Anencephaly, 132
Anesthetic, 101, 108—112, 140
Angiosarcoma, 110
Aniline, 180
Animal feed, polybrominated biphenyl in, 157
Antabuse®, see Disulfiram
Anthelmintic agent, 34, 108
Anticonvulsant, 181
Antifreeze, 80
Antiperspirant, 51
Aroclor 1242®, 156—157
Aroclor 1248®, 157
Aroclor 1254®, 32
Aromatic hydrocarbons, see also specific compounds, 127—133
Aryl esters, 6—7
Astrocytoma, 31—32
At-7, see Hexachlorophene
Ataxia, 5, 8—18, 33—34, 41
ATPase, inhibition of, 161—162
Atropine, 2, 5
Attention span, decreased, 187
Autonomic nervous system,
   damage to, 183
   dysfunction, 112, 133, 141, 170
Autonomic neuropathy, 183
Avicide, 190
Axon
   ballooning, 30
   damage to, 29, 114, 132
   degeneration, 2, 42, 47—48, 146, 189
   giant swelling, 42, 64, 77—78
Axonopathy
   central-peripheral distal, see Central-peripheral distal axonopathy
   cyanide, 26
   ethylene oxide, 86
   n-hexane, 65, 67, 71—73
   isocyanates, 34
   methyl n-butyl ketone, 65, 67, 71—73

nitriles, 29
organic sulfur-containing compounds, 47—48
Axoplasmic flow, inhibition, 79
Axoplasmic transport, deficit in, 54
Azo foaming agent, 186

B

Bactericide, 51, 140
Bacteriostat, 47
Behavioral disturbance, 41, 112, 144, 163, 190
Benzaldehyde, 130
Benzene, 48, 128
Benzenehexachloride, 163
Benzenol, see Phenol
Benzine, 64, 66
Benzoic acid, 130
Benzo(a)-pyrene monooxygenase, 47
Benzyl alcohol, 130
Bilevon, see Hexachlorophene
N,N-Bis acrylamidoacetic acid, 174
Bis-β-aminoproprionitrile, see β,β'-Iminodipropionitrile
Bis(β-cyanoethyl)amine, see β,β'-Iminodipropionitrile
N,N-Bis(2-cyanoethyl)amine, see β,β'-Iminodipropionitrile
Biscyclohexanone oxaldihydrazone, 115, 145, 147, 185—186, 189
  neuropathology, 185—186
  neurotoxicity, 185—186
Bis(cyclohexylidene hydrazide)ethanedioic acid, see Biscyclohexanone oxaldihydrazone
Bis(cyclohexylidene hydrazide)oxalic acid, see Biscyclohexanone oxaldihydrazone
Bis(diethylthiocarbamoyl) disulfide, see Disulfiram
Bis-2-dimethyl-aminoethylether, 27—28
Bis(dimethylthiocarbamoyl)disulfide, see Tetramethylthiuramdisulfide
Bis(2-hydroxy-3,5,6-trichlorophenyl)methane, see Hexachlorophene
Bladder dysfunction, 27—29
Blindness, 30, 82, 142
Blood
  acrylamide level in, 172
  bromide level in, 107
  chlordecone level in, 160
  hypercoagulability, 147
  toluene level in, 130
Brain
  hemorrhage, 108
  increased intracranial pressure, 160
  nucleoprotein synthesis, 26—27
  tumor, 31—32, 110—111
Bromide, inorganic, 104—106
Bromism, 106—107
Bromobenzene, 147
Bromobenzene-4-isothiocyanate, see p-Bromophenylisothiocyanate

Bromobenzene-4-iso thiocyanatobenzene, see p-Bromophenylisothiocyanate
2-Bromobutanoic acid, see 2-Bromobutyric acid
2-Bromobutyric acid, 100, 116—117
α-Bromobutyric acid, see 2-Bromobutyric acid
1-Bromo-4-isothiocyanatobenzene, see p-Bromophenylisothiocyanate
1-Bromopentane, 100, 117
p-Bromophenylacetylurea, 181—182
  mechanism of action, 182
  metabolism, 181—182
  neuropathology, 181
  neurotoxicity, 181
p-Bromophenylisocyanate, 34
4-Bromophenylisothiocyanate, see p-Bromophenylisothiocyanate
p-Bromophenylisothiocyanate, 34
2-Bromopropanoic acid, see 2-Bromopropionic acid
1-Bromopropionic acid, 100
2-Bromopropionic acid, 100, 116
α-Bromopropionic acid, see 2-Bromopropionic acid
Bulbar palsy, 112
Bunina body, 189
1,4-Butanediol, 81
N-t-Butylacrylamide, 174
2-t-Butylazo-2-hydroxy-5-methyl hexane, 186—187
n-Butyl ketone, metabolism of, 74

C

Cancer
  of liver, 110
  respiratory, 31
Capacitor, 156
Carbamates, 3—4
Carbinol, see Methanol
Carbolic acid, see Phenol
Carbon bisulfide, see Carbon disulfide
Carbon chloride, see Carbon tetrachloride
Carbon disulfide, 40—44, 46, 48, 64
  chronic intoxication with, 41
  mechanism of action, 43—44
  neuropathology, 42
  neurotoxicity
    in experimental animals, 41—42
    in humans, 40—41
  toxicity to cutaneous nerves, 42—43
  toxicokinetics, 42—43
Carbonless copying paper, 156
Carbon monoxide, 108, 111
Carbon tetrachloride, 40, 108—109, 162
  neurotoxicity
    in experimental animals, 108—109
    in humans, 108
Cardiac arrhythmia, 183
Cardiovascular damage, 28, 31, 41—42, 44
Cassava, 26
Catechol, 140
Catecholamines, 43, 46, 162

CDTD, see 2'-Chloro-2,4-dinitro-5',6-di(trifluoromethyl)-diphenyl amine
Celestolide, 190
Cellophane manufacture, 40
Central nervous system
   damage to, 183
   stimulation, 140
   tumor of, 31
Central-peripheral distal axonopathy
   acetyl ethyl tetramethyl tetralin, 190
   allyl chloride, 114
   p-bromophenylacetylurea, 181
   2-t-butylazo-2-hydroxy-5-methylhexane, 187
   carbon disulfide, 42
   disulfiram, 46
   dithiocarbamate compounds, 49
   n-hexane, 64, 76—79
   methyl n-butyl ketone, 64, 76—79
   organophosphorus compounds, 2
   pyridinethione derivatives, 51
Central white matter, vacuolation of, 189
Cerebellar granule cell
   degeneration, 50
   necrosis, 100—101, 115—117
   pyknosis, 116
Cerebellum
   ataxia, 129
   atrophy, 129
   damage to, 77, 99—118, 180—181
Cerebral hemorrhage, 46
Cerebrospinal fluid
   elevation of pressure, 146
   increased protein in, 183
   interference with absorption of, 160
Cerebrovascular system, damage to, 42, 185
Ceroid lipofuscinoses, 189
Chelator, 49
Chem-tol, see Pentachlorophenol
Chloracne, 141, 156, 158—159
Chlordane, 163
Chlordecone, 159—162
   metabolism of, 161—162
   neuropathology, 161
   neurotoxicity
      in experimental animals, 160—161
      in humans, 160
   no-effect level, 160
   toxicokinetics, 161
Chlordimeform, 163
Chlorethene, see Vinyl chloride
Chlorethyl, see Ethyl chloride
Chlorethylene, see Vinyl chloride
6-Chloro-6-deoxyfructose, 118
6-Chloro-6-deoxygalactose, 118
6-Chloro-6-deoxyglucitol, 118
6-Chloro-6-deoxyglucose, 118
6-Chloro-6-deoxymannose, 118
Chlorodibenzofuran, 156
2'-Chloro-2,4-dinitro-5',6-di(trifluoromethyl)-diphenyl amine (CDTD), 184
Chloroethane, see Ethyl chloride

2-Chloroethanol, 87
Chloroethylamines, 30
α-Chlorohydrin, 118
Chlorol hydrate, 113
Chloronaphthalene, 156
4-Chlorophenol, 146
Chloropromazine, 147
3-Chloro-1,2-propanediol, see α-Chlorohydrin
2-Chloropropanoic acid, see 2-Chloropropionic acid
3-Chloropropene, see Allyl chloride
3-Chloro-1-propene, see Allyl chloride
2-Chloropropionic acid, 100, 115—116
   metabolism, 116
   neurotoxicity, 115—116
α-Chloropropionic acid, see 2-Chloropropionic acid
Chloropropylene, see Allyl chloride
Chloryl anesthetic, see Ethyl chloride
Cholestyramine, 160—161
Cholinergic effects, 51
Cholinesterase, inhibition of, 2, 47
Choreiform movement, 41
Choroid plexus, degeneration of, 191
Cinnamene, see Styrene monomer
Circulation, impaired, 141
Clioxanide, see 2-Acetoxy-4'-chloro-3,5-diiodobenzanilide
Coal tar distillate, 48
Coating operations, 107, 156
Colbatous chloride, 50
Colonial spirits, see Methanol
Columbian spirits, see Methanol
Compound 1189, see Chlordecone
Convulsion
   biscyclohexane oxaldihydrazone, 186
   2-bromopropionic acid, 116
   carbon tetrachloride, 108
   ethylene glycol, 85—86
   ethylene oxide, 86—88
   mercaptopropionic acid, 117
   methyl chloride, 102
   nitriles, 28, 31
   organophosphorus compounds, 2
   paraquat, 185
   pesticides, 162—163
   phenols, 140—143
   toluene, 129
Corrosion inhibitor, 40
Cosmetics, 80, 85, 188
Cranial nerve
   damage to, 63, 85
   deficit of, 84
   palsy, 112
m-Cresol, 140
p-Cresol, 140
Crops, fumigated, 104
Crotonamide, 174
Cuprinol, see Pentachlorophenol
Cuprizone, see Biscyclohexanone oxaldihydrazone
Cyanide, 26—27
Cyanoacetic acid, 30

2-Cyano-*N*-(2-cyanoethyl)ethanamine, see β,β'-Iminodipropionitrile
Cyanoethylene, see Acrylonitrile
Cyclic hydrocarbons, halogenated, see also specific compounds, 155—163
Cyclic saligenin esters, 17
Cyclohexane, 81
Cyclohexanol, 81
Cyclohexanone, 81
Cycloheximide, 163
Cycloleucine, 186
Cysteine, 32
Cytochrome oxidase, inhibition of, 26, 33, 44
Cytochrome P-450 oxidase, 50, 64, 74—76

D

2,4-D, see 2,4-Dichlorophenoxyacetic acid
DDT, see Dichlorodiphenyl trichloroethane
Decachloroketone, see Chlordecone
1,1a,3,3a,4,5,5a,5b,6-Decachlorooctahydro-1,3,4-metheno-2H-cyclobuta[c,d]pentalen-2-one, see Chlordecone
Defoliant, 141
Degreasing agent, 111
Dementia, 84
Demyelination
  acetyl ethyl tetramethyl tetralin, 189—190
  alkanes, 71
  biscyclohexanone oxaldihydrazone, 185—186
  carbon tetrachloride, 108
  methanol, 84
  nitriles, 26, 30
  organic sulfur-containing compounds, 46—48
  organophosphorus compounds, 2
  paraquat, 185
  phenols, 147
Deodorant, 142, 188
Depilatory, 117
Dermadex, see Hexachlorophene
Dermatitis, 128, 173
Dermatomycosis, treatment of, 34
Detergent, 86
Developmental toxicity, 27
DFP, 3
Diabetes, insulin-dependent, 184
Diacetone acrylamide, 174
o-Diacetyl tetramethyl tetralin, 189—190
Dialkyl amines, 30
Dibutyl ketone, 80
αβ-Dicarbonyls, 79
Dichloroacetic acid, 114—115
  mechanism of action, 115
  neuropathology, 114—115
  neurotoxicity, 114
Dichloroacetylene, 111, 113
Dichlorodiphenyl trichloroethane (DDT), 147, 163
Dichloromethane, see Methylene chloride
Dichlorophene, 146
4,6-Dichlorophenol, 147
2,4-Dichlorophenoxyacetic acid (2,4-D), 147—148, 159
2,4-Dichlorophenoxypropionic acid, 147
2,2-Dichlorovinyl di-*n*-pentyl phosphate, 8
Dichlorvos, 13
Di(2-cyanoethyl)amine, see β,β'-Iminodipropionitrile
Dieldrin, 163
N,N-Diethylacrylamide, 174
Diethylamine, 46
Diethyl dithiocarbamate, 32, 45—46
Diethyl ketone, 81
Diethyl-4-nitrophenyl phosphate, see Paraoxon
N,N-Diethyl-*m*-toluamide, 163
Dihydrooxirene, see Ethylene oxide
2,2'-Dihydroxy-3,3',5,5',6,6'-hexachlorodiphenylmethane, see Hexachlorophene
3,5-Diiodo-4-chlorosalicylanilide, 145
Diisoamyl ketone, 82
Diisobutyl ketone, 82
N,N-Diisopropyl phosphorodiamide fluoridate, see Mipafox
2,5-Diketo-hexanoic acid, 74
γ-Diketone neurotoxicity, 65—73, 77—83
  chemicals tested for, 80—82
N,N-Dimethylacrylamide, 173—174
Dimethylamine, 27, 48
Dimethylamino-hexose-reductone, 30
3-(Dimethylamino)propanenitrile, see β-Dimethylaminopropionitrile
3-(Dimethylamino)propionitrile, see β-Dimethylaminopropionitrile
β-Dimethylaminopropionitrile (DMAPN), 27—29, 170, 183
  neuropathology, 29
  neurotoxicity
    in experimental animals, 28
    in humans, 27—28
β-*N*-Dimethylaminoproprionitrile, see β-Dimethylaminopropionitrile
N,N-Dimethylamino-3-proprionitrile, see β-Dimethylaminopropionitrile
3 (*N,N*-Dimethylamino)proprionitrile, see β-Dimethylaminopropionitrile
Dimethylaminopropylamine, 27
1,1'-Dimethyl-4,4'-bipyridinium, see Paraquat
2,5-Dimethyl-2,3-dihydrofuran, 74
2-(1,1'-Dimethylethyl)azo-5-methyl-2-hexanol, see 2-*t*-Butylazo-2-hydroxy-5-methyl hexane
2,5-Dimethylfuran, 74—76
3,3-Dimethyl-2,5-hexanedione, 71, 77—78, 83
3,4-Dimethyl-2,5-hexanedione, 71, 77—78, 83
Dimethylhydrazine, 186
Dimethyl oxide, see Ethylene oxide
3,3-Dimethyl-1-phenyltriazene, 30
Dimethyl phosphates, 5
Dimethylsulfoxide, 52
2,6-Dinito-3-methoxy-4-*tert*-butyl toluene, 190
2,4-Dinitrophenol, 147
2,4-Dinitrophenyl hydrazone, 186
Dioxan, see 2,3,7,8-Tetrachlorodibenzo-*p*-dioxin

Dioxin, see 2,3,7,8-Tetrachlorodibenzo-*p*-dioxin
2-(*N*,*N*-Dipropyl)amino-5,6-dihydroxytetralin, 190
Di-*n*-propyl ketone, 82
Diquat, 185
Disinfectant, seed, 45, 47
Disodium ethylenebisdithiocarbamate, 49
Disulfiram, 45—47
   mechanism of action, 46, 47
   metabolism, 46
   neuropathology, 46
   neurotoxicity
      in experimental animals, 46
      in humans, 45—46
2,2′-Dithio-bis-pyridine-1-oxide, magnesium sulfate adduct (MDS), 51—54
2,4-Dithiobiuret, 44—45
Dithiocarb, see Sodium diethyldithiocarbamate
Dithiocarbamates, 43, 48—49
Dithiocarbonic anhydride, see Carbon disulfide
Divinylenesulfide, see Thiophene
DMAPN, see β-Dimethylaminopropionitrile
Dopamine, 156—157, 162
Dopamine-β-hydroxylase, inhibition of, 43, 46
Dorsal root ganglion, vacuolated, 49
Dowicide 7, see Pentachlorophenol
Dry cleaning industry, 111
Dyes, 26, 48, 80, 118, 180
Dying-back neuropathy, 42, 114, 171, 181
Dynamite, 118
Dysarthria, 110, 129
Dysdiadochokinesis, 110
Dysmenorrhea, 128

E

Ejaculatory disturbance, 147
Electroencephalogram, abnormal, 133
Electromyogram, abnormal, 131, 163
Electroplating solution, 40
Emotional disturbance, 101
E*n*BK, see Ethyl *n*-butyl ketone
Encephalocoele, 26, 32
Encephalopathy, 41, 104, 129, 183, 185
   hepatic, 109
Endoneurial pressure, increased, 146
Endosulfan, 163
Endrin, 163
Enolase, inhibition of, 79
Ependymoma, malignant, 111
Epilepsy, 104
Epinephrine, 31, 162
1,2-Epoxyethane, see Ethylene oxide
Esthesioneuroepithelioma, 111
1,2-Ethanediol, see Ethylene glycol
Ethanol denaturant, 80
Ethene oxide, see Ethylene oxide
Ethenyl benzene, see Styrene monomer
Ethenyl methyl benzene, see Vinyl toluene
Ether chloratus, see Ethyl chloride
Ether muriatic, see Ethyl chloride

Ethidium bromide, 186
Ethyl *n*-butyl ketone (E*n*BK), 72—73, 80, 83
Ethyl cellulose, production of, 109
Ethyl chloride, 101, 109—110
   metabolism, 110
   neurotoxicity, 109—110
Ethylcrotonate, 174
Ethylene bromide, 47
Ethylene chlorhydrin, 87
Ethylene glycol, 85
Ethylene glycol monomethyl ether, 86
Ethylene oxide, 86—88
   neurotoxicity
      in experimental animals, 87
      in humans, 86—87
Ethylene trichloride, see Trichloroethylene
Ethyl hydrochloric, see Ethyl chloride
N-Ethylmorpholine, 28
3′-Ethyl-5′,6′,7′,8′-tetrahydro-5′,5′,8′,8′-tetramethyl-2′-acetonaphthane, see Acetyl ethyl tetramethyl tetralin
1-(3-Ethyl-5,6,7,8-tetrahydro-5,5,8,8-tetramethyl-3-naphthalenyl)ethanone, see Acetyl ethyl tetramethyl tetralin
Evrisan, see Pentachlorophenol
Exencephaly, 26, 32
Exofene, see Hexachlorophene
Extrapyramidal syndrome, Parkinson-like, 84
Eye, see also Blindness
   corneal opacity, 187
   damage to, 141
   light sensitivity, 48
   pupillary reflex impairment, 64
   retinal degeneration, 54

F

Facial nerve
   damage to, paralysis, 142
Feed, animal, fumigated, 104
Female reproductive toxicity, 162
Ferbam, see Ferric dimethyldithiocarbamate
Ferric dimethyldithiocarbamate, 49
Fetotoxicity, 30, 44, 160—161, 171
Fire extinguisher, 103, 108
Firemaster FF-1®, 157—158
Firemaster PB-6®, 157
Fire retardant, 157
Flavoring agent, 62
Flukicide, 142
Fluorides, organic, 8—10
Foaming agent, 186
Formaldehyde, 84—85
Formic acid, 84—85
Fragrance agent, 62
Freon 113, see Trichlorotrifluoroethane
Fructose-6-phosphate kinase, inhibition of, 79
Fuel, 62, 128
   synthetic, 80
Fumigant

acrylonitrile, 31
aliphatic halogenated hydrocarbons, 102, 104, 108
carbon disulfide, 40
ethylene oxide, 86—87
Fungicide
   aromatic compounds, 186
   organic sulfur-containing compounds, 45, 47, 49, 51
   phenols, 140, 142
2-(2-Furyl)-3-(5-nitro-2-furyl)acrylamide, 174

G

G-11, see Hexachlorophene
Gait, abnormal, 33
Galoxide, 190
Gamophene, see Hexachlorophene
Gasoline, 64
Gastrointestinal hypomobility, 183
Glioblastoma multiforme, 111
Glioma, 31
Glucosides, cyanogenetic, 26
Glue, 63, 128—129, 141
Glutamic decarboxylase, inhibition of, 117—118
Glutaraldehyde, 81
Glutathione, 32, 172
Glutathione-S-transferase, inhibition of, 172
Glyceraldehyde-3-phosphate dehydrogenase, inhibition of, 44, 79, 173
Glycolysis, inhibition of, 43—44, 79, 172—173
Grouting gel, 27

H

Hair care products, 51, 117
Hallucination, 104, 110, 129
HCP, see Hexachlorophene
2,5-HD, see 2,5-Hexanedione
Headache, 31
Hearing disturbance, 158
Heat transfer fluid, 156
Heinz body formation, 181
Hen, neurotoxicity of organophosphorus compounds, 5—8
Hepatotoxicity, 162, 180
2,3,4,2′,4′,5′-Heptabromobiphenyl, 157
Heptachlor, 163
n-Heptane, 81
2,5-Heptanedione, 77—78, 80, 83
2,6-Heptanedione, 77—78
3,5-Heptanedione, 77—78
3-Heptanol, 83
2-Heptanone, see Methyl n-amyl ketone
3-Heptanone, 80
4-Heptanone, see Di-n-propyl ketone
Herbicide, 141, 147, 158, 184
Heterocyclic amines, 30
2,4,5,2′,4′,5′-Hexabromobiphenyl, 157

Hexachlorobenzene, 147
Hexachlorocyclohexane, 147, 163
Hexachlorophene (HCP), 115, 141—147, 184, 186, 189—190
   environmental release, 143
   mechanism of action, 146—147
   metabolism, 146
   neuropathology, 145—146
   neurotoxicity
      in domestic animals, 143
      in experimental animals, 143—145
      in humans, 142—143
      recovery from, 144—145
      single dose, 144
2,4-Hexadiene-1-nitrile, 33
Hexane, practical grade, 64
n-Hexane, 42, 62—80, 83, 187
   mechanism of action, 79—80
   metabolism, 74
   neuropathology, 64
   neurotoxicity
      in experimental animals, 64—73
      in humans, 62—63
   toxicokinetics, 64, 74—76
1,6-Hexanediol, 81
2,5-Hexanediol, 71, 75, 80, 83
2,3-Hexanedione, 77—78, 81
2,4-Hexanedione, 77—78, 81
2,5-Hexanedione (2,5-HD), 64, 69—71, 75—78, 80, 83
   in n-hexane/MnBK neuropathies, 76—77
   metabolism of, 74
3,5-Hexanedione, 81
1-Hexanol, 74—75, 81
2-Hexanol, 72, 74—76, 83
3-Hexanol, 74—75
2-Hexanone, see Methyl n-butyl ketone
Hexosan, see Hexachlorophene
Hippuric acid, 130
Huile H50, see Thiophene
Huile HSO, see Thiophene
Hydraencephaly, 185
Hydraulic fluid, 85, 156
Hydrazine, 186
Hydrocephalus, 132, 144, 185
Hydrocarbons, see also specific compounds
   aliphatic halogenated, 99—126
   aromatic, 127—137
   cyclic halogenated, 155—168
Hydrogen chloride, 111
Hydrogen cyanide, 26
Hydrogen sulfide, 40—41
Hydroquinone, 140
6-Hydroxydopamine, 184
6-Hydroxy-3-heptanone, 83
5-Hydroxy-2-hexanone, 72, 74—76, 80, 83
N-(Hydroxymethyl)acrylamide, 172—174
2-Hydroxy-5-nonanone, 80, 83
4-Hydroxy-pentanoic acid, 74
1-Hydroxy-2-(1H)-pyridinethione, 51
Hyperactivity, 29, 157, 183, 190

Hyperexcitability, 31, 163, 185, 188
Hyperkinesis, 141
Hypokalemia, 129
Hypothalamus, necrosis of, 191

I

IDPN, see β,β'-Iminodipropionitrile
3,3-Imidodipropionamide, 174
3,3'-Imino-*bis*-propanenitrile, see β,β'-Iminodipropionitrile
Imino-β-β-dipropionitrile, see β,β'-Iminodipropionitrile
3,3'-Iminodipropionitrile, see β,β'-Iminodipropionitrile
β,β'-Iminodipropionitrile (IDPN), 29—30, 64
   mechanism of action, 30
   metabolism, 30
   neuropathology, 30
   neurotoxicity, 29—30
Immersion oil, 156
Impotence, 140, 183
Incontinence, 187
Incoordination, 185
Inferior olivary nucleus, degeneration of, 33, 186
Ink, 86, 128, 156
Insecticide, 2, 44, 108, 159
Intramyelinic edema, 185
Iodoacetamide, 174
N-Isobutyloxymethylacrylamide, 174
Isocyanates, see also specific compounds, 34
Isoniazid, 147, 186, 189
Isonicotinic acid, 145
Isopropanol, 82
N-Isopropylacrylamide, 173—174

K

Kelevan®, 160
Kelthane, 163
Kenechlor®, 156
Kepone®, see Chlordecone
β-Keto heptanoic acid, 83
2-Keto-hexanoic acid, 74
2-Keto-1-hexanol, 74
2-Keto-5-hydroxy-hexanoic acid, 74
Ketones, see also specific compounds, 81—82
2-Keto-pentanoic acid, 74
Kidney, damage to, 182

L

Lacquer, 80, 85—86, 128—129, 157
Lacquer thinner, 62
Laetrile, 26
Lafora body, 189
Landry-Guillain-Barré syndrome, 63
Leptophos, 15

Leptophos oxon, 8
Lindane, 163
Liver, cancer of, 110
Lubricant, 156
Lubricating oil, 26
Lucel® 7, see 2-*t*-Butylazo-2-hydroxy-5-methyl hexane
Lysergic acid diethyl-amide, 30

M

Malononitrile, 26—27
Mandelic acid, 133
Mania, 41, 104
MDS, see 2,2'-Dithio-bis-pyridine-1-oxide
Medulla oblongata, necrosis of, 191
Megaesophagus, 171
MEK, see Methyl ethyl ketone
Memory disturbance
   acrylamide, 170
   aromatic compounds, 187
   carbon disulfide, 41
   ethylene glycol, 86
   ethylene oxide, 86
   isocyanates, 34
   methyl bromide, 104
   phenols, 147
   vinyl chloride, 110
Mental retardation, 86
2-Mercaptopropanoic acid, see 2-Mercaptopropionic acid
2-Mercaptopropionic acid, 101, 117
3-Mercaptopropionic acid, 117—118
α-Mercaptopropionic acid, see 2-Mercaptopropionic acid
β-Mercaptopropionic acid, see 3-Mercaptopropionic acid
2-Mercaptopyride, 52—53
2-Mercaptopyridine, 52—53
5-Mercaptothiazolidone, 43
Merex, see Chlordecone
Methacide, see Toluene
Methanol, 80—85, 106
   mechanism of action, 84—85
   neurotoxicity
      in experimental animals, 84
      in humans, 82—84
Methemoglobinemia, 180
Methoxychlor, 163
2-Methoxy ethanol, see Ethylene glycol monomethyl ether
2-Methylacrylamide, 174
N-Methylacrylamide, 173—174
Methyl alcohol, see Methanol
Methyl *n*-amyl ketone, 81
Methyl benzene, see Toluene
Methyl bromide, 101—107
   mechanism of action, 106—107
   metabolism, 106
   neurotoxicity

in experimental animals, 104—106
in humans, 103—104
Methyl n-butyl ketone (MnBK), 42, 62—80, 83, 187
   mechanism of action, 79—80
   neuropathology, 64
   neurotoxicity
      in experimental animals, 64
      in humans, 62—63
   toxicokinetics, 64, 74—76
Methyl chloride, 101—102
   neurotoxicity
      in experimental animals, 102
      in humans, 101—102
   recreational use, 101
2-Methyl-4-chlorophenoxyacetic acid (2,4,5-T), 148
N,N'-Methylene-bis-acrylamide, 173—174
2,2-Methylene bischlorophenol, 146
2,2'-Methylene bis(3,4,6-trichlorophenol), see Hexachlorophene
Methylene chloride, 107—108
Methylene dichloride, see Methylene chloride
Methyl ethyl ketone (MEK), 62, 65, 67, 69, 73, 80—81
Methyl heptyl ketone, see 5-Methyl-2-octanone
Methyl n-heptyl ketone, 82
5-Methyl-2-hexanone, see Methyl isoamyl ketone
Methylhydroxide, see Methanol
3,3'-Methyliminobis-(N-methylpropylamine), 191
Methyl iodide, 101, 107
Methyl isoamyl ketone, 82
Methyl isobutyl ketone, 80—81
Methylmercuric chloride, 172
Methyl methacrylate, 174
5-Methyl-2-octanone, 73, 80, 82
Methylol, see Methanol
Methylolacrylamide, see N-(Hydroxymethyl)acrylamide
2-Methyl pentane, 81
3-Methyl pentane, 81
1-Methyl-4-phenyl-1,2,5,6-tetrahydropyridine, 191
Methyl-n-propyl ketone, 81
4-Methyl-pyrazole, 85
Methyl styrene, see Vinyl toluene
Methyl viologen, see Paraquat
Metronidazole, 118
MiBK, 67—68
Mipafox, 3
Mirbane, 180
Mirex, 159—160, 163
Misonidazole, 118
Miticide, 184
Mixed function oxidase, 172
MnBK, see Methyl n-butyl ketone
Monobromomethane, see Methyl bromide
Monochloroacetaldehyde, 186
Monochloroacetic acid, 113
Monochloroethane, see Ethyl chloride
Monochloromethane, see Methyl chloride
Monohydroxybenzene, see Phenol
Monoiodomethane, see Methyl iodide

Muscle
   atrophy, 42, 131
   necrosis, 2
Musk, 190
Musk 36 A, see Acetyl ethyl tetramethyl tetralin
Musk ambrette, 190
Musk tetraline, see Acetyl ethyl tetramethyl tetralin
Myelin
   damage to, 64
   degeneration, 42, 183
   edema, 114—115
   vacuolization, 114—115, 185
Myelopathy, 47—48
Myocardial damage, 184
Myoclonia, 101
Myoclonic jerk, 163
Myoclonus, 104, 107
Myoglobinuria, 147
Myopathy, 45
Myotonia, 147—148

N

Naphthalene, brominated, 157
Nerve gas, 2
Neural transmission, interference in, 132
Neuritis
   peripheral, 108
   retrobulbar, 112, 131
Neurobehavioral disturbance, 65, 156
Neurofilament
   abnormality, 29—30
   damage to, 79—80
Neurogenic tumor, 47
Neuron
   damage to, 171—172
   degeneration, 42, 46, 113, 185
   lipofuscin deposition, 46
   lipofuscin-like accumulation in, 185
Neuron specific enolase, inhibition of, 173
Neuropathy, 67, 72—73
Neuropsychiatric disturbance, 140, 160
Neurotoxic esterase (NTE), 3
   aging, 3—4
   assay of, 3, 5—6
   inhibition of, 3—6, 8—21
      correlation with clinical neuropathy, 6, 8
      threshold level, 7
   organophosphorylation, 3
   protection, 3—4
   resynthesis, 6
Neurotubules, alteration of, 79
NIAX-calalyst ESN, 27—29
Nicotinamide, 173, 184
Nitriles, see also specific compounds, 27—33
p-Nitroaniline, 184
p-Nitrobenzamide, 181
Nitrobenzene, 180—181
   neurotoxicity
      in experimental animals, 180

in humans, 180
N-(4-Nitrophenyl)-N'-(3-pyridinylmethyl)urea, see N-3-Pyridylmethyl-N'-p-nitrophenylurea
1-Nitrophenyl-3-(3-pyridinylmethyl)urea, see N-3-Pyridylmethyl-N'-p-nitrophenylurea
Nitroprusside, 26
2,5-Nonanediol, 83
2,5-Nonanedione, 80, 83
5-Nonanol, 83
2-Nonanone, see Methyl n-heptyl ketone
5-Nonanone, 73, 80, 83
Norepinephrine, 156—157, 162, 184
Norleucine, see 2-Amino-hexanoic acid
NTE, see Neurotoxic esterase

O

3,6-Octanedione, 77—78, 83
2-Octanone, 82
N-t-Octylacrylamide, 174
Oil disease, 156
Oligodendroglia, degeneration of, 185—186
Omadine®, see 2,2'-Dithio-bis-pyridine-1-oxide
Opsoclonus, 160
Optic nerve, atrophy of, 84—85, 108, 112, 145
Optic neuritis, 46
Organic sulfur-containing compounds, see Sulfur-containing compounds, organic
Organophosphorus compounds, 1—24
  acute effects, 2
  axonopathic inhibitors, 3—4
  axonopathic receptor, 3
  neurotoxicity
    delayed, 2—3
    prediction, 6—7
    testing for, 5—6
  structure-activity relationships, 6—7
Organs, internal, discoloration of, 188
Osteolathyrism, 27, 30
Oxidative phosphorylation, inhibition of
  acids, 115
  aromatic compounds, 184, 189—190
  carbon disulfide, 44
  organic nitrogen compounds, 184
  phenols, 142, 146
Oxirane, see Ethylene oxide

P

Paint, 80, 85—86, 107, 128—129, 141
2-PAM, see Pralidoxime
Panceas, β-islet cel destruction, 182
Papilledema, 142, 160
Paralysis
  acrylamides, 171, 173
  alcohols, 65, 67, 71
  aliphatic chlorinated substances, 118
  alkanes, 65, 67, 71
  allyl chloride, 114

aromatic compounds, 190
cyclic halogenated hydrocarbons, 158
dichloroacetic acid, 114
hypokalemic, 129
methyl bromide, 104
nitriles, 29, 31
nitrobenzene, 180
organic sulfur-containing compounds, 44—50
organophosphorus compounds, 2
phenols, 140—147
Paraoxon, 2—3
Paraplegia, 34
Paraquat, 184—185
Parasiticide, 40
Parkinsonian symptoms, 107, 191
PBB, see Polybrominated biphenyl
PCB, see Polychlorinated biphenyl
PCP, see Pentachlorophenol
Penchlorol, see Pentachlorophenol
d-Penicillamine, 45
Penta, see Pentachlorophenol
Pentabromobiphenyl, 157
Pentachlorophenol (PCP), 140—142, 158—159
  metabolism, 142
  neurotoxicity, 141
N,N-Pentamethyleneacrylamide, 174
n-Pentane, 81
2,4-Pentanedione, 77—78
Pentanoic acid, 74
2-Pentanone, see Methyl n-propyl ketone
1-Pentylbromide, see 1-Bromopentane
n-Pentylbromide, see 1-Bromopentane
Pechloromethane, see Carbon tetrachloride
Perfume, 180, 188
Peripheral nervous system, damage to, 161
Peripheral neuritis, 147
Peripheral neuropathy
  alcohols, 62, 63
  alkanes, 62, 63
  carbon disulfide, 41
  cyclic halogenated hydrocarbons, 159, 163
  methanol, 84
  methyl bromide, 04
  nitriles, 28
  paraquat, 185
  vinyl chloride, 111—112
Permacide, see Pentachlorophenol
Personality disturbance, 34, 85—86, 129
Pesticide
  carbon disulfide, 40
  cyclic halogenated hydrocarbons, 156, 160, 163
  organophosphorus compounds, 5
  phenols, 142
Phantolid, 190
Pharmaceutical industry, 26—27, 29, 48, 114
Phenacemide, 181
Phenobarbital, 32, 45, 50, 75—76
Phenol, and related substances, 139—153
Phenylacetylurea, 181
Phenyl alcohol, see Phenol
Phenylethene, see Styrene monomer

Phenylethylene, see Styrene monomer
Phenyl hydrate, see Phenol
Phenyl hydroxide, see Phenol
Phenylisobutylhydrazine, 186
Phenylisopropylhydrazine, 186
Phenylisothiocyanate, 34
Phenyl phenylacetate, 3
Phenylurea, 181
Phenyl valerate, 3
Phisohex, see Hexachlorophene
Phosgene, 111
Phosphates, 4, 11—13
1-Phospha-2,6,7-trioxabiclyclo[2.2.2]-octane-1-oxide, 2
Phosphinates, 3—4, 16
Phosphinofluoridates, 9
Phosphonates, 4, 6, 14—15
Phosphonofluoridates, 9
Phosphonothioates, 14—15
Phosphoramidates, 4, 6
Phosphorodiamidofluoridates, 9—10
Phosphorofluoridates, 8—9
Phosphothioates, 11—13
Photoreceptor cell, damage to, 146
Photosensitizer, 26
Phthalate esters, 75
Physotigmine, 5
Pigmentation, 188—189
Piperazine, 191
Piperidine, 191
Piperidino-hexose-reductone, 30
Piperonyl butoxide, 50
Plastic industry, 130
Plasticizer, 44, 156
PNU, see N-3-Pyridylmethyl-n-p-nitrophenylurea
Polishes, 85
Polyacrylamide, 170
Polybrominated biphenyl (PBB), 157—158
 neurotoxicity
  in experimental animals, 157—158
  in humans, 157
Polychlorinated biphenyl (PCB), 156—157
 neurotoxicity
  in experimental animals, 156—157
  in humans, 156
Polychlorinated dibenzofuran, 156
Polycyclic musk, see Acetyl ethyl tetramethyl tetralin
Polymer industry, 130
Polyneuritis, 147, 158
Polyneuropathy, 41, 45—46, 114
Polystyrene foam, 101
Polyurethane foam, 27—28, 157
Polyvinylchloride, 110
Postural hypotension, 183—184
Posture, abnormal, 33
Potassium bromide, 106
Potassium conductance, altered, 191
Pralidoxime (2-PAM), 2, 5
Premercapturic acid, 51
Preservative, wood, 141

Preventol P, see Pentachlorophenol
Printing industry, 62
Propellant, 108—110
2-Propenamide, see Acrylamide
2-Propenenitrile, see Acrylonitrile
2-Propenylchloride, see Allyl chloride
S-β-Propionamide glutathione, 174
Propionitrile, 32
Propylene glycol-ethylene oxide copolymer, 87
Propylene oxide, 87
Prostate
 cyst of, 147
 fibrosis of, 147
Pseudobulbar palsy, 84
Psychiatric disturbance
 aromatic hydrocarbons, 131
 carbon disulfide 40—41
 cyclic halogenated hydrocarbons, 159
 disulfiram, 46
 methyl bromide, 104—107
 methyl chloride, 101—102
 trichloroethylene, 112
Psychosis, see Psychiatric disturbance
Purkinje cell, degeneration of, 118, 180
Pyrethroids, 163
Pyridinethione derivatives, 51—54
 mechanism of action, 54
 metabolism, 52—53
 neuropathology, 52
 neurotoxicity, 51—52
Pyridoxamine dithiocarbamic acid, 43
N-3-Pyridylmethyl-N'-p-nitrophenylurea (PNU), 182—184
 mechanism of action, 184
 metabolism, 183—184
 neuropathology, 183
 neurotoxicity
  in experimental animals, 183
  in humans, 182—183
2(Pyridyl-N-oxide)sulfonic acid, 52—53
Pyriform lobe, degeneration of, 186
Pyriminil, see N-3-Pyridylmethyl-N'-p-nitrophenylurea
Pyriminyl, see N-3-Pyridylmethyl-N'-p-nitrophenylurea
Pyrinuron, see N-3-Pyridylmethyl-N'-p-nitrophenylurea
Pyrogallol, 140
Pyrophosphates, 16
Pyruvate metabolism, 115

R

Radicular pain, 108
Raynaud's phenomenon, 110—112
Rayon manufacture, 40
Reactivator oximes, 5
Refrigerant, 101, 103, 108—109
Remyelination, 71, 186
Resins, 31, 40, 44, 48, 86, 140

Resorcinol, 140
Respiratory tract, cancer of, 31
RH 787, see N-3-Pyridylmethyl-N'-p-
   nitrophenylurea
Rigidity, 186
Rodenticide, 44, 182
Rodent repellant, 51
Rubber industry, 31, 40, 44—49, 101, 141

S

Saligenin esters, cyclic, 17
Santobrite, see Pentachlorophenol
Santophen 20, see Pentachlorophenol
Seizure
   acids, 117
   acrylamides, 171
   aromatic hydrocarbons, 128
   disulfiram, 46
   dithiocarbamate compounds, 49
   methyl bromide, 103—104, 107
   organic nitrogen compounds, 183
Senecioic acid amide, 174
Sensorimotor neuropathy
   alkanes, 63
   allyl chloride, 114
   aromatic compounds, 187
   carbon tetrachloride, 108
   dichloroacetic acid, 114
   ethyl chloride, 110
   ketones, 63
   nitriles, 28
   organic nitrogen compounds, 183
   vinyl chloride, 110
Sensorimotor peripheral neuropathy, 86, 112, 148, 170, 183
Sensory disturbance, 141, 147
Sensory neuropathy, 28, 103, 147, 156
Sexual dysfunction, 28
Shampoo, 142
SKF-525A, 172
Skin
   atrophy, 131
   discoloration, 188
Sleep disturbance, 129, 133
Soap, 142, 180, 188
Sodium acetylarsanilate, 30
Sodium bromide, 104, 106
Sodium cyanate, 26
Sodium dichloroacetate, 114—115
Sodium diethyldithiocarbamate, 49
Sodium formate, 85
Sodium pentochlorophenate, 141
Sodium pyridine-2-thiol-1-oxide, see Sodium pyridinethione
Sodium pyridinethione (SPT), 51
Solvent, industrial
   alkanes, 62
   aromatic hydrocarbons, 128
   carbon disulfide, 40
   carbon tetrachloride, 109
   ketones, 62
   methanol, 80
   vinyl chloride, 111
Solvent, abuse, 62—64
Sorbonitrile, see 2,4-Hexadiene-1-nitrile
Speech disturbance, 163
Sperm, abnormal, 116—117
Spermatogenic epithelium, degeneration of, 147
Spinal kyphosis, 52
Spongiosis, of spinal myelin, 109
SPT, see Sodium pyridinethione
Status epilepticus, 103, 107
Status spongiosus, 145, 184—185, 189
Sterilant, 86
Steroidal hormones, 43
Sterologenesis, inhibition of, 79
Styrene monomer, 130—133
   neurotoxicity
      in experimental animals, 132
      in humans, 131—133
Styrene sickness, 131—133
Styrol, see Styrene monomer
Styrolene, see Styrene monomer
Styropol, see Styrene monomer
Substantia nigra, degeneration of, 191
Sugars, chlorinated, 118
Suicidal tendency, 101, 129
Sulfonates, 3
Sulfonyl fluorides, 4
Sulfur-containing compounds, organic, see also specific compounds, 44—54
Surgi-Cen, see Hexachlorophene
Surofene, see Hexachlorophene
Sweat gland excitability, 71
Sympathetic anomaly, 171
Synthetic fibers, 31

T

2,4,5-T, see 2-Methyl-4-chlorophenoxyacetic acid
Taste disturbance, 112
TCB, see 3,4,3',4'-Tetrachlorobiphenyl
TCDBD, see 2,3,7,8-Tetrachlorodibenzo-p-dioxin
TCDD, see 2,3,7,8-Tetrachlorodibenzo-p-dioxin
Tecoram, 49
Temperature sensory loss, 113
Teratogenicity, 27, 32, 47, 147, 156
Testes
   atrophy, 69, 71—72, 88, 180
   degeneration, 102
   dysfunction, 160, 162
Testicular germinal cell, atrophy of, 48, 115, 117, 173, 181—182
Tetrachlorobiphenyl, 156
3,4,3',4'-Tetrachlorobiphenyl (TCB), 156
2,3,6,7-Tetrachlorodibenzo-p-dioxin, see 2,3,7,8-Tetrachlorodibenzo-p-dioxin
2,3,7,8-Tetrachlorodibenzo-1,4-dioxin, see 2,3,7,8-Tetrachlorodibenzo-p-dioxin

2,3,7,8-Tetrachlorodibenzo-*p*-dioxin (TCDD), 141, 147, 158—159
1,1,2,2-Tetrachloroethane, 114
Tetrachlorohydroquinone, 142
Tetrachloromethane, see Carbon tetrachloride
Tetrachlorophenol, 142, 147
Tetraethyl lead, production of, 109
Tetraethylthioperoxydicarbonic diamide, see Disulfiram
Tetraethylthiuram disulfide, see Disulfiram
Tetraethyltin, 147
1,1,4,4-Tetramethyl-6-ethyl-7-acetyl-1,2,3,4-tetrahydronaphthalene, see Acetyl ethyl tetramethyl tetralin
Tetramethylthioperoxydicarbonic acid diamide, see Tetramethylthiuramdisulfide
Tetramethylthiuramdisulfide (TMTD), 47—48
   mechanism of action, 48
   neuropathology, 47—48
   neurotoxicity, 47
Thiacyclopentadiene, see Thiophene
Thiamin, 26, 43, 173
Thiaphene, see Thiophene
2-Thienyl-mercapturic acid, 51
Thioacids, see also specific compounds, 114—118
Thiocyanate, 32
Thiofuram, see Thiophene
Thiofuran, see Thiophene
Thiofurfuran, see Thiophene
Thioglycolic acid, 117
Thioimidocarbonic diamide, see 2,4-Dithiobiuret
2-Thiolactic acid, see 2-Mercaptopropionic acid
Thiole, see Thiophene
Thiophen, see Thiophene
Thiophene, 48—51, 101
   metabolism, 50—51
   neuropathology, 50
   neurotoxicity
      in experimental animals, 50
      in humans, 48—49
Thiotetrole, see Thiophene
Thiourea, 43
   analogues of, 44
Thiram®, see Tetramethylthiuramdisulfide
Thymus
   atrophy, 30, 33, 116—117
   necrosis, 33, 115
Thyroxine, 30
TMTD, see Tetramethylthiuramdisulfide
Toluene, 65, 75, 128—130
   abuse, 129
   metabolism, 130
   neurotoxicity
      in experimental animals, 129—130
      in humans, 128—129
Toluene diisocyanate, 28, 34
Toluol, see Toluene
Tolyethene, see Vinyl toluene
Tonalid, 190
Toxaphene, 163
Transformer, 156

Trans-synaptic degeneration, 50
Tree wound dressing, 51
Tremerad, see 2-Acetoxy-4'-chloro-3,5-diiodobenzanilide
Tremor
   acids, 116—117
   acrylamides, 173
   alkanes, 77
   cyclic halogenated hydrocarbons, 160—163
   ethyl chloride, 110
   ethylene glycol, 86
   ethylene oxide, 86, 88
   ketones, 77
   methyl bromide, 107
   nitriles, 28, 31, 33
   organic nitrogen compounds, 183
   phenols, 141
   thiophene, 50
Tri, see Trichloroethylene
Triaryl phosphates, 7, 19—21
Tribromosalicylanilide, 147
Trichloroacetic acid, 113
Trichloroethanol, 113
Trichloroethylene, 111—113
   decomposition, 113
   metabolism, 113
   neuropathology, 113
   neurotoxicity
      in experimental animals, 112
      in humans, 112
Trichloroethylene epoxide, 113
Trichlorohydroxyphenyl methane, see Hexachlorophene
Trichloronate, see 2,4,5-Trichlorophenyl ethyl ethylphosphonothioate
2,4,5-Trichlorophenol, 158
2,4,5-Trichlorophenoxyacetic acid, 147—148, 158
2,4,5-Trichlorophenyl ethyl ethylphosphonothioate, 8
2,4,5-Trichlorophenyl ethyl phenylphosphonothioate, 5
Trichlorotrifluoroethane, 110
1,1,3-Tricyano-2-amino-1-propene, 27
Triethylphenyl phosphate, 5
Tri-2-ethylphenyl phosphate, 8
Tri-3-ethylphenyl phosphate, 8
Tri-4-ethylphenyl phosphate, 8
Triethyltin, 145, 186, 189
Trigeminal nerve, damage to, 112—113
Trilene, see Trichloroethylene
Tumor
   brain, 31—32, 110—111
   central nervous system, 31
   neurogenic, 47

U

Urine
   mandelic acid level in, 133
   retention, 27—28, 183

V

Vacor®, see N-3-Pyridylmethyl-N'-p-nitrophenylurea
γ-Valerolactone, 74—76, 80
Versalide®, see Acetyl ethyl tetramethyl tetralin
Vinyl benzene, see Styrene monomer
Vinyl chloride, 32, 110—111
  neurotoxicity
    in experimental animals, 111
    in humans, 110—111
Vinyl cyanide, see Acrylonitrile
Vinyl toluene, 132
Vision disturbance
  acrylamides, 170
  alcohols, 63
  alkanes, 63
  aromatic compounds, 187
  aromatic hydrocarbons, 129, 131
  carbon tetrachloride, 108
  ethylene glycol, 85
  ketones, 63
  methanol, 82, 85
  methyl bromide, 104
  methyl chloride, 101
  nitriles, 31
  phenols, 147
Vitamin B6, deficiency, 43

W

Wax, 85
Windshield washer fluid, 80
Wood alcohol, see Methanol
Wood spirits, see Methanol

Y

Yusho, 156

Z

Zinc, chelation of, 54
Zinc dimethyldithiocarbamate, 49
Zinc pyridine-2-thiol-1-oxide, see Zinc pyridinethione
Zinc pyridinethione (ZPT), 51—54
Ziram, see Zinc dimethyldithiocarbamate
ZPT, see Zinc pyridinethione

NO LONGER THE PROPERTY
OF THE
UNIVERSITY OF R.I. LIBRARY